# Acura Automotive Repair Manual

**by Ken Freund
and John H Haynes**
Member of the Guild of Motoring Writers

**Models covered:**
Acura Integra (1986 through 1989) and
Legend (1986 through 1990)

(1B9 – 12020)
(1776)

ABCDE

**Haynes Publishing Group**
Sparkford Nr Yeovil
Somerset BA22 7JJ England

**Haynes North America, Inc**
861 Lawrence Drive
Newbury Park
California 91320 USA

## Acknowledgements

Technical writers who contributed to this project include Mike Stubblefield, Robert Maddox and Larry Warren.

© **Haynes North America, Inc.** 1991

With permission from J.H. Haynes & Co. Ltd.

A book in the **Haynes Automotive Repair Manual Series**

**Printed in the U.S.A.**

All rights reserved. No part of this book may be reproduced or transmitted in any form or by any means, electronic or mechanical, including photocopying, recording or by any information storage or retrieval system, without permission in writing from the copyright holder.

**ISBN 1 85010 776 9**

**Library of Congress Catalog Card Number 91-73626**

While every attempt is made to ensure that the information in this manual is correct, no liability can be accepted by the authors or publishers for loss, damage or injury caused by any errors in, or omissions from, the information given.

# Contents

**Introductory pages**
| | |
|---|---|
| About this manual | 0-6 |
| Introduction to the Acura Integra and Legend | 0-6 |
| Vehicle identification numbers | 0-7 |
| Buying parts | 0-7 |
| Maintenance techniques, tools and working facilities | 0-7 |
| Booster battery (jump) starting | 0-14 |
| Jacking and towing | 0-15 |
| Automotive chemicals and lubricants | 0-16 |
| Safety first! | 0-17 |
| Conversion factors | 0-18 |
| Troubleshooting | 0-19 |

**Chapter 1**
Tune-up and routine maintenance — 1-1

**Chapter 2 Part A**
Integra engine — 2A-1

**Chapter 2 Part B**
Legend engine — 2B-1

**Chapter 2 Part C**
General engine overhaul procedures — 2C-1

**Chapter 3**
Cooling, heating and air conditioning systems — 3-1

**Chapter 4**
Fuel and exhaust systems — 4-1

**Chapter 5**
Engine electrical systems — 5-1

**Chapter 6**
Emissions control systems — 6-1

**Chapter 7 Part A**
Manual transaxle — 7A-1

**Chapter 7 Part B**
Automatic transaxle — 7B-1

**Chapter 8**
Clutch and drivetrain — 8-1

**Chapter 9**
Brakes — 9-1

**Chapter 10**
Suspension and steering systems — 10-1

**Chapter 11**
Body — 11-1

**Chapter 12**
Chassis electrical system — 12-1

**Wiring diagrams** — 12-16

**Index** — IND-1

1989 Acura Integra coupe

1990 Acura Legend sedan

# About this manual

## Its purpose

The purpose of this manual is to help you get the best value from your vehicle. It can do so in several ways. It can help you decide what work must be done, even if you choose to have it done by a dealer service department or a repair shop; it provides information and procedures for routine maintenance and servicing; and it offers diagnostic and repair procedures to follow when trouble occurs.

We hope you use the manual to tackle the work yourself. For many simpler jobs, doing it yourself may be quicker than arranging an appointment to get the vehicle into a shop and making the trips to leave it and pick it up. More importantly, a lot of money can be saved by avoiding the expense the shop must pass on to you to cover its labor and overhead costs. An added benefit is the sense of satisfaction and accomplishment that you feel after doing the job yourself.

## Using the manual

The manual is divided into Chapters. Each Chapter is divided into numbered Sections, which are headed in bold type between horizontal lines. Each Section consists of consecutively numbered paragraphs.

At the beginning of each numbered Section you will be referred to any illustrations which apply to the procedures in that Section. The reference numbers used in illustration captions pinpoint the pertinent Section and the Step within that Section. That is, illustration 3.2 means the illustration refers to Section 3 and Step (or paragraph) 2 within that Section.

Procedures, once described in the text, are not normally repeated. When it's necessary to refer to another Chapter, the reference will be given as Chapter and Section number. Cross references given without use of the word "Chapter" apply to Sections and/or paragraphs in the same Chapter. For example, "see Section 8" means in the same Chapter.

References to the left or right side of the vehicle assume you are sitting in the driver's seat, facing forward.

Even though we have prepared this manual with extreme care, neither the publisher nor the author can accept responsibility for any errors in, or omissions from, the information given.

**NOTE**

A **Note** provides information necessary to properly complete a procedure or information which will make the procedure easier to understand.

**CAUTION**

A **Caution** provides a special procedure or special steps which must be taken while completing the procedure where the **Caution** is found. Not heeding a **Caution** can result in damage to the assembly being worked on.

**WARNING**

A **Warning** provides a special procedure or special steps which must be taken while completing the procedure where the **Warning** is found. Not heeding a **Warning** can result in personal injury.

# Introduction to the Acura Integra and Legend

These models are available in two and four-door liftback and sedan body styles.

The transversely mounted inline four-cylinder or V6 engines used in these models are equipped with electronic fuel injection.

The engine drives the front wheels through either a five-speed manual or four-speed automatic transaxle via independent driveaxles.

Independent suspension, featuring coil spring/damper units, is used on all four wheels on Legend models. Independent front suspension and a solid rear axle is used on Integra models. Integra models are equipped with torsion bars instead of coil springs at the front. The power assisted rack and pinion steering unit is mounted behind the engine.

The brakes are disc at the front with either drum or discs at the rear, depending on model, with power assist standard. Anti-lock brakes are an option.

# Vehicle identification numbers

Modifications are a continuing and unpublicized process in vehicle manufacturing. Since spare parts manuals and lists are compiled on a numerical basis, the individual vehicle numbers are essential to correctly identify the component required.

## Vehicle Identification Number (VIN)

This very important identification number is stamped on the firewall in the engine compartment and on a plate attached to the dashboard inside the windshield on the driver's side of the vehicle. The VIN also appears on the Vehicle Certificate of Title and Registration. It contains information such as where and when the vehicle was manufactured, the model year and the body style.

## Engine numbers

The engine code number, which is commonly needed when ordering engine parts, can be found near the right end of the engine.

## Transaxle number

The transaxle number is commonly needed when ordering transaxle parts. It is on top of the transaxle.

# Buying parts

Replacement parts are available from many sources, which generally fall into one of two categories – authorized dealer parts departments and independent retail auto parts stores. Our advice concerning these parts is as follows:

*Retail auto parts stores:* Good auto parts stores will stock frequently needed components which wear out relatively fast, such as clutch components, exhaust systems, brake parts, tune-up parts, etc. These stores often supply new or reconditioned parts on an exchange basis, which can save a considerable amount of money. Discount auto parts stores are often very good places to buy materials and parts needed for general vehicle maintenance such as oil, grease, filters, spark plugs, belts, touch-up paint, bulbs, etc. They also usually sell tools and general accessories, have convenient hours, charge lower prices and can often be found not far from home.

*Authorized dealer parts department:* This is the best source for parts which are unique to the vehicle and not generally available elsewhere (such as major engine parts, transaxle parts, trim pieces, etc.).

*Warranty information:* If the vehicle is still covered under warranty, be sure that any replacement parts purchased – regardless of the source – do not invalidate the warranty!

To be sure of obtaining the correct parts, have engine and chassis numbers available and, if possible, take the old parts along for positive identification.

# Maintenance techniques, tools and working facilities

## Maintenance techniques

There are a number of techniques involved in maintenance and repair that will be referred to throughout this manual. Application of these techniques will enable the home mechanic to be more efficient, better organized and capable of performing the various tasks properly, which will ensure that the repair job is thorough and complete.

## Fasteners

Fasteners are nuts, bolts, studs and screws used to hold two or more parts together. There are a few things to keep in mind when working with fasteners. Almost all of them use a locking device of some type, either a lockwasher, locknut, locking tab or thread adhesive. All threaded fasteners should be clean and straight, with undamaged threads and undamaged corners on the hex head where the wrench fits. Develop the habit of replacing all damaged nuts and bolts with new ones. Special locknuts with nylon or fiber inserts can only be used once. If they are removed, they lose their locking ability and must be replaced with new ones.

Rusted nuts and bolts should be treated with a penetrating fluid to ease removal and prevent breakage. Some mechanics use turpentine in a spout-type oil can, which works quite well. After applying the rust penetrant, let it work for a few minutes before trying to loosen the nut or bolt. Badly rusted fasteners may have to be chiseled or sawed off or removed with a special nut breaker, available at tool stores.

If a bolt or stud breaks off in an assembly, it can be drilled and removed with a special tool commonly available for this purpose. Most automotive machine shops can perform this task, as well as other repair procedures, such as the repair of threaded holes that have been stripped out.

Flat washers and lockwashers, when removed from an assembly, should always be replaced exactly as removed. Replace any damaged washers with new ones. Never use a lockwasher on any soft metal surface (such as aluminum), thin sheet metal or plastic.

## Fastener sizes

For a number of reasons, automobile manufacturers are making wider and wider use of metric fasteners. Therefore, it is important to be able to tell the difference between standard (sometimes called U.S. or SAE) and metric hardware, since they cannot be interchanged.

All bolts, whether standard or metric, are sized according to diameter, thread pitch and length. For example, a standard 1/2 – 13 x 1 bolt is 1/2 inch in diameter, has 13 threads per inch and is 1 inch long. An M12 – 1.75 x 25 metric bolt is 12 mm in diameter, has a thread pitch of 1.75 mm (the distance between threads) and is 25 mm long. The two bolts are nearly identical, and easily confused, but they are not interchangeable.

In addition to the differences in diameter, thread pitch and length, metric and standard bolts can also be distinguished by examining the bolt heads. To begin with, the distance across the flats on a standard bolt head is measured in inches, while the same dimension on a metric bolt is sized in millimeters (the same is true for nuts). As a result, a standard wrench should not be used on a metric bolt and a metric wrench should not be used on a standard bolt. Also, most standard bolts have slashes radiating out from the center of the head to denote the grade or strength of the bolt, which is an indication of the amount of torque that can be applied to it. The greater the number of slashes, the greater the strength of the bolt. Grades 0 through 5 are commonly used on automobiles. Metric bolts have a property class (grade) number, rather than a slash, molded into their heads to indicate bolt strength. In this case, the higher the number, the stronger the bolt. Property class numbers 8.8, 9.8 and 10.9 are commonly used on automobiles.

Strength markings can also be used to distinguish standard hex nuts from metric hex nuts. Many standard nuts have dots stamped into one side, while metric nuts are marked with a number. The greater the number of dots, or the higher the number, the greater the strength of the nut.

Metric studs are also marked on their ends according to property class (grade). Larger studs are numbered (the same as metric bolts), while smaller studs carry a geometric code to denote grade.

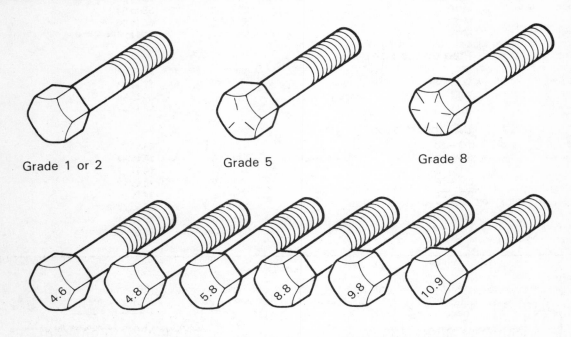

Bolt strength markings (top – standard/SAE/USS; bottom – metric)

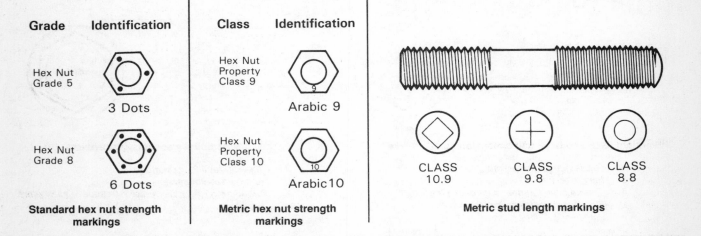

Standard hex nut strength markings

Metric hex nut strength markings

Metric stud length markings

# Maintenance techniques, tools and working facilities

It should be noted that many fasteners, especially Grades 0 through 2, have no distinguishing marks on them. When such is the case, the only way to determine whether it is standard or metric is to measure the thread pitch or compare it to a known fastener of the same size.

Standard fasteners are often referred to as SAE, as opposed to metric. However, it should be noted that SAE technically refers to a non-metric *fine thread* fastener only. Coarse thread non-metric fasteners are referred to as USS sizes.

Since fasteners of the same size (both standard and metric) may have different strength ratings, be sure to reinstall any bolts, studs or nuts removed from your vehicle in their original locations. Also, when replacing a fastener with a new one, make sure that the new one has a strength rating equal to or greater than the original.

## Tightening sequences and procedures

Most threaded fasteners should be tightened to a specific torque value (torque is the twisting force applied to a threaded component such as a nut or bolt). Overtightening the fastener can weaken it and cause it to break, while undertightening can cause it to eventually come loose. Bolts, screws and studs, depending on the material they are made of and their thread diameters, have specific torque values, many of which are noted in the Specifications at the beginning of each Chapter. Be sure to follow the torque recommendations closely. For fasteners not assigned a specific torque, a general torque value chart is presented here as a guide. These torque values are for dry (unlubricated) fasteners threaded into steel or cast iron (not aluminum). As was previously mentioned, the size and grade of a fastener determine the amount of torque that can safely be

| | Ft-lbs | Nm |
|---|---|---|
| **Metric thread sizes** | | |
| M-6 | 6 to 9 | 9 to 12 |
| M-8 | 14 to 21 | 19 to 28 |
| M-10 | 28 to 40 | 38 to 54 |
| M-12 | 50 to 71 | 68 to 96 |
| M-14 | 80 to 140 | 109 to 154 |
| **Pipe thread sizes** | | |
| 1/8 | 5 to 8 | 7 to 10 |
| 1/4 | 12 to 18 | 17 to 24 |
| 3/8 | 22 to 33 | 30 to 44 |
| 1/2 | 25 to 35 | 34 to 47 |
| **U.S. thread sizes** | | |
| 1/4 – 20 | 6 to 9 | 9 to 12 |
| 5/16 – 18 | 12 to 18 | 17 to 24 |
| 5/16 – 24 | 14 to 20 | 19 to 27 |
| 3/8 – 16 | 22 to 32 | 30 to 43 |
| 3/8 – 24 | 27 to 38 | 37 to 51 |
| 7/16 – 14 | 40 to 55 | 55 to 74 |
| 7/16 – 20 | 40 to 60 | 55 to 81 |
| 1/2 – 13 | 55 to 80 | 75 to 108 |

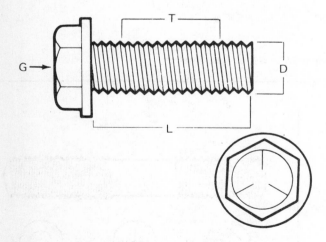

**Standard (SAE and USS) bolt dimensions/grade marks**

- G  Grade marks (bolt length)
- L  Length (in inches)
- T  Thread pitch (number of threads per inch)
- D  Nominal diameter (in inches)

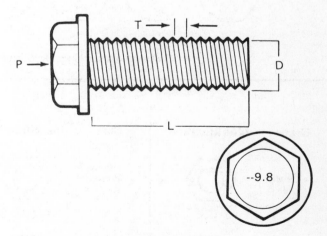

**Metric bolt dimensions/grade marks**

- P  Property class (bolt strength)
- L  Length (in millimeters)
- T  Thread pitch (distance between threads in millimeters)
- D  Diameter

applied to it. The figures listed here are approximate for Grade 2 and Grade 3 fasteners. Higher grades can tolerate higher torque values.

Fasteners laid out in a pattern, such as cylinder head bolts, oil pan bolts, differential cover bolts, etc., must be loosened or tightened in sequence to avoid warping the component. This sequence will normally be shown in the appropriate Chapter. If a specific pattern is not given, the following procedures can be used to prevent warping.

Initially, the bolts or nuts should be assembled finger-tight only. Next, they should be tightened one full turn each, in a criss-cross or diagonal pattern. After each one has been tightened one full turn, return to the first one and tighten them all one-half turn, following the same pattern. Finally, tighten each of them one-quarter turn at a time until each fastener has been tightened to the proper torque. To loosen and remove the fasteners, the procedure would be reversed.

### Component disassembly

Component disassembly should be done with care and purpose to help ensure that the parts go back together properly. Always keep track of the sequence in which parts are removed. Make note of special characteristics or marks on parts that can be installed more than one way, such as a grooved thrust washer on a shaft. It is a good idea to lay the disassembled parts out on a clean surface in the order that they were removed. It may also be helpful to make sketches or take instant photos of components before removal.

When removing fasteners from a component, keep track of their locations. Sometimes threading a bolt back in a part, or putting the washers and nut back on a stud, can prevent mix-ups later. If nuts and bolts cannot be returned to their original locations, they should be kept in a compartmented box or a series of small boxes. A cupcake or muffin tin is ideal for this purpose, since each cavity can hold the bolts and nuts from a particular area (i.e. oil pan bolts, valve cover bolts, engine mount bolts, etc.). A pan of this type is especially helpful when working on assemblies with very small parts, such as the carburetor, alternator, valve train or interior dash and trim pieces. The cavities can be marked with paint or tape to identify the contents.

Whenever wiring looms, harnesses or connectors are separated, it is a good idea to identify the two halves with numbered pieces of masking tape so they can be easily reconnected.

### Gasket sealing surfaces

Throughout any vehicle, gaskets are used to seal the mating surfaces between two parts and keep lubricants, fluids, vacuum or pressure contained in an assembly.

Many times these gaskets are coated with a liquid or paste-type gasket sealing compound before assembly. Age, heat and pressure can sometimes cause the two parts to stick together so tightly that they are very difficult to separate. Often, the assembly can be loosened by striking it with a soft-face hammer near the mating surfaces. A regular hammer can be used if a block of wood is placed between the hammer and the part. Do not hammer on cast parts or parts that could be easily damaged. With any particularly stubborn part, always recheck to make sure that every fastener has been removed.

Avoid using a screwdriver or bar to pry apart an assembly, as they can easily mar the gasket sealing surfaces of the parts, which must remain smooth. If prying is absolutely necessary, use an old broom handle, but keep in mind that extra clean up will be necessary if the wood splinters.

After the parts are separated, the old gasket must be carefully scraped off and the gasket surfaces cleaned. Stubborn gasket material can be soaked with rust penetrant or treated with a special chemical to soften it so it can be easily scraped off. A scraper can be fashioned from a piece of copper tubing by flattening and sharpening one end. Copper is recommended because it is usually softer than the surfaces to be scraped, which reduces the chance of gouging the part. Some gaskets can be removed with a wire brush, but regardless of the method used, the mating surfaces must be left clean and smooth. If for some reason the gasket surface is gouged, then a gasket sealer thick enough to fill scratches will have to be used during reassembly of the components. For most applications, a non-drying (or semi-drying) gasket sealer should be used.

### Hose removal tips

**Warning:** *If the vehicle is equipped with air conditioning, do not disconnect any of the A/C hoses without first having the system depressurized by a dealer service department or a service station.*

Hose removal precautions closely parallel gasket removal precautions. Avoid scratching or gouging the surface that the hose mates against or the connection may leak. This is especially true for radiator hoses. Because of various chemical reactions, the rubber in hoses can bond itself to the metal spigot that the hose fits over. To remove a hose, first loosen the hose clamps that secure it to the spigot. Then, with slip-joint pliers, grab the hose at the clamp and rotate it around the spigot. Work it back and forth until it is completely free, then pull it off. Silicone or other lubricants will ease removal if they can be applied between the hose and the outside of the spigot. Apply the same lubricant to the inside of the hose and the outside of the spigot to simplify installation.

As a last resort (and if the hose is to be replaced with a new one anyway), the rubber can be slit with a knife and the hose peeled from the spigot. If this must be done, be careful that the metal connection is not damaged.

If a hose clamp is broken or damaged, do not reuse it. Wire-type clamps usually weaken with age, so it is a good idea to replace them with screw-type clamps whenever a hose is removed.

## *Tools*

A selection of good tools is a basic requirement for anyone who plans to maintain and repair his or her own vehicle. For the owner who has few tools, the initial investment might seem high, but when compared to the spiraling costs of professional auto maintenance and repair, it is a wise one.

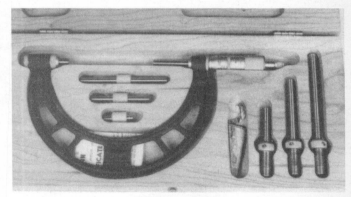

**Micrometer set**

**Dial indicator set**

## Maintenance techniques, tools and working facilities

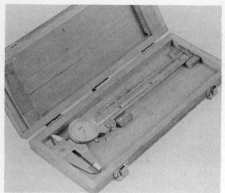
Dial caliper

Hand-operated vacuum pump

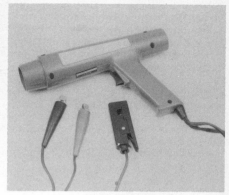
Timing light

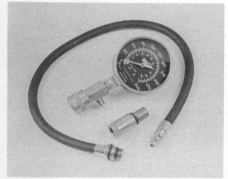
Compression gauge with spark plug hole adapter

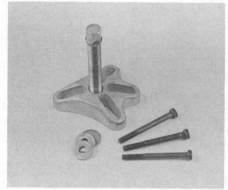
Damper/steering wheel puller

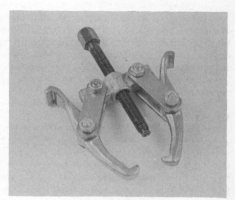
General purpose puller

Hydraulic lifter removal tool

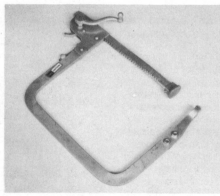
Valve spring compressor

Valve spring compressor

Ridge reamer

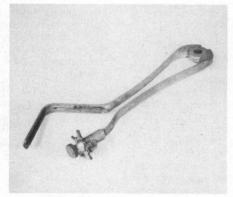
Piston ring groove cleaning tool

Ring removal/installation tool

Ring compressor

Cylinder hone

Brake hold-down spring tool

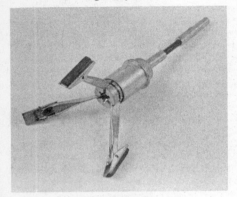
Brake cylinder hone

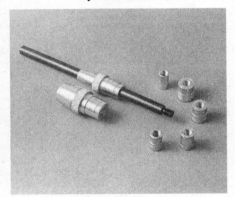
Clutch plate alignment tool

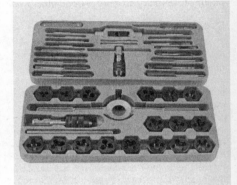
Tap and die set

To help the owner decide which tools are needed to perform the tasks detailed in this manual, the following tool lists are offered: *Maintenance and minor repair, Repair/overhaul* and *Special*.

The newcomer to practical mechanics should start off with the maintenance and minor repair tool kit, which is adequate for the simpler jobs performed on a vehicle. Then, as confidence and experience grow, the owner can tackle more difficult tasks, buying additional tools as they are needed. Eventually the basic kit will be expanded into the repair and overhaul tool set. Over a period of time, the experienced do-it-yourselfer will assemble a tool set complete enough for most repair and overhaul procedures and will add tools from the special category when it is felt that the expense is justified by the frequency of use.

### Maintenance and minor repair tool kit

The tools in this list should be considered the minimum required for performance of routine maintenance, servicing and minor repair work. We recommend the purchase of combination wrenches (box-end and open-end combined in one wrench). While more expensive than open end wrenches, they offer the advantages of both types of wrench.

*Combination wrench set (1/4-inch to 1 inch or 6 mm to 19 mm)*
*Adjustable wrench, 8 inch*
*Spark plug wrench with rubber insert*
*Spark plug gap adjusting tool*
*Feeler gauge set*
*Brake bleeder wrench*
*Standard screwdriver (5/16-inch x 6 inch)*
*Phillips screwdriver (No. 2 x 6 inch)*
*Combination pliers – 6 inch*
*Hacksaw and assortment of blades*
*Tire pressure gauge*
*Grease gun*
*Oil can*
*Fine emery cloth*
*Wire brush*
*Battery post and cable cleaning tool*
*Oil filter wrench*
*Funnel (medium size)*
*Safety goggles*
*Jackstands(2)*
*Drain pan*

**Note:** *If basic tune-ups are going to be part of routine maintenance, it will be necessary to purchase a good quality stroboscopic timing light and combination tachometer/dwell meter. Although they are included in the list of special tools, it is mentioned here because they are absolutely necessary for tuning most vehicles properly.*

### Repair and overhaul tool set

These tools are essential for anyone who plans to perform major repairs and are in addition to those in the maintenance and minor repair tool kit. Included is a comprehensive set of sockets which, though expensive, are invaluable because of their versatility, especially when various extensions and drives are available. We recommend the 1/2-inch drive over the 3/8-inch drive. Although the larger drive is bulky and more expensive, it has the capacity of accepting a very wide range of large sockets. Ideally, however, the mechanic should have a 3/8-inch drive set and a 1/2-inch drive set.

*Socket set(s)*
*Reversible ratchet*
*Extension – 10 inch*
*Universal joint*
*Torque wrench (same size drive as sockets)*
*Ball peen hammer – 8 ounce*
*Soft-face hammer (plastic/rubber)*
*Standard screwdriver (1/4-inch x 6 inch)*
*Standard screwdriver (stubby – 5/16-inch)*
*Phillips screwdriver (No. 3 x 8 inch)*
*Phillips screwdriver (stubby – No. 2)*

# Maintenance techniques, tools and working facilities 0 – 13

*Pliers – vise grip*
*Pliers – lineman's*
*Pliers – needle nose*
*Pliers – snap-ring (internal and external)*
*Cold chisel – 1/2-inch*
*Scribe*
*Scraper (made from flattened copper tubing)*
*Centerpunch*
*Pin punches (1/16, 1/8, 3/16-inch)*
*Steel rule/straightedge – 12 inch*
*Allen wrench set (1/8 to 3/8-inch or 4 mm to 10 mm)*
*A selection of files*
*Wire brush (large)*
*Jackstands (second set)*
*Jack (scissor or hydraulic type)*

**Note:** *Another tool which is often useful is an electric drill with a chuck capacity of 3/8-inch and a set of good quality drill bits.*

## Special tools

The tools in this list include those which are not used regularly, are expensive to buy, or which need to be used in accordance with their manufacturer's instructions. Unless these tools will be used frequently, it is not very economical to purchase many of them. A consideration would be to split the cost and use between yourself and a friend or friends. In addition, most of these tools can be obtained from a tool rental shop on a temporary basis.

This list primarily contains only those tools and instruments widely available to the public, and not those special tools produced by the vehicle manufacturer for distribution to dealer service departments. Occasionally, references to the manufacturer's special tools are included in the text of this manual. Generally, an alternative method of doing the job without the special tool is offered. However, sometimes there is no alternative to their use. Where this is the case, and the tool cannot be purchased or borrowed, the work should be turned over to the dealer service department or an automotive repair shop.

*Valve spring compressor*
*Piston ring groove cleaning tool*
*Piston ring compressor*
*Piston ring installation tool*
*Cylinder compression gauge*
*Cylinder ridge reamer*
*Cylinder surfacing hone*
*Cylinder bore gauge*
*Micrometers and/or dial calipers*
*Hydraulic lifter removal tool*
*Balljoint separator*
*Universal-type puller*
*Impact screwdriver*
*Dial indicator set*
*Stroboscopic timing light (inductive pick-up)*
*Hand operated vacuum/pressure pump*
*Tachometer/dwell meter*
*Universal electrical multimeter*
*Cable hoist*
*Brake spring removal and installation tools*
*Floor jack*

## Buying tools

For the do-it-yourselfer who is just starting to get involved in vehicle maintenance and repair, there are a number of options available when purchasing tools. If maintenance and minor repair is the extent of the work to be done, the purchase of individual tools is satisfactory. If, on the other hand, extensive work is planned, it would be a good idea to purchase a modest tool set from one of the large retail chain stores. A set can usually be bought at a substantial savings over the individual tool prices, and they often come with a tool box. As additional tools are needed, add–on sets, individual tools and a larger tool box can be purchased to expand the tool selection. Building a tool set gradually allows the cost of the tools to be spread over a longer period of time and gives the mechanic the freedom to choose only those tools that will actually be used.

Tool stores will often be the only source of some of the special tools that are needed, but regardless of where tools are bought, try to avoid cheap ones, especially when buying screwdrivers and sockets, because they won't last very long. The expense involved in replacing cheap tools will eventually be greater than the initial cost of quality tools.

## Care and maintenance of tools

Good tools are expensive, so it makes sense to treat them with respect. Keep them clean and in usable condition and store them properly when not in use. Always wipe off any dirt, grease or metal chips before putting them away. Never leave tools lying around in the work area. Upon completion of a job, always check closely under the hood for tools that may have been left there so they won't get lost during a test drive.

Some tools, such as screwdrivers, pliers, wrenches and sockets, can be hung on a panel mounted on the garage or workshop wall, while others should be kept in a tool box or tray. Measuring instruments, gauges, meters, etc. must be carefully stored where they cannot be damaged by weather or impact from other tools.

When tools are used with care and stored properly, they will last a very long time. Even with the best of care, though, tools will wear out if used frequently. When a tool is damaged or worn out, replace it. Subsequent jobs will be safer and more enjoyable if you do.

## *Working facilities*

Not to be overlooked when discussing tools is the workshop. If anything more than routine maintenance is to be carried out, some sort of suitable work area is essential.

It is understood, and appreciated, that many home mechanics do not have a good workshop or garage available, and end up removing an engine or doing major repairs outside. It is recommended, however, that the overhaul or repair be completed under the cover of a roof.

A clean, flat workbench or table of comfortable working height is an absolute necessity. The workbench should be equipped with a vise that has a jaw opening of at least four inches.

As mentioned previously, some clean, dry storage space is also required for tools, as well as the lubricants, fluids, cleaning solvents, etc. which soon become necessary.

Sometimes waste oil and fluids, drained from the engine or cooling system during normal maintenance or repairs, present a disposal problem. To avoid pouring them on the ground or into a sewage system, pour the used fluids into large containers, seal them with caps and take them to an authorized disposal site or recycling center. Plastic jugs, such as old antifreeze containers, are ideal for this purpose.

Always keep a supply of old newspapers and clean rags available. Old towels are excellent for mopping up spills. Many mechanics use rolls of paper towels for most work because they are readily available and disposable. To help keep the area under the vehicle clean, a large cardboard box can be cut open and flattened to protect the garage or shop floor.

Whenever working over a painted surface, such as when leaning over a fender to service something under the hood, always cover it with an old blanket or bedspread to protect the finish. Vinyl covered pads, made especially for this purpose, are available at auto parts stores.

# Booster battery (jump) starting

Observe these precautions when using a booster battery to start a vehicle:
a) Before connecting the booster battery, make sure the ignition switch is in the Off position.
b) Turn off the lights, heater and other electrical loads.
c) Your eyes should be shielded. Safety goggles are a good idea.
d) Make sure the booster battery is the same voltage as the dead one in the vehicle.
e) The two vehicles MUST NOT TOUCH each other!
f) Make sure the transmission is in Neutral (manual) or Park (automatic).
g) If the booster battery is not a maintenance-free type, remove the vent caps and lay a cloth over the vent holes.

Connect the red jumper cable to the positive (+) terminals of each battery.

Connect one end of the black jumper cable to the negative (–) terminal of the booster battery. The other end of this cable should be connected to a good ground on the vehicle to be started, such as a bolt or bracket on the engine block **(see illustration)**. Make sure the cable will not come into contact with the fan, drivebelts or other moving parts of the engine.

Start the engine using the booster battery, then, with the engine running at idle speed, disconnect the jumper cables in the reverse order of connection.

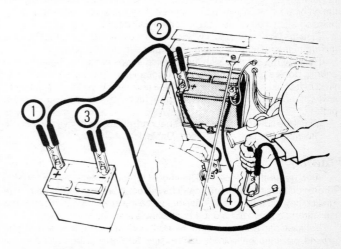

**Make the booster battery cable connections in the numerical order shown (note that the negative cable of the booster battery is NOT attached to the negative terminal of the dead battery)**

# Jacking and towing

## Jacking

**Warning:** *The jack supplied with the vehicle should only be used for changing a tire or placing jackstands under the frame. Never work under the vehicle or start the engine while this jack is being used as the only means of support.*

The vehicle should be on level ground. Place the shift lever in Park, if you have an automatic, or Reverse if you have a manual transaxle. Block the wheel diagonally opposite the wheel being changed. Set the parking brake.

Remove the spare tire and jack from stowage. Remove the wheel cover with a screwdriver or the tapered end of the lug nut wrench by inserting and twisting the handle and then prying against the back of the wheel cover. Loosen the lug nuts about one-half turn.

Place the scissors-type jack under the side of the vehicle and adjust the jack height until it fits between the notches in the vertical rocker panel flange nearest the wheel to be changed. There is a front and rear jacking point on each side of the vehicle **(see illustration)**.

Turn the jack handle clockwise until the tire clears the ground. Remove the lug nuts and pull the wheel off. Replace it with the spare.

Replace the lug nuts with the beveled edges facing in. Tighten them snugly. Don't attempt to tighten them completely until the vehicle is lowered or it could slip off the jack. Turn the jack handle counterclockwise to lower the vehicle. Remove the jack and tighten the lug nuts in a criss-cross pattern.

Install the cover and be sure it's snapped into place all the way around. Stow the tire, jack and wrench. Unblock the wheels.

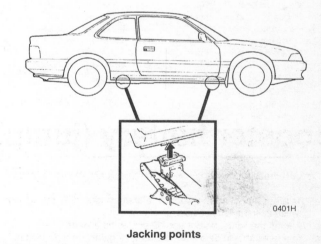

**Jacking points**

## Towing

As a general rule, the vehicle should be towed with the front (drive) wheels off the ground. If they can't be raised, place them on a dolly. The ignition key must be in the I position, since the steering lock mechanism isn't strong enough to hold the front wheels straight while towing.

Vehicles can be towed from the front only with all four wheels on the ground, provided that speeds don't exceed 30 mph and the distance is not over 50 miles. Before towing, check the transmission fluid level (see Chapter 1). If the level is below the HOT line on the dipstick, add fluid or use a towing dolly. Release the parking brake, put the transaxle in Neutral and place the ignition key in the I position. **Caution:** *Never tow a vehicle with an automatic transaxle from the rear with the front wheels on the ground. On 1990 Legend models, to avoid serious damage to the automatic transaxle, first start the engine and shift to D, then to N and shut the engine off. If the engine does not run or the transmission cannot be shifted while the engine is running, the vehicle must be transported on flat bed equipment.*

Equipment specifically designed for towing should be used. It should be attached to the main structural members of the vehicle, not the bumpers or brackets.

Safety is a major consideration when towing and all applicable state and local laws must be obeyed. A safety chain system must be used at all times. Remember that power steering and power brakes will not work with the engine off.

# Automotive chemicals and lubricants

A number of automotive chemicals and lubricants are available for use during vehicle maintenance and repair. They include a wide variety of products ranging from cleaning solvents and degreasers to lubricants and protective sprays for rubber, plastic and vinyl.

## Cleaners

**Carburetor cleaner and choke cleaner** is a strong solvent for gum, varnish and carbon. Most carburetor cleaners leave a dry-type lubricant film which will not harden or gum up. Because of this film it is not recommended for use on electrical components.

**Brake system cleaner** is used to remove grease and brake fluid from the brake system, where clean surfaces are absolutely necessary. It leaves no residue and often eliminates brake squeal caused by contaminants.

**Electrical cleaner** removes oxidation, corrosion and carbon deposits from electrical contacts, restoring full current flow. It can also be used to clean spark plugs, carburetor jets, voltage regulators and other parts where an oil-free surface is desired.

**Demoisturants** remove water and moisture from electrical components such as alternators, voltage regulators, electrical connectors and fuse blocks. They are non-conductive, non-corrosive and non-flammable.

**Degreasers** are heavy-duty solvents used to remove grease from the outside of the engine and from chassis components. They can be sprayed or brushed on and, depending on the type, are rinsed off either with water or solvent.

## Lubricants

**Motor oil** is the lubricant formulated for use in engines. It normally contains a wide variety of additives to prevent corrosion and reduce foaming and wear. Motor oil comes in various weights (viscosity ratings) from 5 to 80. The recommended weight of the oil depends on the season, temperature and the demands on the engine. Light oil is used in cold climates and under light load conditions. Heavy oil is used in hot climates and where high loads are encountered. Multi-viscosity oils are designed to have characteristics of both light and heavy oils and are available in a number of weights from 5W-20 to 20W-50.

**Gear oil** is designed to be used in differentials, manual transmissions and other areas where high-temperature lubrication is required.

**Chassis and wheel bearing grease** is a heavy grease used where increased loads and friction are encountered, such as for wheel bearings, balljoints, tie-rod ends and universal joints.

**High-temperature wheel bearing grease** is designed to withstand the extreme temperatures encountered by wheel bearings in disc brake equipped vehicles. It usually contains molybdenum disulfide (moly), which is a dry-type lubricant.

**White grease** is a heavy grease for metal-to-metal applications where water is a problem. White grease stays soft under both low and high temperatures (usually from –100 to +190-degrees F), and will not wash off or dilute in the presence of water.

**Assembly lube** is a special extreme pressure lubricant, usually containing moly, used to lubricate high-load parts (such as main and rod bearings and cam lobes) for initial start-up of a new engine. The assembly lube lubricates the parts without being squeezed out or washed away until the engine oiling system begins to function.

**Silicone lubricants** are used to protect rubber, plastic, vinyl and nylon parts.

**Graphite lubricants** are used where oils cannot be used due to contamination problems, such as in locks. The dry graphite will lubricate metal parts while remaining uncontaminated by dirt, water, oil or acids. It is electrically conductive and will not foul electrical contacts in locks such as the ignition switch.

**Moly penetrants** loosen and lubricate frozen, rusted and corroded fasteners and prevent future rusting or freezing.

**Heat-sink grease** is a special electrically non-conductive grease that is used for mounting electronic ignition modules where it is essential that heat is transferred away from the module.

## Sealants

**RTV sealant** is one of the most widely used gasket compounds. Made from silicone, RTV is air curing, it seals, bonds, waterproofs, fills surface irregularities, remains flexible, doesn't shrink, is relatively easy to remove, and is used as a supplementary sealer with almost all low and medium temperature gaskets.

**Anaerobic sealant** is much like RTV in that it can be used either to seal gaskets or to form gaskets by itself. It remains flexible, is solvent resistant and fills surface imperfections. The difference between an anaerobic sealant and an RTV-type sealant is in the curing. RTV cures when exposed to air, while an anaerobic sealant cures only in the absence of air. This means that an anaerobic sealant cures only after the assembly of parts, sealing them together.

**Thread and pipe sealant** is used for sealing hydraulic and pneumatic fittings and vacuum lines. It is usually made from a Teflon compound, and comes in a spray, a paint-on liquid and as a wrap-around tape.

## Chemicals

**Anti-seize compound** prevents seizing, galling, cold welding, rust and corrosion in fasteners. High-temperature anti-seize, usually made with copper and graphite lubricants, is used for exhaust system and exhaust manifold bolts.

**Anaerobic locking compounds** are used to keep fasteners from vibrating or working loose and cure only after installation, in the absence of air. Medium strength locking compound is used for small nuts, bolts and screws that may be removed later. High-strength locking compound is for large nuts, bolts and studs which aren't removed on a regular basis.

**Oil additives** range from viscosity index improvers to chemical treatments that claim to reduce internal engine friction. It should be noted that most oil manufacturers caution against using additives with their oils.

**Gas additives** perform several functions, depending on their chemical makeup. They usually contain solvents that help dissolve gum and varnish that build up on carburetor, fuel injection and intake parts. They also serve to break down carbon deposits that form on the inside surfaces of the combustion chambers. Some additives contain upper cylinder lubricants for valves and piston rings, and others contain chemicals to remove condensation from the gas tank.

## Miscellaneous

**Brake fluid** is specially formulated hydraulic fluid that can withstand the heat and pressure encountered in brake systems. Care must be taken so this fluid does not come in contact with painted surfaces or plastics. An opened container should always be resealed to prevent contamination by water or dirt.

**Weatherstrip adhesive** is used to bond weatherstripping around doors, windows and trunk lids. It is sometimes used to attach trim pieces.

**Undercoating** is a petroleum-based, tar-like substance that is designed to protect metal surfaces on the underside of the vehicle from corrosion. It also acts as a sound-deadening agent by insulating the bottom of the vehicle.

**Waxes and polishes** are used to help protect painted and plated surfaces from the weather. Different types of paint may require the use of different types of wax and polish. Some polishes utilize a chemical or abrasive cleaner to help remove the top layer of oxidized (dull) paint on older vehicles. In recent years many non-wax polishes that contain a wide variety of chemicals such as polymers and silicones have been introduced. These non-wax polishes are usually easier to apply and last longer than conventional waxes and polishes.

# Safety first!

Regardless of how enthusiastic you may be about getting on with the job at hand, take the time to ensure that your safety is not jeopardized. A moment's lack of attention can result in an accident, as can failure to observe certain simple safety precautions. The possibility of an accident will always exist, and the following points should not be considered a comprehensive list of all dangers. Rather, they are intended to make you aware of the risks and to encourage a safety conscious approach to all work you carry out on your vehicle.

## Essential DOs and DON'Ts

**DON'T** rely on a jack when working under the vehicle. Always use approved jackstands to support the weight of the vehicle and place them under the recommended lift or support points.

**DON'T** attempt to loosen extremely tight fasteners (i.e. wheel lug nuts) while the vehicle is on a jack – it may fall.

**DON'T** start the engine without first making sure that the transmission is in Neutral (or Park where applicable) and the parking brake is set.

**DON'T** remove the radiator cap from a hot cooling system – let it cool or cover it with a cloth and release the pressure gradually.

**DON'T** attempt to drain the engine oil until you are sure it has cooled to the point that it will not burn you.

**DON'T** touch any part of the engine or exhaust system until it has cooled sufficiently to avoid burns.

**DON'T** siphon toxic liquids such as gasoline, antifreeze and brake fluid by mouth, or allow them to remain on your skin.

**DON'T** inhale brake lining dust – it is potentially hazardous (see *Asbestos* below).

**DON'T** allow spilled oil or grease to remain on the floor – wipe it up before someone slips on it.

**DON'T** use loose fitting wrenches or other tools which may slip and cause injury.

**DON'T** push on wrenches when loosening or tightening nuts or bolts. Always try to pull the wrench toward you. If the situation calls for pushing the wrench away, push with an open hand to avoid scraped knuckles if the wrench should slip.

**DON'T** attempt to lift a heavy component alone – get someone to help you.

**DON'T** rush or take unsafe shortcuts to finish a job.

**DON'T** allow children or animals in or around the vehicle while you are working on it.

**DO** wear eye protection when using power tools such as a drill, sander, bench grinder, etc. and when working under a vehicle.

**DO** keep loose clothing and long hair well out of the way of moving parts.

**DO** make sure that any hoist used has a safe working load rating adequate for the job.

**DO** get someone to check on you periodically when working alone on a vehicle.

**DO** carry out work in a logical sequence and make sure that everything is correctly assembled and tightened.

**DO** keep chemicals and fluids tightly capped and out of the reach of children and pets.

**DO** remember that your vehicle's safety affects that of yourself and others. If in doubt on any point, get professional advice.

## Asbestos

Certain friction, insulating, sealing, and other products – such as brake linings, brake bands, clutch linings, torque converters, gaskets, etc. – contain asbestos. *Extreme care must be taken to avoid inhalation of dust from such products, since it is hazardous to health.* If in doubt, assume that they *do* contain asbestos.

## Fire

Remember at all times that gasoline is highly flammable. Never smoke or have any kind of open flame around when working on a vehicle. But the risk does not end there. A spark caused by an electrical short circuit, by two metal surfaces contacting each other, or even by static electricity built up in your body under certain conditions, can ignite gasoline vapors, which in a confined space are highly explosive. Do not, under any circumstances, use gasoline for cleaning parts. Use an approved safety solvent.

Always disconnect the battery ground (–) cable *at the battery* before working on any part of the fuel system or electrical system. Never risk spilling fuel on a hot engine or exhaust component.

It is strongly recommended that a fire extinguisher suitable for use on fuel and electrical fires be kept handy in the garage or workshop at all times. Never try to extinguish a fuel or electrical fire with water.

## Fumes

Certain fumes are highly toxic and can quickly cause unconsciousness and even death if inhaled to any extent. Gasoline vapor falls into this category, as do the vapors from some cleaning solvents. Any draining or pouring of such volatile fluids should be done in a well ventilated area.

When using cleaning fluids and solvents, read the instructions on the container carefully. Never use materials from unmarked containers.

Never run the engine in an enclosed space, such as a garage. Exhaust fumes contain carbon monoxide, which is extremely poisonous. If you need to run the engine, always do so in the open air, or at least have the rear of the vehicle outside the work area.

If you are fortunate enough to have the use of an inspection pit, never drain or pour gasoline and never run the engine while the vehicle is over the pit. The fumes, being heavier than air, will concentrate in the pit with possibly lethal results.

## The battery

Never create a spark or allow a bare light bulb near a battery. They normally give off a certain amount of hydrogen gas, which is highly explosive.

Always disconnect the battery ground (–) cable *at the battery* before working on the fuel or electrical systems.

If possible, loosen the filler caps or cover when charging the battery from an external source (this does not apply to sealed or maintenance-free batteries). Do not charge at an excessive rate or the battery may burst.

Take care when adding water to a non maintenance-free battery and when carrying a battery. The electrolyte, even when diluted, is very corrosive and should not be allowed to contact clothing or skin.

Always wear eye protection when cleaning the battery to prevent the caustic deposits from entering your eyes.

## Household current

When using an electric power tool, inspection light, etc., which operates on household current, always make sure that the tool is correctly connected to its plug and that, where necessary, it is properly grounded. Do not use such items in damp conditions and, again, do not create a spark or apply excessive heat in the vicinity of fuel or fuel vapor.

## Secondary ignition system voltage

A severe electric shock can result from touching certain parts of the ignition system (such as the spark plug wires) when the engine is running or being cranked, particularly if components are damp or the insulation is defective. In the case of an electronic ignition system, the secondary system voltage is much higher and could prove fatal.

# Conversion factors

### Length (distance)
| | | | | | |
|---|---|---|---|---|---|
| Inches (in) | X | 25.4 = Millimetres (mm) | X | 0.0394 | = Inches (in) |
| Feet (ft) | X | 0.305 = Metres (m) | X | 3.281 | = Feet (ft) |
| Miles | X | 1.609 = Kilometres (km) | X | 0.621 | = Miles |

### Volume (capacity)
| | | | | | |
|---|---|---|---|---|---|
| Cubic inches (cu in; in$^3$) | X | 16.387 = Cubic centimetres (cc; cm$^3$) | X | 0.061 | = Cubic inches (cu in; in$^3$) |
| Imperial pints (Imp pt) | X | 0.568 = Litres (l) | X | 1.76 | = Imperial pints (Imp pt) |
| Imperial quarts (Imp qt) | X | 1.137 = Litres (l) | X | 0.88 | = Imperial quarts (Imp qt) |
| Imperial quarts (Imp qt) | X | 1.201 = US quarts (US qt) | X | 0.833 | = Imperial quarts (Imp qt) |
| US quarts (US qt) | X | 0.946 = Litres (l) | X | 1.057 | = US quarts (US qt) |
| Imperial gallons (Imp gal) | X | 4.546 = Litres (l) | X | 0.22 | = Imperial gallons (Imp gal) |
| Imperial gallons (Imp gal) | X | 1.201 = US gallons (US gal) | X | 0.833 | = Imperial gallons (Imp gal) |
| US gallons (US gal) | X | 3.785 = Litres (l) | X | 0.264 | = US gallons (US gal) |

### Mass (weight)
| | | | | | |
|---|---|---|---|---|---|
| Ounces (oz) | X | 28.35 = Grams (g) | X | 0.035 | = Ounces (oz) |
| Pounds (lb) | X | 0.454 = Kilograms (kg) | X | 2.205 | = Pounds (lb) |

### Force
| | | | | | |
|---|---|---|---|---|---|
| Ounces-force (ozf; oz) | X | 0.278 = Newtons (N) | X | 3.6 | = Ounces-force (ozf; oz) |
| Pounds-force (lbf; lb) | X | 4.448 = Newtons (N) | X | 0.225 | = Pounds-force (lbf; lb) |
| Newtons (N) | X | 0.1 = Kilograms-force (kgf; kg) | X | 9.81 | = Newtons (N) |

### Pressure
| | | | | | |
|---|---|---|---|---|---|
| Pounds-force per square inch (psi; lbf/in$^2$; lb/in$^2$) | X | 0.070 = Kilograms-force per square centimetre (kgf/cm$^2$; kg/cm$^2$) | X | 14.223 | = Pounds-force per square inch (psi; lbf/in$^2$; lb/in$^2$) |
| Pounds-force per square inch (psi; lbf/in$^2$; lb/in$^2$) | X | 0.068 = Atmospheres (atm) | X | 14.696 | = Pounds-force per square inch (psi; lbf/in$^2$; lb/in$^2$) |
| Pounds-force per square inch (psi; lbf/in$^2$; lb/in$^2$) | X | 0.069 = Bars | X | 14.5 | = Pounds-force per square inch (psi; lbf/in$^2$; lb/in$^2$) |
| Pounds-force per square inch (psi; lbf/in$^2$; lb/in$^2$) | X | 6.895 = Kilopascals (kPa) | X | 0.145 | = Pounds-force per square inch (psi; lbf/in$^2$; lb/in$^2$) |
| Kilopascals (kPa) | X | 0.01 = Kilograms-force per square centimetre (kgf/cm$^2$; kg/cm$^2$) | X | 98.1 | = Kilopascals (kPa) |

### Torque (moment of force)
| | | | | | |
|---|---|---|---|---|---|
| Pounds-force inches (lbf in; lb in) | X | 1.152 = Kilograms-force centimetre (kgf cm; kg cm) | X | 0.868 | = Pounds-force inches (lbf in; lb in) |
| Pounds-force inches (lbf in; lb in) | X | 0.113 = Newton metres (Nm) | X | 8.85 | = Pounds-force inches (lbf in; lb in) |
| Pounds-force inches (lbf in; lb in) | X | 0.083 = Pounds-force feet (lbf ft; lb ft) | X | 12 | = Pounds-force inches (lbf in; lb in) |
| Pounds-force feet (lbf ft; lb ft) | X | 0.138 = Kilograms-force metres (kgf m; kg m) | X | 7.233 | = Pounds-force feet (lbf ft; lb ft) |
| Pounds-force feet (lbf ft; lb ft) | X | 1.356 = Newton metres (Nm) | X | 0.738 | = Pounds-force feet (lbf ft; lb ft) |
| Newton metres (Nm) | X | 0.102 = Kilograms-force metres (kgf m; kg m) | X | 9.804 | = Newton metres (Nm) |

### Power
| | | | | | |
|---|---|---|---|---|---|
| Horsepower (hp) | X | 745.7 = Watts (W) | X | 0.0013 | = Horsepower (hp) |

### Velocity (speed)
| | | | | | |
|---|---|---|---|---|---|
| Miles per hour (miles/hr; mph) | X | 1.609 = Kilometres per hour (km/hr; kph) | X | 0.621 | = Miles per hour (miles/hr; mph) |

### Fuel consumption*
| | | | | | |
|---|---|---|---|---|---|
| Miles per gallon, Imperial (mpg) | X | 0.354 = Kilometres per litre (km/l) | X | 2.825 | = Miles per gallon, Imperial (mpg) |
| Miles per gallon, US (mpg) | X | 0.425 = Kilometres per litre (km/l) | X | 2.352 | = Miles per gallon, US (mpg) |

### Temperature
Degrees Fahrenheit = (°C x 1.8) + 32    Degrees Celsius (Degrees Centigrade; °C) = (°F − 32) x 0.56

*It is common practice to convert from miles per gallon (mpg) to litres/100 kilometres (l/100km), where mpg (Imperial) x l/100 km = 282 and mpg (US) x l/100 km = 235

# Troubleshooting

## Contents

| Symptom | Section |
|---|---|
| **Engine** | |
| Engine backfires | 15 |
| Engine diesels (continues to run) after switching off | 18 |
| Engine hard to start when cold | 3 |
| Engine hard to start when hot | 4 |
| Engine lacks power | 14 |
| Engine lopes while idling or idles erratically | 8 |
| Engine misses at idle speed | 9 |
| Engine misses throughout driving speed range | 10 |
| Engine rotates but will not start | 2 |
| Engine runs with oil pressure light on | 17 |
| Engine stalls | 13 |
| Engine starts but stops immediately | 6 |
| Engine stumbles on acceleration | 11 |
| Engine surges while holding accelerator steady | 12 |
| Engine will not rotate when attempting to start | 1 |
| Oil puddle under engine | 7 |
| Pinging or knocking engine sounds during acceleration or uphill | 16 |
| Starter motor noisy or excessively rough in engagement | 5 |
| **Engine electrical system** | |
| Alternator light fails to go out | 20 |
| Battery will not hold a charge | 19 |
| Alternator light fails to come on when key is turned on | 21 |
| **Fuel system** | |
| Excessive fuel consumption | 22 |
| Fuel leakage and/or fuel odor | 23 |
| **Cooling system** | |
| Coolant loss | 28 |
| External coolant leakage | 26 |
| Internal coolant leakage | 27 |
| Overcooling | 25 |
| Overheating | 24 |
| Poor coolant circulation | 29 |
| **Clutch** | |
| Clutch pedal stays on floor | 39 |
| Clutch slips (engine speed increases with no increase in vehicle speed) | 35 |
| Fluid in area of master cylinder dust cover and on pedal | 31 |
| Fluid on release cylinder | 32 |
| Grabbing (chattering) as clutch is engaged | 36 |
| High pedal effort | 40 |
| Noise in clutch area | 38 |
| Pedal feels spongy when depressed | 33 |
| Pedal travels to floor – no pressure or very little resistance | 30 |
| Transaxle rattling (clicking) | 37 |
| Unable to select gears | 34 |
| **Manual transaxle** | |
| Clicking noise in turns | 44 |
| Clunk on acceleration or deceleration | 43 |

| Symptom | Section |
|---|---|
| Knocking noise at low speeds | 41 |
| Leaks lubricant | 50 |
| Locked in second gear | 51 |
| Noise most pronounced when turning | 42 |
| Noisy in all gears | 48 |
| Noisy in neutral with engine running | 46 |
| Noisy in one particular gear | 47 |
| Slips out of gear | 49 |
| Vibration | 45 |
| **Automatic transaxle** | |
| Engine will start in gears other than Park or Neutral | 56 |
| Fluid leakage | 52 |
| General shift mechanism problems | 54 |
| Transaxle fluid brown or has burned smell | 53 |
| Transaxle slips, shifts roughly, is noisy or has no drive in forward or reverse gears | 57 |
| Transaxle will not downshift with accelerator pedal pressed to the floor | 55 |
| **Driveaxles** | |
| Clicking noise in turns | 58 |
| Shudder or vibration during acceleration | 59 |
| Vibration at highway speeds | 60 |
| **Brakes** | |
| Brake pedal feels spongy when depressed | 68 |
| Brake pedal travels to the floor with little resistance | 69 |
| Brake roughness or chatter (pedal pulsates) | 63 |
| Dragging brakes | 66 |
| Excessive brake pedal travel | 65 |
| Excessive pedal effort required to stop vehicle | 64 |
| Grabbing or uneven braking action | 67 |
| Noise (high-pitched squeal when the brakes are applied) | 62 |
| Parking brake does not hold | 70 |
| Vehicle pulls to one side during braking | 61 |
| **Suspension and steering systems** | |
| Abnormal or excessive tire wear | 72 |
| Abnormal noise at the front end | 77 |
| Cupped tires | 82 |
| Erratic steering when braking | 79 |
| Excessive pitching and/or rolling around corners or during braking | 80 |
| Excessive play or looseness in steering system | 86 |
| Excessive tire wear on inside edge | 84 |
| Excessive tire wear on outside edge | 83 |
| Hard steering | 75 |
| Poor returnability of steering to center | 76 |
| Rattling or clicking noise in rack and pinion | 87 |
| Shimmy, shake or vibration | 74 |
| Suspension bottoms | 81 |
| Tire tread worn in one place | 85 |
| Vehicle pulls to one side | 71 |
| Wander or poor steering stability | 78 |
| Wheel makes a thumping noise | 73 |

# Troubleshooting

This section provides an easy reference guide to the more common problems which may occur during the operation of your vehicle. These problems and their possible causes are grouped under headings denoting various components or systems, such as Engine, Cooling system, etc. They also refer you to the chapter and/or section which deals with the problem.

Remember that successful troubleshooting is not a mysterious black art practiced only by professional mechanics. It is simply the result of the right knowledge combined with an intelligent, systematic approach to the problem. Always work by a process of elimination, starting with the simplest solution and working through to the most complex – and never overlook the obvious. Anyone can run the gas tank dry or leave the lights on overnight, so don't assume that you are exempt from such oversights.

Finally, always establish a clear idea of why a problem has occurred and take steps to ensure that it doesn't happen again. If the electrical system fails because of a poor connection, check the other connections in the system to make sure that they don't fail as well. If a particular fuse continues to blow, find out why – don't just replace one fuse after another. Remember, failure of a small component can often be indicative of potential failure or incorrect functioning of a more important component or system.

## Engine

### 1 Engine will not rotate when attempting to start

1 Battery terminal connections loose or corroded (Chapter 1).
2 Battery discharged or faulty (Chapter 1).
3 Automatic transaxle not completely engaged in Park (Chapter 7) or clutch not completely depressed (Chapter 8).
4 Broken, loose or disconnected wiring in the starting circuit (Chapters 5 and 12).
5 Starter motor pinion jammed in flywheel ring gear (Chapter 5).
6 Starter solenoid faulty (Chapter 5).
7 Starter motor faulty (Chapter 5).
8 Ignition switch faulty (Chapter 12).
9 Starter pinion or flywheel teeth worn or broken (Chapter 5).

### 2 Engine rotates but will not start

1 Fuel tank empty.
2 Battery discharged (engine rotates slowly) (Chapter 5).
3 Battery terminal connections loose or corroded (Chapter 1).
4 Leaking fuel injector(s), fuel pump, pressure regulator, etc. (Chapter 4).
5 Fuel not reaching fuel rail (Chapter 4).
6 Ignition components damp or damaged (Chapter 5).
7 Worn, faulty or incorrectly gapped spark plugs (Chapter 1).
8 Broken, loose or disconnected wiring in the starting circuit (Chapter 5).
9 Loose distributor is changing ignition timing (Chapter 5).
10 Broken, loose or disconnected wires at the ignition coil or faulty coil (Chapter 5).

### 3 Engine hard to start when cold

1 Battery discharged or low (Chapter 1).
2 Malfunctioning fuel system (Chapter 4).
3 Faulty cold start injector (Chapter 4).
4 Injector(s) leaking (Chapter 4).
5 Distributor rotor carbon tracked (Chapter 5).

### 4 Engine hard to start when hot

1 Air filter clogged (Chapter 1).
2 Fuel not reaching the fuel injection system (Chapter 4).
3 Corroded battery connections, especially ground (Chapter 1).

### 5 Starter motor noisy or excessively rough in engagement

1 Pinion or flywheel gear teeth worn or broken (Chapter 5).
2 Starter motor mounting bolts loose or missing (Chapter 5).

### 6 Engine starts but stops immediately

1 Loose or faulty electrical connections at distributor, coil or alternator (Chapter 5).
2 Insufficient fuel reaching the fuel injector(s) (Chapters 1 and 4).
3 Vacuum leak at the gasket between the intake manifold/plenum and throttle body (Chapters 1 and 4).

### 7 Oil puddle under engine

1 Oil pan gasket and/or oil pan drain bolt washer leaking (Chapter 2).
2 Oil pressure sending unit leaking (Chapter 2).
3 Valve covers leaking (Chapter 2).
4 Engine oil seals leaking (Chapter 2).
5 Oil pump housing leaking (Chapter 2).

### 8 Engine lopes while idling or idles erratically

1 Vacuum leakage (Chapters 2 and 4).
2 Leaking EGR valve (Chapter 6).
3 Air filter clogged (Chapter 1).
4 Fuel pump not delivering sufficient fuel to the fuel injection system (Chapter 4).
5 Leaking head gasket (Chapter 2).
6 Timing belt and/or pulleys worn (Chapter 2).
7 Camshaft lobes worn (Chapter 2).

### 9 Engine misses at idle speed

1 Spark plugs worn or not gapped properly (Chapter 1).
2 Faulty spark plug wires (Chapter 1).
3 Vacuum leaks (Chapter 1).
4 Incorrect ignition timing (Chapter 1).
5 Uneven or low compression (Chapter 2).

### 10 Engine misses throughout driving speed range

1 Fuel filter clogged and/or impurities in the fuel system (Chapter 1).
2 Low fuel output at the injector(s) (Chapter 4).
3 Faulty or incorrectly gapped spark plugs (Chapter 1).
4 Incorrect ignition timing (Chapter 5).
5 Cracked distributor cap, disconnected distributor wires or damaged distributor components (Chapters 1 and 5).
6 Leaking spark plug wires (Chapters 1 or 5).
7 Faulty emission system components (Chapter 6).
8 Low or uneven cylinder compression pressures (Chapter 2).
9 Weak or faulty ignition system (Chapter 5).
10 Vacuum leak in fuel injection system, intake manifold, air control valve or vacuum hoses (Chapter 4).

# Troubleshooting 0-21

### 11 Engine stumbles on acceleration

1. Spark plugs fouled (Chapter 1).
2. Spark plug wire(s) faulty
3. Fuel injection system needs adjustment or repair (Chapter 4).
4. Fuel filter clogged (Chapters 1 and 4).
5. Incorrect ignition timing (Chapter 5).
6. Intake manifold air leak (Chapters 2 and 4).

### 12 Engine surges while holding accelerator steady

1. Intake air leak (Chapter 4).
2. Fuel pump faulty (Chapter 4).
3. Loose fuel injector wire harness connectors (Chapter 4).
4. Defective ECU (Chapter 6).

### 13 Engine stalls

1. Fuel filter clogged and/or water and impurities in the fuel system (Chapters 1 and 4).
2. Distributor components damp or damaged (Chapter 5).
3. Faulty emissions system components (Chapter 6).
4. Faulty or incorrectly gapped spark plugs (Chapter 1).
5. Faulty spark plug wires (Chapter 1).
6. Vacuum leak in the fuel injection system, intake manifold or vacuum hoses (Chapters 2 and 4).
7. Valve clearances incorrectly set (Chapter 1).
8. Idle speed too low (Chapter 1).

### 14 Engine lacks power

1. Incorrect ignition timing (Chapter 5).
2. Excessive play in distributor shaft (Chapter 5).
3. Worn rotor, distributor cap or wires (Chapters 1 and 5).
4. Faulty or incorrectly gapped spark plugs (Chapter 1).
5. Fuel injection system out of adjustment or excessively worn (Chapter 4).
6. Faulty coil (Chapter 5).
7. Brakes binding (Chapter 9).
8. Automatic transaxle fluid level incorrect (Chapter 1).
9. Clutch slipping (Chapter 8).
10. Fuel filter clogged and/or impurities in the fuel system (Chapters 1 and 4).
11. Emission control system not functioning properly (Chapter 6).
12. Low or uneven cylinder compression pressures (Chapter 2).

### 15 Engine backfires

1. Emission control system not functioning properly (Chapter 6).
2. Ignition timing incorrect (Chapter 5).
3. Faulty secondary ignition system (cracked spark plug insulator, faulty plug wires, distributor cap and/or rotor) (Chapters 1 and 5).
4. Fuel injection system in need of adjustment or worn excessively (Chapter 4).
5. Vacuum leak at fuel injector(s), intake manifold, air control valve or vacuum hoses (Chapters 2 and 4).
6. Valve clearances incorrectly set and/or valves sticking (Chapter 1).

### 16 Pinging or knocking engine sounds during acceleration or uphill

1. Incorrect grade of fuel.
2. Ignition timing incorrect (Chapter 5).
3. Fuel injection system in need of adjustment (Chapter 4).
4. Improper or damaged spark plugs or wires (Chapter 1).
5. Worn or damaged distributor components (Chapter 5).
6. Faulty emission system (Chapter 6).
7. Vacuum leak (Chapters 2 and 4).

### 17 Engine runs with oil pressure light on

1. Low oil level (Chapter 1).
2. Idle rpm below specification (Chapter 1).
3. Short in wiring circuit (Chapter 12).
4. Faulty oil pressure sender (Chapter 2).
5. Worn engine bearings and/or oil pump (Chapter 2).

### 18 Engine diesels (continues to run) after switching off

1. Excessive engine operating temperature (Chapter 3).
2. Idle speed too high (Chapter 1).

## Engine electrical system

### 19 Battery will not hold a charge

1. Alternator drivebelt defective or not adjusted properly (Chapter 1).
2. Battery electrolyte level low (Chapter 1).
3. Battery terminals loose or corroded (Chapter 1).
4. Alternator not charging properly (Chapter 5).
5. Loose, broken or faulty wiring in the charging circuit (Chapter 5).
6. Short in vehicle wiring (Chapter 12).
7. Internally defective battery (Chapters 1 and 5).

### 20 Alternator light fails to go out

1. Faulty alternator or charging circuit (Chapter 5).
2. Alternator drivebelt defective or out of adjustment (Chapter 1).
3. Alternator voltage regulator inoperative (Chapter 5).

### 21 Alternator light fails to come on when key is turned on

1. Warning light bulb defective (Chapter 12).
2. Fault in the printed circuit, dash wiring or bulb holder (Chapter 12).

## Fuel system

### 22 Excessive fuel consumption

1. Dirty or clogged air filter element (Chapter 1).
2. Incorrectly set ignition timing (Chapter 5).

## Troubleshooting

3   Emissions system not functioning properly (Chapter 6).
4   Fuel injection internal parts excessively worn or damaged (Chapter 4).
5   Low tire pressure or incorrect tire size (Chapter 1).

### 23   Fuel leakage and/or fuel odor

1   Leaking fuel feed or return line (Chapters 1 and 4).
2   Tank overfilled.
3   Evaporative canister filter clogged (Chapters 1 and 6).
4   Fuel injector internal parts excessively worn (Chapter 4).

## Cooling system

### 24   Overheating

1   Insufficient coolant in system (Chapter 1).
2   Water pump drivebelt defective or out of adjustment (Chapter 1).
3   Radiator core blocked or grille restricted (Chapter 3).
4   Thermostat faulty (Chapter 3).
5   Electric coolant fan blades broken or cracked (Chapter 3).
6   Radiator cap not maintaining proper pressure (Chapter 3).
7   Ignition timing incorrect (Chapter 5).

### 25   Overcooling

1   Faulty thermostat (Chapter 3).
2   Inaccurate temperature gauge sending unit (Chapter 3)

### 26   External coolant leakage

1   Deteriorated/damaged hoses; loose clamps (Chapters 1 and 3).
2   Water pump seal defective (Chapter 3).
3   Leakage from radiator core or coolant reservoir bottle (Chapter 3).
4   Engine drain or water jacket core plugs leaking (Chapter 2).

### 27   Internal coolant leakage

1   Leaking cylinder head gasket (Chapter 2).
2   Cracked cylinder bore or cylinder head (Chapter 2).

### 28   Coolant loss

1   Too much coolant in system (Chapter 1).
2   Coolant boiling away because of overheating (Chapter 3).
3   Internal or external leakage (Chapter 3).
4   Faulty radiator cap (Chapter 3).

### 29   Poor coolant circulation

1   Inoperative water pump (Chapter 3).
2   Restriction in cooling system (Chapters 1 and 3).
3   Water pump drivebelt defective/out of adjustment (Chapter 1).
4   Thermostat sticking (Chapter 3).

## Clutch

### 30   Pedal travels to floor – no pressure or very little resistance

1   Master or release cylinder faulty – Legend models – (Chapter 8).
2   Clutch cable broken – Integra models – (Chapter 8).
3   Hose/pipe burst or leaking – Legend models – (Chapter 8).
4   Connections leaking – Legend models – (Chapter 8).
5   No fluid in reservoir – Legend models – (Chapter 8).
6   If fluid level in reservoir rises as pedal is depressed, master cylinder center valve seal is faulty – Legend models – (Chapter 8).
7   If there is fluid on dust seal at master cylinder, piston primary seal is leaking – Legend models – (Chapter 8).
8   Broken release bearing or fork (Chapter 8).

### 31   Fluid in area of master cylinder dust cover and on pedal

Rear seal failure in master cylinder (Chapter 8).

### 32   Fluid on release cylinder

Release cylinder plunger seal faulty (Chapter 8).

### 33   Pedal feels spongy when depressed

Air in system (Chapter 8).

### 34   Unable to select gears

1   Faulty transaxle (Chapter 7).
2   Faulty clutch disc (Chapter 8).
3   Fork and bearing not assembled properly (Chapter 8).
4   Faulty pressure plate (Chapter 8).
5   Pressure plate-to-flywheel bolts loose (Chapter 8).

### 35   Clutch slips (engine speed increases with no increase in vehicle speed)

1   Clutch plate worn (Chapter 8).
2   Clutch plate is oil soaked by leaking rear main seal (Chapter 8).
3   Clutch plate not seated. It may take 30 or 40 normal starts for a new one to seat.
4   Warped pressure plate or flywheel (Chapter 8).
5   Weak diaphragm spring (Chapter 8).
6   Clutch plate overheated. Allow to cool.

### 36   Grabbing (chattering) as clutch is engaged

1   Oil on clutch plate lining, burned or glazed facings (Chapter 8).
2   Worn or loose engine or transaxle mounts (Chapters 2 and 7).
3   Worn splines on clutch plate hub (Chapter 8).
4   Warped pressure plate or flywheel (Chapter 8).
5   Burned or smeared resin on flywheel or pressure plate (Chapter 8).

### 37   Transaxle rattling (clicking)

1   Release fork loose (Chapter 8).
2   Clutch plate damper spring failure (Chapter 8).

## Troubleshooting 0-23

### 38 Noise in clutch area

1 Fork shaft improperly installed (Chapter 8).
2 Faulty bearing (Chapter 8).

### 39 Clutch pedal stays on floor

1 Piston binding in bore – Legend models – (Chapter 8).
2 Broken release bearing or fork (Chapter 8).
3 Clutch cable broken – Integra models – (Chapter 8)

### 40 High pedal effort

1 Piston binding in bore – Legend models – Chapter 8).
2 Pressure plate faulty (Chapter 8).
3 Incorrect size master or release cylinder – Legend models – (Chapter 8).
4 Damaged or binding clutch cable – Integra models – (Chapter 8).

## Manual transaxle

### 41 Knocking noise at low speeds

1 Worn driveaxle constant velocity (CV) joints (Chapter 8).
2 Worn side gear shaft counterbore in differential case (Chapter 7A).*

### 42 Noise most pronounced when turning

Differential gear noise (Chapter 7A).*

### 43 Clunk on acceleration or deceleration

1 Loose engine or transaxle mounts (Chapters 2 and 7A).
2 Worn differential pinion shaft in case.*
3 Worn side gear shaft counterbore in differential case (Chapter 7A).*
4 Worn or damaged driveaxle inboard CV joints (Chapter 8).

### 44 Clicking noise in turns

Worn or damaged outboard CV joint (Chapter 8).

### 45 Vibration

1 Rough wheel bearing (Chapters 1 and 10).
2 Damaged driveaxle (Chapter 8).
3 Out-of-round tires (Chapter 1).
4 Tire out of balance (Chapters 1 and 10).
5 Worn CV joint (Chapter 8).

### 46 Noisy in neutral with engine running

1 Damaged input gear bearing (Chapter 7A).*
2 Damaged clutch release bearing (Chapter 8).

### 47 Noisy in one particular gear

1 Damaged or worn constant-mesh gears (Chapter 7A).*
2 Damaged or worn synchronizers (Chapter 7A).*
3 Bent reverse fork (Chapter 7A).*
4 Damaged fourth speed gear or output gear (Chapter 7A).*
5 Worn or damaged reverse idler gear or idler bushing (Chapter 7A).*

### 48 Noisy in all gears

1 Insufficient lubricant (Chapter 7A).
2 Damaged or worn bearings (Chapter 7A).*
3 Worn or damaged input gear shaft and/or output gear shaft (Chapter 7A).*

### 49 Slips out of gear

1 Worn or improperly adjusted linkage (Chapter 7A).
2 Transaxle loose on engine (Chapter 7A).
3 Shift linkage does not work freely, binds (Chapter 7A).
4 Input gear bearing retainer broken or loose (Chapter 7A).*
5 Dirt between clutch cover and engine housing (Chapter 7A).
6 Worn shift fork (Chapter 7A).*

### 50 Leaks lubricant

1 Side gear shaft seals worn (Chapter 8).
2 Excessive amount of lubricant in transaxle (Chapters 1 and 7A).
3 Loose or broken input gear shaft bearing retainer (Chapter 7A).*
4 Input gear bearing retainer O-ring and/or lip seal damaged (Chapter 7A).*

### 51 Locked in second gear

Lock pin or interlock pin missing (Chapter 7A).*

* Although the corrective action necessary to remedy the symptoms described is beyond the scope of the home mechanic, the above information should be helpful in isolating the cause of the condition so that the owner can communicate clearly with a professional mechanic.

## Automatic transaxle

**Note:** *Due to the complexity of the automatic transaxle, it is difficult for the home mechanic to properly diagnose and service this component. For problems other than the following, the vehicle should be taken to a dealer or transmission shop.*

### 52 Fluid leakage

1 Automatic transmission fluid is a deep red color. Fluid leaks should not be confused with engine oil, which can easily be blown onto the transaxle by air flow.
2 To pinpoint a leak, first remove all built-up dirt and grime from the transaxle housing with degreasing agents and/or steam cleaning. Then drive the vehicle at low speeds so air flow will not blow the leak far from its source. Raise the vehicle and determine where the leak is coming from. Common areas of leakage are:
   a) Pan (Chapters 1 and 7)
   b) Dipstick tube (Chapters 1 and 7)

# Troubleshooting

c) Transaxle oil lines (Chapter 7)
d) Speed sensor (Chapter 7)

## 53 Transaxle fluid brown or has a burned smell

Transaxle fluid burned (Chapter 1).

## 54 General shift mechanism problems

1   Chapter 7, Part B, deals with checking and adjusting the shift linkage on automatic transaxles. Common problems which may be attributed to poorly adjusted linkage are:
   a) Engine starting in gears other than Park or Neutral.
   b) Indicator on shifter pointing to a gear other than the one actually being used.
   c) Vehicle moves when in Park.
2   Refer to Chapter 7B for the shift linkage adjustment procedure.

## 55 Transaxle will not downshift with accelerator pedal pressed to the floor

Throttle Valve (TV) cable out of adjustment (Chapter 7B).

## 56 Engine will start in gears other than Park or Neutral

Neutral start switch malfunctioning (Chapter 7B).

## 57 Transaxle slips, shifts roughly, is noisy or has no drive in forward or reverse gears

There are many probable causes for the above problems, but the home mechanic should be concerned with only one possibility — fluid level. Before taking the vehicle to a repair shop, check the level and condition of the fluid as described in Chapter 1. Correct the fluid level as necessary or change the fluid and filter if needed. If the problem persists, have a professional diagnose the cause.

# Driveaxles

## 58 Clicking noise in turns

Worn or damaged outboard CV joint (Chapter 8).

## 59 Shudder or vibration during acceleration

1   Excessive toe-in (Chapter 10).
2   Incorrect spring or torsion bar heights (Chapter 10).
3   Worn or damaged inboard or outboard CV joints (Chapter 8).
4   Sticking inboard CV joint assembly (Chapter 8).

## 60 Vibration at highway speeds

1   Out-of-balance front wheels and/or tires (Chapters 1 and 10).
2   Out-of-round front tires (Chapters 1 and 10).
3   Worn CV joint(s) (Chapter 8).

# Brakes

**Note:** *Before assuming that a brake problem exists, make sure that:*
   a) The tires are in good condition and properly inflated (Chapter 1).
   b) The front end alignment is correct (Chapter 10).
   c) The vehicle is not loaded with weight in an unequal manner.

## 61 Vehicle pulls to one side during braking

1   Incorrect tire pressures (Chapter 1).
2   Front end out of line (have the front end aligned).
3   Front or rear tires not matched to one another.
4   Restricted brake lines or hoses (Chapter 9).
5   Malfunctioning caliper or drum brake assembly (Chapter 9).
6   Loose suspension parts (Chapter 10).
7   Loose calipers (Chapter 9).
8   Excessive wear of brake shoe or pad material or disc/drum on one side.

## 62 Noise (high-pitched squeal when the brakes are applied)

1   Front and/or rear disc brake pads worn out. The noise comes from the wear sensor rubbing against the disc (does not apply to all vehicles). Replace pads with new ones immediately (Chapter 9).
2   incorrectly installed new pads (many require an anti-squeal compound on the backing plates).

## 63 Brake roughness or chatter (pedal pulsates)

1   Excessive lateral runout (Chapter 9).
2   Uneven pad wear (Chapter 9).
3   Defective rotor (Chapter 9).

## 64 Excessive brake pedal effort required to stop vehicle

1   Malfunctioning power brake booster (Chapter 9).
2   Partial system failure (Chapter 9).
3   Excessively worn pads or shoes (Chapter 9).
4   Piston in caliper or wheel cylinder stuck or sluggish (Chapter 9).
5   Brake pads or shoes contaminated with oil or grease (Chapter 9).
6   New pads or shoes installed and not yet seated. It will take a while for the new material to seat against the rotor or drum.

## 65 Excessive brake pedal travel

1   Partial brake system failure (Chapter 9).
2   Insufficient fluid in master cylinder (Chapters 1 and 9).
3   Air trapped in system (Chapters 1 and 9).

## 66 Dragging brakes

1   Incorrect adjustment of brake light switch (Chapter 9).
2   Master cylinder pistons not returning correctly (Chapter 9).
3   Restricted brake lines or hoses (Chapters 1 and 9).
4   Incorrect parking brake adjustment (Chapter 9).

# Troubleshooting    0-25

## 67  Grabbing or uneven braking action

1. Malfunction of proportioning valve (Chapter 9).
2. Malfunction of power brake booster unit (Chapter 9).
3. Binding brake pedal mechanism (Chapter 9).

## 68  Brake pedal feels spongy when depressed

1. Air in hydraulic lines (Chapter 9).
2. Master cylinder mounting bolts loose (Chapter 9).
3. Master cylinder defective (Chapter 9).

## 69  Brake pedal travels to the floor with little resistance

1. Little or no fluid in the master cylinder reservoir caused by leaking caliper piston(s) (Chapter 9).
2. Loose, damaged or disconnected brake lines (Chapter 9).

## 70  Parking brake does not hold

Parking brake linkage improperly adjusted (Chapters 1 and 9).

## Suspension and steering systems

**Note:** *Before attempting to diagnose the suspension and steering systems, perform the following preliminary checks:*
   a) Tires for wrong pressure and uneven wear.
   b) Steering universal joints from the column to the steering gear for loose connectors or wear.
   c) Front and rear suspension and the steering gear assembly for loose or damaged parts.
   d) Out-of-round or out-of-balance tires, bent rims and loose and/or rough wheel bearings.

## 71  Vehicle pulls to one side

1. Mismatched or uneven tires (Chapter 10).
2. Broken or sagging springs or torsion bars (Chapter 10).
3. Wheel alignment (Chapter 10).
4. Front brake dragging (Chapter 9).

## 72  Abnormal or excessive tire wear

1. Wheel alignment (Chapter 10).
2. Sagging springs or torsion bars or broken springs (Chapter 10).
3. Tire out of balance (Chapter 10).
4. Worn strut damper (Chapter 10).
5. Overloaded vehicle.
6. Tires not rotated regularly.

## 73  Wheel makes a thumping noise

1. Blister or bump on tire (Chapter 10).
2. Improper strut damper action (Chapter 10).

## 74  Shimmy, shake or vibration

1. Tire or wheel out-of-balance or out-of-round (Chapter 10).
2. Loose or worn wheel bearings (Chapters 1, 8 and 10).
3. Worn tie-rod ends (Chapter 10).
4. Worn lower balljoints (Chapters 1 and 10).
5. Excessive wheel runout (Chapter 10).
6. Blister or bump on tire (Chapter 10).

## 75  Hard steering

1. Lack of lubrication at balljoints, tie-rod ends and steering gear assembly (Chapter 10).
2. Front wheel alignment (Chapter 10).
3. Low tire pressure(s) (Chapters 1 and 10).

## 76  Poor returnability of steering to center

1. Lack of lubrication at balljoints and tie-rod ends (Chapter 10).
2. Binding in balljoints (Chapter 10).
3. Binding in steering column (Chapter 10).
4. Lack of lubricant in steering gear assembly (Chapter 10).
5. Front wheel alignment (Chapter 10).

## 77  Abnormal noise at the front end

1. Lack of lubrication at balljoints and tie-rod ends (Chapters 1 and 10).
2. Damaged strut mounting (Chapter 10).
3. Worn control arm bushings or tie-rod ends (Chapter 10).
4. Loose stabilizer bar (Chapter 10).
5. Loose wheel nuts (Chapters 1 and 10).
6. Loose suspension bolts (Chapter 10).

## 78  Wander or poor steering stability

1. Mismatched or uneven tires (Chapter 10).
2. Lack of lubrication at balljoints and tie-rod ends (Chapters 1 and 10).
3. Worn strut assemblies (Chapter 10).
4. Loose stabilizer bar (Chapter 10).
5. Broken or sagging springs (Chapter 10).
6. Wheel alignment (Chapter 10).

## 79  Erratic steering when braking

1. Wheel bearings worn (Chapter 10).
2. Broken or sagging springs (Chapter 10).
3. Leaking wheel cylinder or caliper (Chapter 10).
4. Warped rotors or drums (Chapter 10).

## 80  Excessive pitching and/or rolling around corners or during braking

1. Loose stabilizer bar (Chapter 10).
2. Worn strut dampers or mountings (Chapter 10).
3. Broken or sagging springs (Chapter 10).
4. Overloaded vehicle.

## 81 Suspension bottoms

1. Overloaded vehicle.
2. Worn strut dampers (Chapter 10).
3. Incorrect, broken or sagging springs (Chapter 10).

## 82 Cupped tires

1. Front wheel or rear wheel alignment (Chapter 10).
2. Worn strut dampers (Chapter 10).
3. Wheel bearings worn (Chapter 10).
4. Excessive tire or wheel runout (Chapter 10).
5. Worn balljoints (Chapter 10).

## 83 Excessive tire wear on outside edge

1. Inflation pressures incorrect (Chapter 1).
2. Excessive speed in turns.
3. Front end alignment incorrect (excessive toe-in). Have professionally aligned.
4. Suspension arm bent or twisted (Chapter 10).

## 84 Excessive tire wear on inside edge

1. Inflation pressures incorrect (Chapter 1).
2. Front end alignment incorrect (toe-out). Have professionally aligned.
3. Loose or damaged steering components (Chapter 10).

## 85 Tire tread worn in one place

1. Tires out of balance.
2. Damaged or buckled wheel. Inspect and replace if necessary.
3. Defective tire (Chapter 1).

## 86 Excessive play or looseness in steering system

1. Wheel bearing(s) worn (Chapter 10).
2. Tie-rod end loose (Chapter 10).
3. Steering gear loose (Chapter 10).
4. Worn or loose steering intermediate shaft (Chapter 10).

## 87 Rattling or clicking noise in steering gear

1. Insufficient or improper lubricant in steering gear assembly (Chapter 10).
2. Steering gear attachment loose (Chapter 10).

# Chapter 1  Tune-up and routine maintenance

## Contents

| | |
|---|---|
| Air filter replacement | 17 |
| Automatic transaxle fluid change | 28 |
| Automatic transaxle fluid level check | 7 |
| Battery check, maintenance and charging | 10 |
| Brake check | 15 |
| Clutch release arm freeplay check and adjustment (Integra models only) | 16 |
| Cooling system check | 13 |
| Cooling system servicing (draining, flushing and refilling) | 26 |
| Driveaxle boot check | 24 |
| Drivebelt check, adjustment and replacement | 11 |
| Engine oil and oil filter change | 8 |
| Evaporative emissions control system check | 33 |
| Exhaust Gas Recirculation (EGR) system check (Legend models only) | 34 |
| Exhaust system check | 27 |
| Fluid level checks | 4 |
| Fuel filter replacement | 35 |
| Fuel system check | 21 |
| Idle speed check and adjustment | 32 |
| Ignition timing check and adjustment | 30 |
| Introduction | 1 |
| Maintenance schedule | 2 |
| Manual transaxle lubricant change | 29 |
| Manual transaxle lubricant level check | 22 |
| Positive Crankcase Ventilation (PCV) valve check and replacement | 25 |
| Power steering fluid level check | 6 |
| Spark plug check and replacement | 18 |
| Spark plug wire, distributor cap and rotor check and replacement | 19 |
| Steering and suspension check | 23 |
| Throttle linkage inspection | 31 |
| Tire and tire pressure checks | 5 |
| Tire rotation | 14 |
| Tune-up general information | 3 |
| Underhood hose check and replacement | 12 |
| Valve clearance check and adjustment (Integra models only) | 20 |
| Windshield wiper blade inspection and replacement | 9 |

## Specifications

### Recommended lubricants and fluids

**Note:** *Listed here are manufacturer recommendations at the time this manual was written. Manufacturers occasionally upgrade their fluid and lubricant specifications, so check with your local auto parts store for current recommendations.*

| | |
|---|---|
| Engine oil type | API grade SG or SG/CC multigrade and fuel efficient oil |
| Viscosity | See accompanying chart |
| Fuel | Unleaded gasoline, 87 octane or higher |
| Automatic transaxle fluid type | Dexron II automatic transmission fluid |
| Manual transaxle | |
|   Lubricant type | API grade SF or SE engine oil |
|   Viscosity | See accompanying chart |
| Brake fluid type | DOT 3 brake fluid |
| Power steering system fluid | Honda power steering fluid |

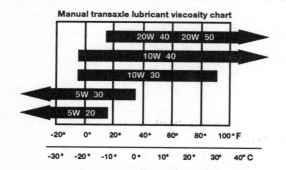
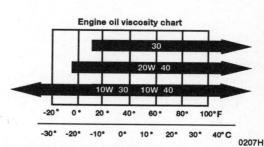

Recommended SAE viscosity grades for engine oils and manual transaxle lubricants
For best fuel economy and cold starting, select the lowest SAE viscosity grade oil for the expected temperature range

### Ignition system

| | |
|---|---|
| Spark plug | |
|   Type | Champion RC9YCN4, Nippondenso Q20 PR-U11 |
|   Gap | 0.044 inch |
| Spark plug wire resistance | Less than 25000 ohms |
| Ignition timing | Refer to the *Vehicle Emission Control Information* label in the engine compartment |
| Engine firing order | |
|   Legend | 1-4-2-5-3-6 |
|   Integra | 1-3-4-2 |

## Chapter 1 Tune-up and routine maintenance

### Cooling system
Thermostat rating
    Starts to kopen ......................................... 173-degrees F (78-degrees C)
    Fully open ............................................. 196-degrees F (91-degrees C)

### Accessory drivebelt deflection
Power steering pump
    Integra ................................................ 3/4 to 7/8-inch (18 to 22 mm)
    Legend ................................................ 3/4 to 15/16-inch (19 to 24 mm)
Alternator ............................................... 9/32 to 11/32-inch (7 to 9 mm)
Air conditioning compressor
    Integra ................................................ 9/32 to 11/32-inch (7 to 9 mm)
    Legend ................................................ 3/4 to 7/8-inch (18 to 22 mm)

### Clutch cable release arm freeplay
(Integra models only) .................................... 3/32 to 13/64-inch (4 to 5 mm)

### Brakes
Disc brake pad lining thickness (minimum) .................. 1/16-inch (1.6 mm)
Drum brake shoe lining thickness (minimum) ................ 3/32-inch (2 mm)
Parking brake adjustment
    Integra ................................................ 7 to 11 clicks
    Legend ................................................ 4 to 8 clicks

### General
Valve clearances – engine cold – (Integra models only)
    Intake valve .......................................... 0.005 to 0.007-inch (0.13 to 0.17 mm)
    Exhaust valve ........................................ 0.006 to 0.008-inch (0.15 to 0.19 mm)
Throttle cable deflection limit ............................. 3/8 to 1/2-inch (10 to 12 mm)

### Capacities
Engine oil
    Integra ................................................ 3.7 qt (3.5 liter)
    Legend ................................................ 4.8 qt (4.5 liter)
Automatic transaxle
    Integra ................................................ 2.5 qt (2.4 liter)
    Legend ................................................ 3.4 qt (3.2 liter)
Manual transaxle
    Integra ................................................ 2.4 qt (2.3 liter)
    Legend ................................................ 2.3 qt (2.2 liter)
Coolant ................................................ 6.0 qt (5.7 liter) (approximate)

### Torque specifications      Ft-lbs
Automatic transaxle drain plug ............................ 29
Manual transaxle drain and filler plugs ..................... 29 to 33
Wheel lug nuts .......................................... 80
Fuel filter
    Banjo bolt ............................................ 16
    Service bolt .......................................... 9
    Clamp bolt ........................................... 9
Spark plugs
    Legend ............................................... 16
    Integra ............................................... 13

**1986 and 1987**

**1988 and 1989**

*The blackened terminal shown on the distributor cap indicates the Number One spark plug wire position*

**Cylinder location and distributor rotation**

### 1 Introduction

This chapter is designed to help the home mechanic maintain the Acura Legend and Integra for peak performance, economy, safety and long life.

On the following pages is a master maintenance schedule, followed by sections dealing specifically with each item on the schedule. Visual checks, adjustments, component replacement and other helpful items are included. Refer to the accompanying photos of the engine compartment and the underside of the vehicle for the location of various components.

Servicing your Legend/Integra in accordance with the mileage/time maintenance schedule and the following Sections will provide it with a planned maintenance program that should result in a long and reliable service life. This is a comprehensive plan, so maintaining some items but not others at the specified service intervals will not produce the same results.

As you service your Legend or Integra, you will discover that many of the procedures can – and should – be grouped together because of the nature of the particular procedure you're performing or because of the close proximity of two otherwise unrelated components to one another.

For example, if the vehicle is raised for chassis lubrication, you should inspect the exhaust, suspension, steering and fuel systems while you're under the vehicle. When you're rotating the tires, it makes good sense to check the brakes and wheel bearings since the wheels are already removed.

Finally, let's suppose you have to borrow or rent a torque wrench. Even if you only need to tighten the spark plugs, you might as well check the torque of as many critical fasteners as time allows.

The first step of this maintenance program is to prepare yourself before the actual work begins. Read through all sections pertinent to the procedures you're planning to do, then make a list of and gather together all the parts and tools you will need to do the job. If it looks as if you might run into problems during a particular segment of some procedure, seek advice from your local parts man or dealer service department.

# Chapter 1 Tune-up and routine maintenance

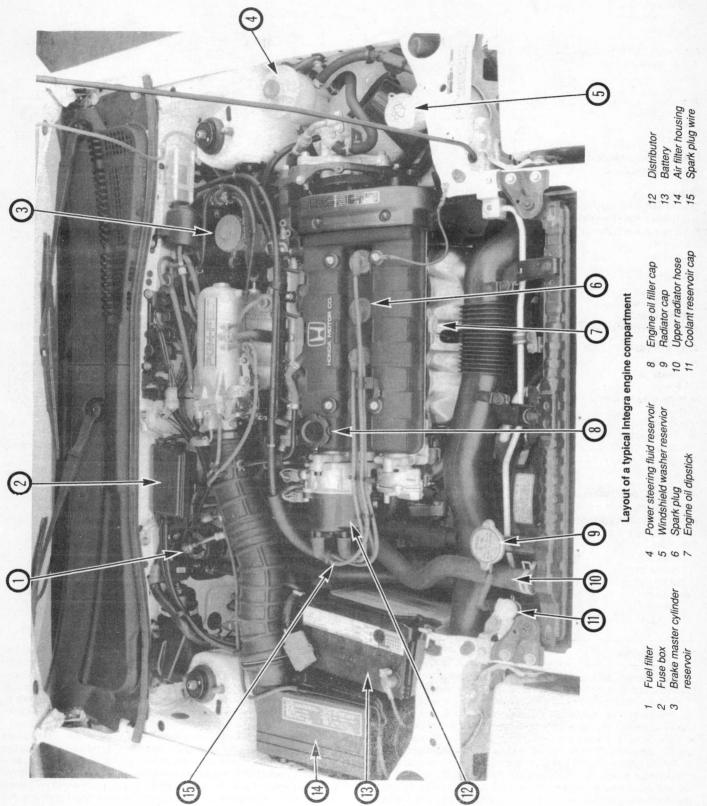

Layout of a typical Integra engine compartment

1 Fuel filter
2 Fuse box
3 Brake master cylinder reservoir
4 Power steering fluid reservoir
5 Windshield washer reservoir
6 Spark plug
7 Engine oil dipstick
8 Engine oil filler cap
9 Radiator cap
10 Upper radiator hose
11 Coolant reservoir cap
12 Distributor
13 Battery
14 Air filter housing
15 Spark plug wire

# Chapter 1 Tune-up and routine maintenance

Layout of the underside components of a typical Integra engine compartment

1. Exhaust pipe
2. Brake caliper
3. Driveaxle boot
4. Engine oil drain plug
5. Manual transaxle lubricant drain plug

# Chapter 1 Tune-up and routine maintenance

Layout of the rear underside components of a typical Integra

1. Muffler
2. Brake caliper
3. Exhaust system hanger
4. Exhaust pipe
5. Fuel tank

# Chapter 1 Tune-up and routine maintenance

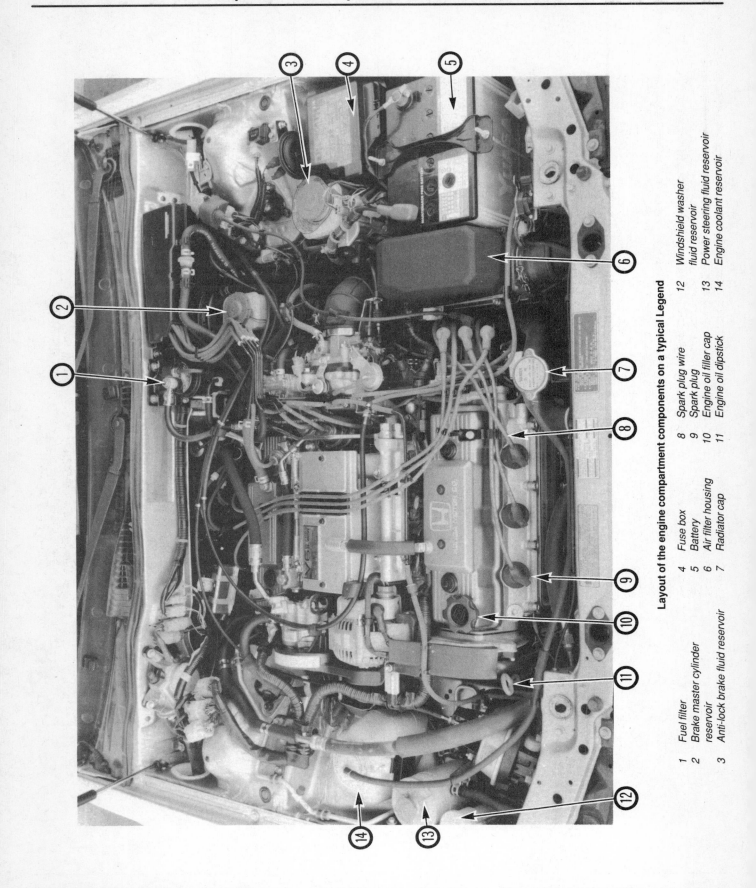

Layout of the engine compartment components on a typical Legend

1. Fuel filter
2. Brake master cylinder reservoir
3. Anti-lock brake fluid reservoir
4. Fuse box
5. Battery
6. Air filter housing
7. Radiator cap
8. Spark plug wire
9. Spark plug
10. Engine oil filler cap
11. Engine oil dipstick
12. Windshield washer fluid reservoir
13. Power steering fluid reservoir
14. Engine coolant reservoir

# Chapter 1 Tune-up and routine maintenance 1-7

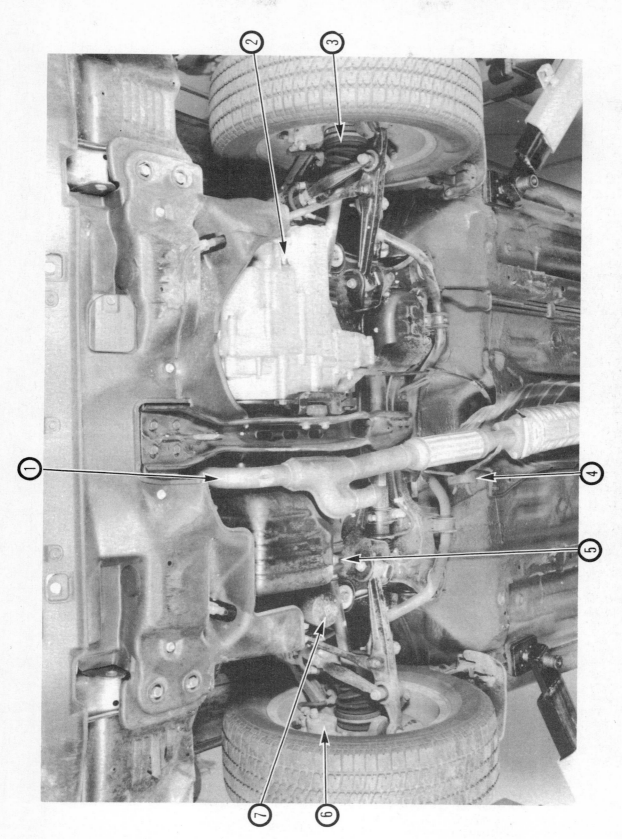

Layout of the underside components of a typical Legend engine compartment

1 Exhaust pipe
2 Manual transaxle drain plug
3 Driveaxle boot
4 Exhaust system hanger
5 Engine oil drain plug
6 Brake caliper
7 Engine oil filter

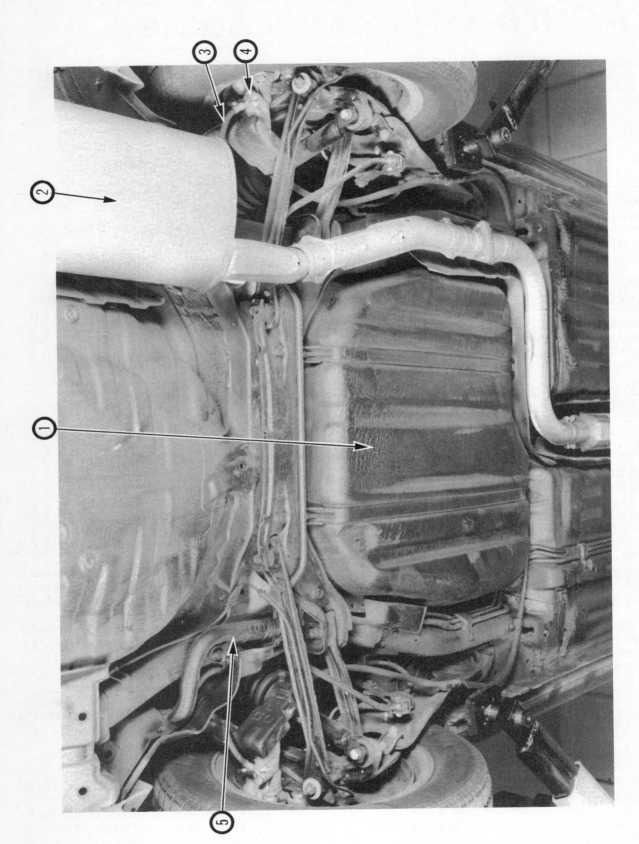

Layout of the rear underside components of a typical Legend

1 Fuel tank
2 Muffler
3 Brake hose
4 Brake caliper
5 Fuel filler pipe

## 2 Acura maintenance schedule

The maintenance intervals in this manual are provided with the assumption that you, not the dealer, will be doing the work. These are the minimum maintenance intervals recommended by the factory for Legends and Integras that are driven daily. If you wish to keep your vehicle in peak condition at all times, you may wish to perform some of these procedures even more often. Because frequent maintenance enhances the efficiency, performance and resale value of your car, we encourage you to do so. If you drive in dusty areas, tow a trailer, idle or drive at low speeds for extended periods or drive for short distances (less than four miles) in below freezing temperatures, shorter intervals are also recommended.

When your vehicle is new, it should be serviced by a factory authorized dealer service department to protect the factory warranty. In many cases, the initial maintenance check is done at no cost to the owner.

### Every 250 miles or weekly, whichever comes first

- Check the engine oil level (Section 4)
- Check the engine coolant level (Section 4)
- Check the windshield washer fluid level (Section 4)
- Check the brake fluid level (Section 4)
- Check the tires and tire pressures (Section 5)

### Every 3000 miles or 3 months, whichever comes first

*All items listed above plus:*
- Check the power steering fluid level (Section 6)
- Check the automatic transaxle fluid level (Section 7)
- Change the engine oil and oil filter (Section 8)

### Every 7500 miles or 6 months, whichever comes first

*All items listed above plus:*
- Inspect and replace, if necessary, the windshield wiper blades (Section 9)
- Check and adjust, if necessary, the clutch release arm freeplay (Integra models only) (Section 16)
- Check and service the battery (Section 10)
- Check and adjust, if necessary, the engine drivebelts (Section 11)
- Inspect and replace, if necessary, all underhood hoses (Section 12)
- Check the cooling system (Section 13)
- Rotate the tires (Section 14)
- Check the front disc brake pads (Section 15)

### Every 15,000 miles or 12 months, whichever comes first

*All items listed above plus:*
- Adjust the valve clearances (Integra models only) (Section 20)
- Inspect the brake system (Section 15)*
- Replace the air filter (Section 17)
- Inspect the fuel system (Section 21)
- Check and replace, if necessary, the spark plugs (Section 18)
- Inspect and replace, if necessary, the spark plug wires, distributor cap and rotor (Section 19)
- Check the manual transaxle lubricant level (Section 22)*
- Inspect the suspension and steering components (Section 23)*
- Check the driveaxle boots (Section 24)

### Every 30,000 miles or 24 months, whichever comes first

*All items listed above plus:*
- Check and replace, if necessary, the PCV valve (Section 25)
- Service the cooling system (drain, flush and refill) (Section 26)
- Inspect the exhaust system (Section 27)
- Change the automatic transaxle fluid (Section 28)**
- Change the manual transaxle lubricant (Section 29)

### Every 60,000 miles or 24 months, whichever comes first

*All items listed above plus:*
- Replace the fuel filter (Section 35)
- Check and adjust, if necessary, the engine ignition timing (Section 30)
- Check and adjust, if necessary, the engine idle speed (Section 32)
- Inspect the evaporative emissions control system (Section 33)
- Check the Exhaust Gas Recirculation (EGR) system (Section 34)
- Check the operation of the throttle linkage (Section 31)

*This item is affected by "severe" operating conditions as described below. If your vehicle is operated under "severe" conditions, perform all maintenance indicated with a * at 3000 mile/3 month intervals.
Severe conditions are indicated if you mainly operate your vehicle under one or more of the following conditions:*
- Operating in dusty areas
- Towing a trailer
- Idling for extended periods and/or low speed operation
- Operating when outside temperatures remain below freezing and when most trips are less than four miles

**If operated under one or more of the following conditions, change the automatic transaxle fluid every 15,000 miles:*
- In heavy city traffic where the outside temperature regularly reaches 90-degrees F (32-degrees C) or higher
- In hilly or mountainous terrain

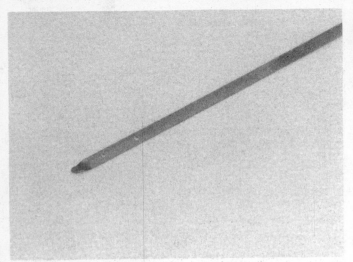

**4.4 The oil level should be between the two holes in the dipstick – if it isn't, add enough oil to bring the level to or near the upper hole (it takes one quart to raise the level from the lower to the upper hole)**

## 3 Tune-up general information

The term tune-up is used in this manual to represent a combination of individual operations rather than one specific procedure.

If, from the time the vehicle is new, the routine maintenance schedule is followed closely and frequent checks are made of fluid levels and high wear items, as suggested throughout this manual, the engine will be kept in relatively good running condition and the need for additional work will be minimized.

More likely than not, however, there will be times when the engine is running poorly due to lack of regular maintenance. This is even more likely if a used vehicle, which has not received regular and frequent maintenance checks, is purchased. In such cases, an engine tune-up will be needed outside of the regular routine maintenance intervals.

The first step in any tune-up or engine diagnosis to help correct a poor running engine would be a cylinder compression check. A check of the engine compression (Chapter 2 Part B) will give valuable information regarding the overall performance of many internal components and should be used as a basis for tune-up and repair procedures. If, for instance, a compression check indicates serious internal engine wear, a conventional tune-up will not help the running condition of the engine and would be a waste of time and money. Because of its importance, compression checking should be performed by someone with the proper compression testing gauge and the knowledge to use it properly.

The following series of operations are those most often needed to bring a generally poor running engine back into a proper state of tune.

### Minor tune-up

Clean, inspect and test the battery (Section 10)
Check all engine related fluids (Section 4)
Check and adjust the drivebelts (Section 11)
Replace the spark plugs (Section 18)
Inspect the distributor cap and rotor (Section 19)
Inspect the spark plug and coil wires (Section 19)
Check and adjust the idle speed (Section 32)
Check the air filter (Section 17)
Check the cooling system (Section 13)
Check all underhood hoses (Section 12)

### Major tune-up

All items listed under minor tune-up, plus . . .
Check the EGR system (Legend models only) (Section 34)
Check the ignition timing (Section 30)
Check the charging system (Chapter 5)

**4.6 The threaded oil filler cap is located on the camshaft cover – to prevent dirt from contaminating the engine, always make sure the area around this opening is clean before unscrewing the cap**

Check the fuel system (Section 21)
Replace the air filter (Section 17)
Replace the distributor cap and rotor (Section 19)
Replace the spark plug wires (Section 19)

## 4 Fluid level checks

1 Fluids are an essential part of the lubrication, cooling, brake, clutch and other systems. Because these fluids gradually become depleted and/or contaminated during normal operation of the vehicle, they must be periodically replenished. See Recommended lubricants, fluids and capacities at the beginning of this Chapter before adding fluid to any of the following components. **Note:** *The vehicle must be on level ground before fluid levels can be checked.*

### Engine oil

*Refer to illustrations 4.4 and 4.6*

2 The engine oil level is checked with a dipstick located at the front side of the engine. The dipstick extends through a metal tube from which it protrudes down into the engine oil pan.

3 The oil level should be checked before the vehicle has been driven, or about 15 minutes after the engine has been shut off. If the oil is checked immediately after driving the vehicle, some of the oil will remain in the upper engine components, producing an inaccurate reading on the dipstick.

4 Pull the dipstick from the tube and wipe all the oil from the end with a clean rag or paper towel. Insert the clean dipstick all the way back into its metal tube and pull it out again. Observe the oil at the end of the dipstick. At its highest point, the level should be between the upper and lower holes **(see illustration)**.

5 It takes one quart of oil to raise the level from the lower hole to the upper hole on the dipstick. Do not allow the level to drop below the lower hole or oil starvation may cause engine damage. Conversely, overfilling the engine (adding oil above the upper hole) may cause oil fouled spark plugs, oil leaks or oil seal failures.

6 Remove the threaded cap from the camshaft cover to add oil **(see illustration)**. Use an oil can spout or funnel to prevent spills. After adding the oil, install the filler cap hand tight. Start the engine and look carefully for any small leaks around the oil filter or drain plug. Stop the engine and check the oil level again after it has had sufficient time to drain from the upper block and cylinder head galleys.

7 Checking the oil level is an important preventive maintenance step. A continually dropping oil level indicates oil leakage through damaged seals, from loose connections, or past worn rings or valve guides. If the oil

# Chapter 1 Tune-up and routine maintenance

1–11

4.9 Make sure the coolant level in the reservoir is between the Max and Min lines (which can be seen using a flashlight) – if it's below the Min line, add a sufficient quantity of the specified mixture of antifreeze and water

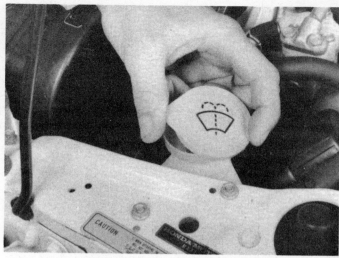

4.14 The windshield washer fluid reservoir is located at the left front corner of the engine compartment on Integra models – fluid can be added after flipping up the cap

looks milky in color or has water droplets in it, a cylinder head gasket may be blown. The engine should be checked immediately. The condition of the oil should also be checked. Each time you check the oil level, slide your thumb and index finger up the dipstick before wiping off the oil. If you see small dirt or metal particles clinging to the dipstick, the oil should be changed (see Section 8).

### Engine coolant

*Refer to illustration 4.9*

8  All vehicles covered by this manual are equipped with a pressurized coolant recovery system. A white coolant reservoir located on the right side of the engine compartment is connected by a hose to the base of the coolant filler cap. If the coolant heats up during engine operation, coolant can escape through a pressurized filler cap, then through a connecting hose into the reservoir. As the engine cools, the coolant is automatically drawn back into the cooling system to maintain the correct level.

9  The coolant level should be checked regularly. It must be between the Max and Min lines on the tank. The level will vary with the temperature of the engine. When the engine is cold, the coolant level should be at or slightly above the Min mark on the tank. Once the engine has warmed up, the level should be at or near the Max mark. If it isn't, allow the fluid in the tank to cool, then remove the cap from the reservoir **(see illustration)** and add coolant to bring the level up to the Max line. Use only ethylene/glycol type coolant and water in the mixture ratio recommended by your owner's manual. Do not use supplemental inhibitors or additives. If only a small amount of coolant is required to bring the system up to the proper level, water can be used. However, repeated additions of water will dilute the recommended antifreeze and water solution. In order to maintain the proper ratio of antifreeze and water, it is advisable to top up the coolant level with the correct mixture. Refer to your owner's manual for the recommended ratio.

10  If the coolant level drops within a short time after replenishment, there may be a leak in the system. Inspect the radiator, hoses, engine coolant filler cap, drain plugs, air bleeder plugs and water pump. If no leak is evident, have the radiator cap pressure tested by your dealer. **Warning:** *Never remove the radiator cap or the coolant recovery reservoir cap when the engine is running or has just been shut down, because the cooling system is hot. Escaping steam and scalding liquid could cause serious injury.*

11  If it is necessary to open the radiator cap, wait until the system has cooled completely, then wrap a thick cloth around the cap and turn it to the first stop. If any steam escapes, wait until the system has cooled further, then remove the cap.

12  When checking the coolant level, always note its condition. It should be relatively clear. If it is brown or rust colored, the system should be

4.15 Remove the cell caps to check the water level in the battery – if the level is low, add distilled water only

drained, flushed and refilled. Even if the coolant appears to be normal, the corrosion inhibitors wear out with use, so it must be replaced at the specified intervals.

13  Do not allow antifreeze to come in contact with your skin or painted surfaces of the vehicle. Flush contacted areas immediately with plenty of water.

### Windshield washer fluid

*Refer to illustration 4.14*

14  Fluid for the windshield washer system is stored in a plastic reservoir which is located at the left (Integra) or right (Legend) front corner of the engine compartment **(see illustration)**. In milder climates, plain water can be used to top up the reservoir, but the reservoir should be kept no more than 2/3 full to allow for expansion should the water freeze. In colder climates, the use of a specially designed windshield washer fluid, available at your dealer and any auto parts store, will help lower the freezing point of the fluid. Mix the solution with water in accordance with the manufacturer's directions on the container. Do not use regular antifreeze. It will damage the vehicle's paint.

### Battery electrolyte

*Refer to illustration 4.15*

15  The vehicles covered by this manual are equipped with a battery which is permanently sealed (except for vent holes) and has no filler caps. Water doesn't have to be added to these batteries at any time. If a conven-

# Chapter 1  Tune-up and routine maintenance

**4.17a**  The brake fluid level should be kept between the MIN and MAX marks on the translucent plastic reservoir – lift up the cap to add fluid

**4.17b**  The Anti-Lock Brake system fluid level should be between the MIN and MAX lines

**4.17c**  Keep the level between the MIN and MAX lines on the clutch fluid reservoir (Legend models)

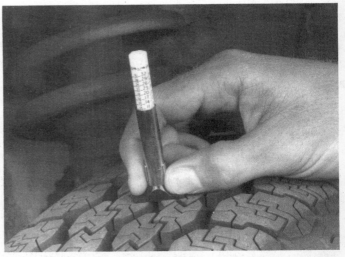

**5.2**  Use a tire tread depth indicator to monitor tire wear – they are available at auto parts stores and service stations and cost very little

tional battery is installed on your vehicle, check the electrolyte level of all six battery cells. It must be between the upper and lower levels – normally filled up to the bottom of the split-ring indicator in each cell – **(see illustration)**. If the level is low, unsnap or unscrew the filler/vent cap and add distilled water. Install and securely retighten the cap. **Caution:** *Overfilling the cells may cause electrolyte to spill over during periods of heavy charging, causing corrosion or damage.*

### Brake and clutch fluid

*Refer to illustrations 4.17a, 4.17b and 4.17c*

16  The brake master cylinder is mounted on the front of the power booster unit and the clutch master cylinder (Legend models) next to it on the firewall within the engine compartment. Later Legend models have a separate reservoir for the Anti-Lock Brake (ALB) system located in front of the brake master cylinder. ALB-equipped vehicles should be driven for a few minutes to equalize the fluid in the system before checking the fluid level in the reservoir. If the level rises significantly above the MAX mark, have the system checked by a dealer because this could indicate a malfunction in the ALB system.

17  To check the fluid level of the brake, clutch master cylinder or ALB reservoir, simply look at the MAX and MIN marks on the reservoir **(see illustrations)**. The level should be between the two marks.

18  If the level is low, wipe the top of the reservoir cover with a clean rag to prevent contamination of the brake system before lifting the cap.

19  Add only the specified brake fluid to the brake, clutch or ALB reservoir (refer to Recommended lubricants and fluids at the front of this chapter or to your owner's manual). Mixing different types of brake fluid can damage the system. Fill the brake master cylinder reservoir only to about 3/4-inch below the Max line – this brings the fluid to the correct level when you put the cap back on. **Warning:** *Use caution when filling the reservoir – brake fluid can harm your eyes and damage painted surfaces. Do not use brake fluid that has been opened for more than one year or has been left open. Brake fluid absorbs moisture from the air. Excess moisture can cause a dangerous loss of braking.*

20  While the reservoir cap is removed, inspect the master cylinder reservoir for contamination. If deposits, dirt particles or water droplets are present, the system should be drained and refilled (see Chapters 8 and 9).

# Chapter 1  Tune-up and routine maintenance

| Condition | Probable cause | Corrective action | Condition | Probable cause | Corrective action |
|---|---|---|---|---|---|
| Shoulder wear | • Underinflation (both sides wear)<br>• Incorrect wheel camber (one side wear)<br>• Hard cornering<br>• Lack of rotation | • Measure and adjust pressure.<br>• Repair or replace axle and suspension parts.<br>• Reduce speed.<br>• Rotate tires. | Toe wear (Feathered edge) | • Incorrect toe | • Adjust toe-in. |
| Center wear | • Overinflation<br>• Lack of rotation | • Measure and adjust pressure.<br>• Rotate tires. | Uneven wear | • Incorrect camber or caster<br>• Malfunctioning suspension<br>• Unbalanced wheel<br>• Out-of-round brake drum<br>• Lack of rotation | • Repair or replace axle and suspension parts.<br>• Repair or replace suspension parts.<br>• Balance or replace.<br>• Turn or replace.<br>• Rotate tires. |

5.3  This chart will help you determine the condition of the tires, the probable cause(s) of abnormal wear and the corrective action necessary

5.4a  If a tire loses air on a steady basis, check the valve core first to make sure it's snug (special inexpensive wrenches are commonly available at auto parts stores)

21  After filling the reservoir to the proper level, make sure the lid is properly seated to prevent fluid leakage and/or system pressure loss.
22  The brake fluid in the master cylinder will drop slightly as the brake pads at each wheel wear down during normal operation. If the master cylinder requires repeated replenishing to keep it at the proper level, this is an indication of leakage in the brake system, which should be corrected immediately. Check all brake lines and connections, along with the wheel cylinders and booster (see Section 15 for more information). A drop in the clutch reservoir level indicates a leak in the clutch hydraulic system (Chapter 8).
23  If, upon checking the brake master cylinder fluid level, you discover an empty or nearly empty reservoir, the brake system should be bled (see Chapter 9).

## 5  Tire and tire pressure checks

*Refer to illustrations 5.2, 5.3, 5.4a, 5.4b and 5.8*

1  Periodic inspection of the tires may spare you from the inconvenience of being stranded with a flat tire. It can also provide you with vital information regarding possible problems in the steering and suspension systems before major damage occurs.
2  Normal tread wear can be monitored with a simple, inexpensive device known as a tread depth indicator **(see illustration)**. When the tread depth reaches the specified minimum, replace the tire(s).
3  Note any abnormal tread wear **(see illustration)**. Tread pattern irregularities such as cupping, flat spots and more wear on one side than the other are indications of front end alignment and/or balance problems. If any of these conditions are noted, take the vehicle to a tire shop or service station to correct the problem.
4  Look closely for cuts, punctures and embedded nails or tacks. Sometimes a tire will hold its air pressure for a short time or leak down very slowly even after a nail has embedded itself into the tread. If a slow leak persists, check the valve core to make sure it is tight **(see illustration)**. Examine the tread for an object that may have embedded itself into the tire or for a "plug" that may have begun to leak (radial tire punctures are repaired with a plug that is installed in a puncture). If a puncture is suspected, it can be easily verified by spraying a solution of soapy water onto the puncture area

# Chapter 1 Tune-up and routine maintenance

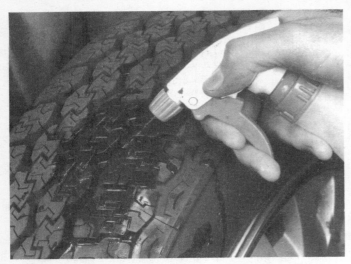

5.4b  If the valve core is tight, raise the corner of the vehicle with the low tire and spray a soapy water solution onto the tread as the tire is turned slowly – leaks will cause small bubbles to appear

5.8  To extend the life of the tires, check the air pressure at least once a week with an accurate gauge (don't forget the spare!)

6.4  On Integra models, the power steering fluid reservoir is translucent so the fluid level can be checked without removing the cap – keep the level between the two lines

6.5a  On Legend models the power steering fluid cap has a dipstick attached – twist the cap in the appropriate direction when removing or installing

(see illustration). The soapy solution will bubble if there is a leak. Unless the puncture is inordinately large, a tire shop or gas station can usually repair the punctured tire.

5  Carefully inspect the inboard sidewall of each tire for evidence of brake fluid leakage. If you see any, inspect the brakes immediately.

6  Correct tire air pressure adds miles to the lifespan of the tires, improves mileage and enhances overall ride quality. Tire pressure cannot be accurately estimated by looking at a tire, particularly if it is a radial. A tire pressure gauge is therefore essential. Keep an accurate gauge in the glovebox. The pressure gauges fitted to the nozzles of air hoses at gas stations are often inaccurate.

7  Always check tire pressure when the tires are cold. "Cold," in this case, means the vehicle has not been driven over a mile in the three hours preceding a tire pressure check. A pressure rise of four to eight pounds is not uncommon once the tires are warm.

8  Unscrew the valve cap protruding from the wheel or hubcap and push the gauge firmly onto the valve (see illustration). Note the reading on the gauge and compare this figure to the recommended tire pressure shown on the tire placard on the left door jamb. Be sure to reinstall the valve cap to keep dirt and moisture out of the valve stem mechanism. Check all four tires and, if necessary, add enough air to bring them up to the recommended pressure levels.

9  Don't forget to keep the spare tire inflated to the specified pressure (consult your owner's manual). Note that the air pressure specified for the compact spare is significantly higher than the pressure of the regular tires.

## 6  Power steering fluid level check

*Refer to illustrations 6.4, 6.5a and 6.5b*

1  Unlike manual steering, the power steering system relies on fluid which may, over a period of time, require replenishing.

2  The fluid reservoir for the power steering pump is located on the inner fender panel near the left (Integra) or right (Legend) front of the engine compartment.

3  For the check, the front wheels should be pointed straight ahead and the engine should be off. The fluid should be cold when checking the Integra power steering fluid level while the Legend can be checked with the fluid hot or cold.

# Chapter 1  Tune-up and routine maintenance

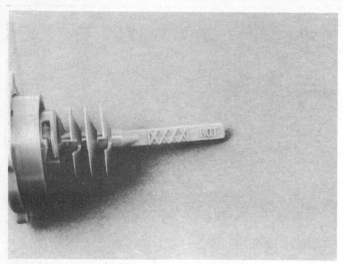

6.5b  The dipstick is marked on both sides so the fluid can be checked hot or cold – the level should be in the crosshatched area

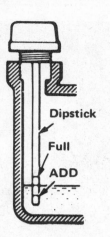

7.3a  On Integra models, the dipstick screws into the transaxle case on the passenger's side of the engine compartment

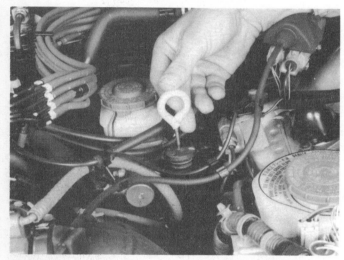

7.3b  On Legend models, the automatic transaxle dipstick inserts into the case on the driver's side of the engine compartment

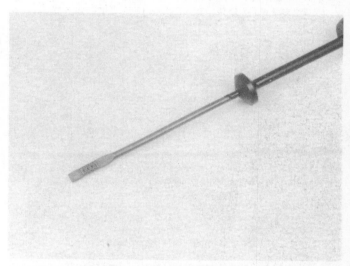

7.5  The automatic transaxle fluid level should be in the cross-hatched area on the dipstick

4   On Integra models, the reservoir is translucent plastic and the fluid level can be checked visually **(see illustration)**.
5   On Legend models, the level is checked with a dipstick attached to the inside of the filler cap **(see illustrations)**.
6   If additional fluid is required, pour the specified type directly into the reservoir, using a funnel to prevent spills.
7   If the reservoir requires frequent fluid additions, all power steering hoses, hose connections, the power steering pump and the rack and pinion assembly should be carefully checked for leaks.

## 7  Automatic transaxle fluid level check

*Refer to illustrations 7.3a, 7.3b and 7.5*

1   The level of the automatic transaxle fluid should be carefully maintained. Low fluid level can lead to slipping or loss of drive, while overfilling can cause foaming, loss of fluid and transaxle damage.
2   The transaxle fluid level should only be checked when the engine is off.
3   Remove the dipstick **(see illustrations)**. Check the level of the fluid on the dipstick and note its condition.

4   Wipe the fluid from the dipstick with a clean rag and reinsert it, but don't screw it in.
5   Pull the dipstick out again and note the fluid level **(see illustration)**. The level should be between the upper and lower marks on the dipstick. If the level is low, add the specified automatic transmission fluid through the dipstick opening with a funnel.
6   Add just enough of the specified fluid to fill the transaxle to the proper level. It takes about one pint to raise the level from the upper mark to the lower mark, so add the fluid a little at a time and keep checking the level until it is correct.
7   The condition of the fluid should also be checked along with the level. If the fluid at the end of the dipstick is black or a dark reddish brown color, or if it emits a burned smell, the fluid should be changed (see Section 28). If you are in doubt about the condition of the fluid, purchase some new fluid and compare the two for color and smell.

## 8  Engine oil and oil filter change

*Refer to illustrations 8.2, 8.7, 8.12, 8.14, 8.15 and 8.19*

1   Frequent oil changes are the best preventive maintenance the home

# Chapter 1 Tune-up and routine maintenance

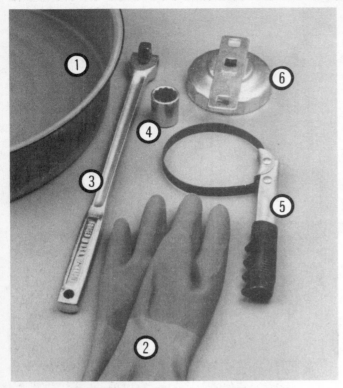

**8.2 These tools are required when changing the engine oil and filter**

1. **Drain pan** – It should be fairly shallow in depth, but wide to prevent spills
2. **Rubber gloves** – When removing the drain plug and filter, you will get oil on your hands (the gloves will prevent burns)
3. **Breaker bar** – Sometimes the oil drain plug is tight and a long breaker bar is needed to loosen it
4. **Socket** – To be used with the breaker bar or a ratchet (must be the correct size to fit the drain plug – six-point preferred)
5. **Filter wrench** – This is a metal band-type wrench, which requires clearance around the filter to be effective
6. **Filter wrench** – This type fits on the bottom of the filter and can be turned with a ratchet or breaker bar (different size wrenches are available for different types of filters)

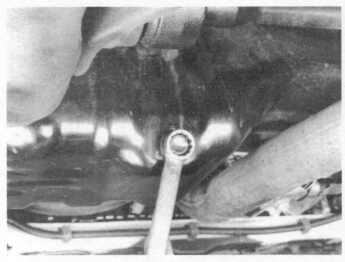

**8.7 Use the proper size box-end wrench or six-point socket to remove the oil drain plug without rounding off the corners**

**8.12 The oil filter is usually on very tight and will require a special wrench for removal – DO NOT use the wrench to tighten the new filter**

mechanic can give the engine, because aging oil becomes diluted and contaminated, which leads to premature engine wear.

2  Make sure you have all the necessary tools before you begin this procedure **(see illustration)**. You should also have plenty of rags or newspapers handy for mopping up any spills.

3  Access to the underside of the vehicle is greatly improved if the vehicle can be lifted on a hoist, driven onto ramps or supported by jackstands. **Warning:** *Do not work under a vehicle which is supported only by a bumper, hydraulic or scissors-type jack.*

4  If this is your first oil change, get under the vehicle and familiarize yourself with the locations of the oil drain plug and the oil filter. The engine and exhaust components will be warm during the actual work, so try to anticipate any potential problems before the engine and accessories are hot.

5  Park the vehicle on a level spot. Start the engine and allow it to reach its normal operating temperature. Warm oil and sludge will flow out more easily. Turn off the engine when it's warmed up. Remove the filler cap from the camshaft cover.

6  Raise the vehicle and support it securely on jackstands. **Warning:** *To avoid personal injury, never get beneath the vehicle when it is supported by only by a jack. The jack provided with your vehicle is designed solely for raising the vehicle to remove and replace the wheels. Always use jackstands to support the vehicle when it becomes necessary to place your body underneath the vehicle.*

7  Being careful not to touch the hot exhaust components, place the drain pan under the drain plug in the bottom of the pan and remove the plug **(see illustration)**. You may want to wear gloves while unscrewing the plug the final few turns if the engine is hot.

8  Allow the old oil to drain into the pan. It may be necessary to move the pan farther under the engine as the oil flow slows to a trickle. Inspect the old oil for the presence of metal shavings and chips.

9  After all the oil has drained, wipe off the drain plug with a clean rag. Even minute metal particles clinging to the plug would immediately contaminate the new oil.

10  Clean the area around the drain plug opening, reinstall the plug and tighten it securely, but do not strip the threads.

11  Move the drain pan into position under the oil filter.

## Integra and 1988 and later Legend models

12  Loosen the oil filter **(see illustration)** by turning it counterclockwise with the filter wrench. Any standard filter wrench will work. Sometimes the

# Chapter 1  Tune-up and routine maintenance

8.14  Lubricate the oil filter gasket with clean engine oil before installing the filter on the engine

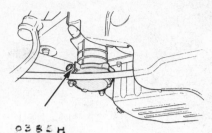

8.15  Remove the oil filter housing drain bolt (arrow) with a socket or box-end wrench

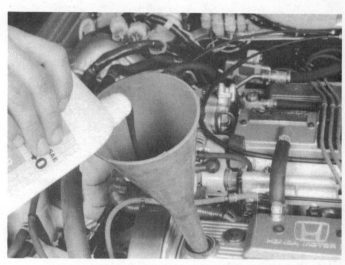

8.19  By using a funnel you'll avoid spilling oil on top of the engine

oil filter is screwed on so tightly that it cannot be loosened. If this situation occurs, punch a metal bar or long screwdriver directly through the side of the canister and use it as a T-bar to turn the filter. Be prepared for oil to spurt out of the canister as it is punctured. Once the filter is loose, use your hands to unscrew it from the block. Just as the filter is detached from the block, immediately tilt the open end up to prevent the oil inside the filter from spilling out. **Warning:** *The engine exhaust manifold may still be hot, so be careful.*

13  With a clean rag, wipe off the mounting surface on the block. If a residue of old oil is allowed to remain, it will smoke when the block is heated up. It will also prevent the new filter from seating properly. Also make sure that the none of the old gasket remains stuck to the mounting surface. It can be removed with a scraper if necessary.

14  Compare the old filter with the new one to make sure they are the same type. Smear some clean engine oil on the rubber gasket of the new filter and screw it into place **(see illustration)**. Because overtightening the filter will damage the gasket, do not use a filter wrench to tighten the filter. Tighten it by hand until the gasket contacts the seating surface. Then seat the filter by giving it an additional 3/4-turn.

### *1986 and 1987 Legend models*

15  Remove the oil filter drain bolt and allow the residual oil to drain out **(see illustration)**.
16  Remove the three cap nuts below the drain bolt and lower the cover, O-ring and filter from the housing.
17  Insert the new filter, then place the cover and O-ring in position and install the cap nuts and the filter drain bolt. Tighten the cap nuts and drain bolt securely.

### *All models*

18  Remove all tools, rags, etc. from under the vehicle, being careful not to spill the oil in the drain pan, then lower the vehicle.
19  Add new oil to the engine through the oil filler cap in the camshaft cover. Use a spout or funnel to prevent oil from spilling onto the top of the engine **(see illustration)**. Pour three quarts of fresh oil into the engine. Wait a few minutes to allow the oil to drain into the pan, then check the level on the oil dipstick (see Section 4 if necessary). If the oil level is at or near the upper hole on the dipstick, install the filler cap hand tight, start the engine and allow the new oil to circulate.
20  Allow the engine to run for about a minute. While the engine is running, look under the vehicle and check for leaks at the oil pan drain plug and around the oil filter. If either is leaking, stop the engine and tighten the plug or filter or, on the early Legend model, the filter cap nuts slightly.

21  Wait a few minutes to allow the oil to trickle down into the pan, then recheck the level on the dipstick and, if necessary, add enough oil to bring the level to the upper hole.
22  During the first few trips after an oil change, make it a point to check frequently for leaks and proper oil level.
23  The old oil drained from the engine cannot be reused in its present state and should be discarded. Oil reclamation centers, auto repair shops and gas stations will normally accept the oil, which can be refined and used again. After the oil has cooled, it can be drained into a suitable container (capped plastic jugs, topped bottles, milk cartons, etc.) for transport to one of these disposal sites.

## 9  Windshield wiper blade inspection and replacement

*Refer to illustrations 9.6, 9.7 and 9.8*

1  The windshield wiper and blade assembly should be inspected periodically for damage, loose components and cracked or worn blade elements.
2  Road film can build up on the wiper blades and affect their efficiency, so they should be washed regularly with a mild detergent solution.
3  The action of the wiping mechanism can loosen bolts, nuts and fasteners, so they should be checked and tightened, as necessary, at the same time the wiper blades are checked.
4  If the wiper blade elements are cracked, worn or warped, or no longer clean adequately, they should be replaced with new ones.
5  Lift the arm assembly away from the glass for clearance.

# Chapter 1  Tune-up and routine maintenance

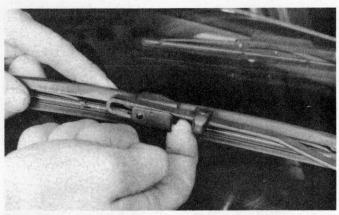

**9.6  Press in on the lock tab and push the blade assembly out of the hook at the end to remove it**

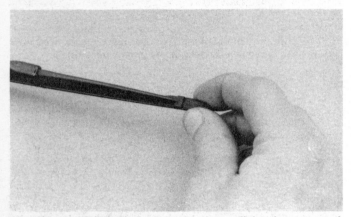

**9.7  Squeeze the blade element tabs, then pull the element out of the metal frame and remove it**

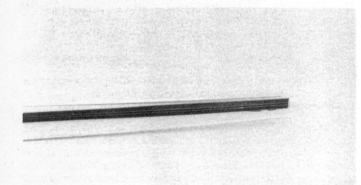

**9.8  The metal retainers must be inserted into the slots in the rubber element before installation**

6   Press in on the lock tab and push the blade assembly down the wiper arm, out of the hook at the end **(see illustration)**.
7   Squeeze the blade element tabs tightly and pull the element out of the metal frame **(see illustration)**.
8   Remove the metal retainers from the element and install them in the new element **(see illustration)**.
9   Insert the element into the frame and push it until the element tabs lock.
10  Place the metal arm assembly in the hook on the wiper arm and press it into place until the lock tab snaps into place.

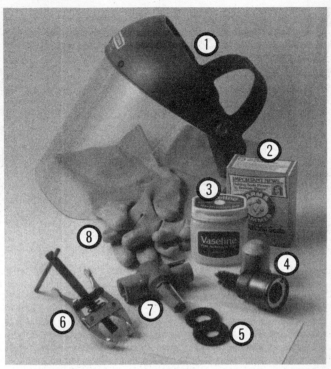

**10.1  Tools and materials required for battery maintenance**

1   **Face shield/safety goggles** – When removing corrosion with a brush, the acidic particles can easily fly up into your eyes
2   **Baking soda** – A solution of baking soda and water can be used to neutralize corrosion
3   **Petroleum jelly** – A layer of this on the battery posts will help prevent corrosion
4   **Battery post/cable cleaner** – This wire brush cleaning tool will remove all traces of corrosion from the battery posts and cable clamps
5   **Treated felt washers** – Placing one of these on each post, directly under the cable clamps, will help prevent corrosion
6   **Puller** – Sometimes the cable clamps are very difficult to pull off the posts, even after the nut/bolt has been completely loosened. This tool pulls the clamp straight up and off the post without damage.
7   **Battery post/cable cleaner** – Here is another cleaning tool which is a slightly different version of number 4 above, but it does the same thing
8   **Rubber gloves** – Another safety item to consider when servicing the battery; remember that's acid inside the battery!

## 10  Battery check, maintenance and charging

*Refer to illustrations 10.1, 10.6a, 10.6b, 10.7a and 10.7b*

### Check and maintenance

1   A routine preventive maintenance program for the battery in your vehicle is the only way to ensure quick and reliable starts. But before performing any battery maintenance, make sure that you have the proper equipment necessary to work safely around the battery **(see illustration)**.
2   There are also several precautions that should be taken whenever battery maintenance is performed. Before servicing the battery, always turn the engine and all accessories off and disconnect the cable from the negative terminal of the battery.
3   The battery produces hydrogen gas, which is both flammable and explosive. Never create a spark, smoke or light a match around the battery.

# Chapter 1  Tune-up and routine maintenance

10.6a  Battery terminal corrosion usually appears as light, fluffy powder

10.6b  Removing the cable from a battery post with a wrench – sometimes a special battery pliers is required for this procedure if corrosion has caused deterioration of the nut hex (always remove the negative cable first and hook it up last!)

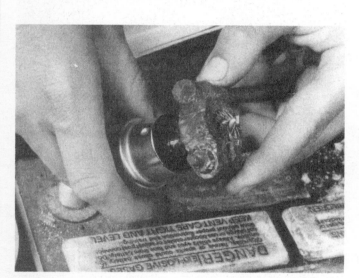

10.7a  When cleaning the cable clamps, all corrosion must be removed (the inside of the clamp is tapered to match the taper on the post, so don't remove too much material)

10.7b  Regardless of the type of tool used on the battery posts, a clean, shiny surface should be the result

Always charge the battery in a ventilated area.

4   Electrolyte contains poisonous and corrosive sulfuric acid. Do not allow it to get in your eyes, on your skin on on your clothes. Never ingest it. Wear protective safety glasses when working near the battery. Keep children away from the battery.

5   Note the external condition of the battery. If the positive terminal and cable clamp on your vehicle's battery is equipped with a rubber protector, make sure that it's not torn or damaged. It should completely cover the terminal. Look for any corroded or loose connections, cracks in the case or cover or loose hold-down clamps. Also check the entire length of each cable for cracks and frayed conductors.

6   If corrosion, which looks like white, fluffy deposits (**see illustration**) is evident, particularly around the terminals, the battery should be removed for cleaning. Loosen the cable clamp nuts with a wrench, being careful to remove the negative cable first, and slide them off the terminals (**see illustration**). Then disconnect the hold-down clamp nuts, remove the clamp and lift the battery from the engine compartment.

7   Clean the cable clamps thoroughly with a battery brush or a terminal cleaner and a solution of warm water and baking soda (**see illustration**). Wash the terminals and the top of the battery case with the same solution but make sure that the solution doesn't get into the battery. When cleaning the cables, terminals and battery top, wear safety goggles and rubber gloves to prevent any solution from coming in contact with your eyes or hands. Wear old clothes too – even diluted, sulfuric acid splashed onto clothes will burn holes in them. If the terminals have been extensively corroded, clean them up with a terminal cleaner (**see illustration**). Thoroughly wash all cleaned areas with plain water.

8   Before reinstalling the battery in the engine compartment, inspect the plastic battery carrier. If it's dirty or covered with corrosion, remove it and clean it in the same solution of warm water and baking soda. Inspect the metal brackets which support the carrier to make sure that they are not covered with corrosion. If they are, wash them off. If corrosion is extensive, sand the brackets down to bare metal and spray them with a zinc-based primer (available in spray cans at auto paint and body supply stores).

9   Reinstall the battery carrier and the battery back into the engine compartment. Make sure that no parts or wires are laying on the carrier during installation of the battery.

10   Install a pair of specially treated felt washers around the terminals (available at auto parts stores), then coat the terminals and the cable clamps with petroleum jelly or grease to prevent further corrosion. Install

# Chapter 1 Tune-up and routine maintenance

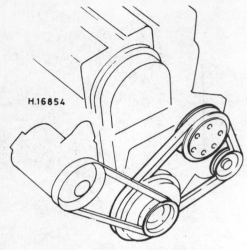

11.2  A typical drivebelt layout

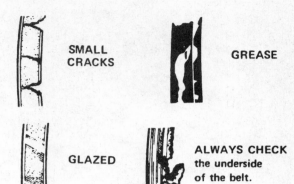

11.3a  Here are some of the more common problems associated with drivebelts (check the belts very carefully to prevent an untimely breakdown)

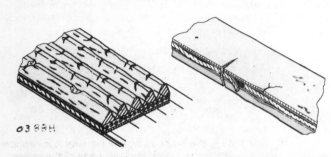

11.3b  Check V-ribbed belts for signs of wear like these – if the belt looks worn, replace it

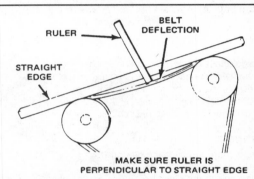

11.4  Measuring drivebelt deflection with a straightedge and ruler

the cable clamps and tighten the nuts, being careful to install the negative cable last.

11  Install the hold-down clamp and nuts. Tighten the nuts only enough to hold the battery firmly in place. Overtightening these nuts can crack the battery case.

## Charging

12  Remove all of the cell caps (if equipped) and cover the holes with a clean cloth to prevent spattering electrolyte. Disconnect the negative battery cable and hook the battery charger leads to the battery posts (positive to positive, negative to negative), then plug in the charger. Make sure it is set at 12 volts if it has a selector switch.

13  If you're using a charger with a rate higher than two amps, check the battery regularly during charging to make sure it doesn't overheat. If you're using a trickle charger, you can safely let the battery charge overnight after you've checked it regularly for the first couple of hours.

14  If the battery has removeable cell caps, measure the specific gravity with a hydrometer every hour during the last few hours of the charging cycle. Hydrometers are available inexpensively from auto parts stores – follow the instructions that come with the hydrometer. Consider the battery charged when there's no change in the specific gravity reading for two hours and the electrolyte in the cells is gassing (bubbling) freely. The specific gravity reading from each cell should be very close to the others. If not, the battery probably has a bad cell(s).

15  Some batteries with sealed tops have built-in hydrometers on the top that indicate the state of charge by the color displayed in the hydrometer window. Normally, a bright-colored hydrometer indicates a full charge and a dark hydrometer indicates the battery still needs charging. Check the battery manufacturer's instructions to be sure you know what the colors mean.

16  If the battery has a sealed top and no built-in hydrometer, you can hook up a digital voltmeter across the battery terminals to check the charge. A fully charged battery should read 12.6 volts or higher.

17  Further information on the battery and jump starting can be found in Chapter 5 and at the front of this manual.

## 11  Drivebelt check, adjustment and replacement

*Refer to illustrations 11.2, 11.3a, 11.3b, 11.4, 11.6, 11.7 and 11.10*

### Check

1  The alternator and air conditioning compressor drivebelts are either V-belts or V-ribbed belts. Sometimes referred to as "fan" belts, the drivebelts are located at the left end of the engine. The good condition and proper adjustment of the alternator belt is critical to the operation of the engine. Because of their composition and the high stresses to which they are subjected, drivebelts stretch and deteriorate as they get older. They must therefore be periodically inspected.

2  The number of belts used on a particular vehicle depends on the accessories installed **(see illustration)**.

3  With the engine off, open the hood and locate the drivebelts at the left or right end of the engine. With a flashlight, check each belt: On V-belts, check for cracks and separation of the belt plies **(see illustration)**. On V-ribbed belts, check for separation of the adhesive rubber on both sides of the core, core separation from the belt side, a severed core, separation of the ribs from the adhesive rubber, cracking or separation of the ribs, and torn or worn ribs or cracks in the inner ridges of the ribs **(see illustration)**. On both belt types, check for fraying and glazing, which gives the belt a shiny appearance. Both sides of the belt should be inspected, which means you will have to twist the belt to check the underside. Use your fingers to feel the belt where you can't see it. If any of the above conditions are evident, replace the belt (go to Step 8).

# Chapter 1  Tune-up and routine maintenance

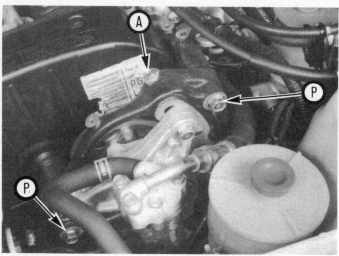

11.6  On Integra models, loosen the pivot bolts (P) and turn the adjusting bolt (A) to rotate the power steering pump – tighten the lower pivot bolt last

11.7  Turn the idler adjusting bolt (arrow) in or out to adjust the drivebelt tension

4  The tightness of each belt is checked by pushing on it at a distance halfway between the pulleys **(see illustration)**. Apply about 10 pounds of force with your thumb and see how much the belt moves down (deflects). Refer to the Specifications section at the beginning of this chapter for the amount of deflection allowed in each belt.

## Adjustment

5  If adjustment is necessary, it is done by moving the belt-driven accessory on the bracket.
6  For some components, there will be an adjusting bolt and a pivot bolt **(see illustration)**. Both must be loosened slightly to enable you to move the component. After the two bolts have been loosened, move the component away from the engine (to tighten the belt) or toward the engine (to loosen the belt). After adjustment, tighten the bolts securely.
7  On some components, loosen the pivot bolt and locknut on the adjusting bolt. Turn the adjusting bolt to tension the belt. On other components, the idler pulley is moved to tension the drivebelt **(see illustration)**.

## Replacement

8  To replace a belt, follow the above procedures for drivebelt adjustment but slip the belt off the crankshaft pulley and remove it. If you are replacing the alternator belt, you will have to remove the air conditioning compressor belt first because of the way they are arranged on the crankshaft pulley. Because of this and because belts tend to wear out more or less together, it is a good idea to replace both belts at the same time. Mark each belt and its appropriate pulley groove so the replacement belts can be installed in their proper positions.
9  Take the old belts to the parts store in order to make a direct comparison for length, width and design.
10  After replacing a V-ribbed drivebelt, make sure it fits properly in the ribbed grooves in the pulleys **(see illustration)**. It is essential that the belt be properly centered.
11  Adjust the belt(s) in accordance with the procedure outlined above.

## 12  Underhood hose check and replacement

**Caution:** *Replacement of air conditioning hoses must be left to a dealer service department or air conditioning shop that has the equipment to depressurize the system safely. Never remove air conditioning components or hoses until the system has been depressurized.*

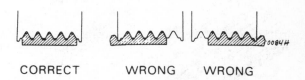

11.10  When installing the V-ribbed belt, make sure it is centered on the pulley – it must not overlap either edge of the pulley

### General

1  High temperatures in the engine compartment can cause the deterioration of the rubber and plastic hoses used for engine, accessory and emission systems operation. Periodic inspection should be made for cracks, loose clamps, material hardening and leaks.
2  Information specific to the cooling system hoses can be found in Section 13.
3  Some, but not all, hoses are secured to the fittings with clamps. Where clamps are used, check to be sure they haven't lost their tension, allowing the hose to leak. If clamps aren't used, make sure the hose has not expanded and/or hardened where it slips over the fitting, allowing it to leak.

### Vacuum hoses

4  It's quite common for vacuum hoses, especially those in the emissions system, to be color coded or identified by colored stripes molded into them. Various systems require hoses with different wall thicknesses, collapse resistance and temperature resistance. When replacing hoses, be sure the new ones are made of the same material.
5  Often the only effective way to check a hose is to remove it completely from the vehicle. If more than one hose is removed, be sure to label the hoses and fittings to ensure correct installation.
6  When checking vacuum hoses, be sure to include any plastic T-fittings in the check. Inspect the fittings for cracks and the hose where it fits over the fitting for distortion, which could cause leakage.
7  A small piece of vacuum hose (1/4-inch inside diameter) can be used as a stethoscope to detect vacuum leaks. Hold one end of the hose to your ear and probe around vacuum hoses and fittings, listening for the "hissing" sound characteristic of a vacuum leak. **Warning:** *When probing with the vacuum hose stethoscope, be very careful not to come into contact with moving engine components such as the drivebelts, cooling fan, etc.*

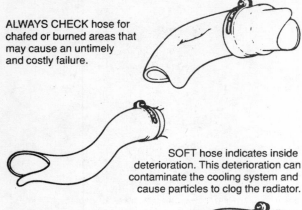

**13.4 Hoses, like drivebelts, have a habit of failing at the worst possible time – to prevent the inconvenience of a blown radiator or heater hose, inspect them carefully as shown here**

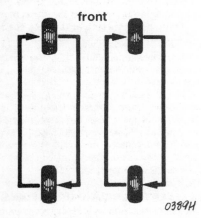

**14.2 The recommended rotation pattern for these models**

## Fuel hose

**Warning:** *There are certain precautions which must be taken when inspecting or servicing fuel system components. Work in a well ventilated area and do not allow open flames (cigarettes, appliance pilot lights, etc.) or bare light bulbs near the work area. Mop up any spills immediately and do not store fuel soaked rags where they could ignite. The fuel system is under pressure, so if any fuel lines are to be disconnected, the pressure in the system must be relieved first (see Chapter 4 for more information).*

8   Check all rubber fuel lines for deterioration and chafing. Check especially for cracks in areas where the hose bends and just before fittings, such as where a hose attaches to the fuel filter.

9   When replacing hose, use only hose that is specifically designed for your fuel injection system.

## Metal lines

10   Sections of metal line are often used for fuel line between the fuel pump and fuel injection unit. Check carefully to be sure the line has not been bent or crimped and that cracks have not started in the line.
11   If a section of metal fuel line must be replaced, only seamless steel tubing should be used, since copper and aluminum tubing don't have the strength necessary to withstand normal engine vibration.
12   Check the metal brake lines where they enter the master cylinder and brake proportioning unit (if used) for cracks in the lines or loose fittings. Any sign of brake fluid leakage calls for an immediate thorough inspection of the brake system.

## 13   Cooling system check

*Refer to illustration 13.4*

1   Many major engine failures can be attributed to a faulty cooling system. If the vehicle is equipped with an automatic transmission, the cooling system also cools the transmission fluid and thus plays an important role in prolonging transmission life.
2   The cooling system should be checked with the engine cold. Do this before the vehicle is driven for the day or after the engine has been shut off for at least three hours.
3   Remove the radiator cap by turning it to the left until it reaches a stop. If you hear a hissing sound (indicating there is still pressure in the system), wait until it stops. Now press down on the cap with the palm of your hand and continue turning to the left until the cap can be removed. Thoroughly clean the cap, inside and out, with clean water. Also clean the filler neck on the radiator. All traces of corrosion should be removed. The coolant inside the radiator should be relatively transparent. If it's rust colored, the system should be drained and refilled (Section 26). If the coolant level isn't up to the top, add additional antifreeze/coolant mixture (see Section 4).
4   Carefully check the large upper and lower radiator hoses along with the smaller diameter heater hoses which run from the engine to the firewall. Inspect each hose along its entire length, replacing any hose which is cracked, swollen or shows signs of deterioration. Cracks may become more apparent if the hose is squeezed **(see illustration)**. Regardless of condition, it's a good idea to replace hoses with new ones every two years.
5   Make sure that all hose connections are tight. A leak in the cooling system will usually show up as white or rust colored deposits on the areas adjoining the leak. If wire-type clamps are used at the ends of the hoses, it may be a good idea to replace them with more secure screw-type clamps.
6   Use compressed air or a soft brush to remove bugs, leaves, etc. from the front of the radiator or air conditioning condenser. Be careful not to damage the delicate cooling fins or cut yourself on them.
7   Every other inspection, or at the first indication of cooling system problems, have the cap and system pressure tested. If you don't have a pressure tester, most gas stations and repair shops will do this for a minimal charge.

## 14   Tire rotation

*Refer to illustration 14.2*

1   The tires should be rotated at the specified intervals and whenever uneven wear is noticed. Since the vehicle will be raised and the tires removed anyway, check the brakes (Section 15) at this time.
2   Radial tires must be rotated in a specific pattern **(see illustration)**.
3   Refer to the information in Jacking and towing at the front of this manual for the proper procedures to follow when raising the vehicle and changing a tire. If the brakes are to be checked, do not apply the parking brake as stated. Make sure the tires are blocked to prevent the vehicle from rolling.
4   Preferably, the entire vehicle should be raised at the same time. This can be done on a hoist or by jacking up each corner and then lowering the vehicle onto jackstands placed under the frame rails. Always use four jackstands and make sure the vehicle is firmly supported.

# Chapter 1  Tune-up and routine maintenance

15.6  You will find an inspection hole like this in each caliper – placing a steel ruler across the hole should enable you to determine the thickness of the remaining pad material

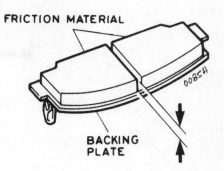

15.9  If a more precise measurement of pad thickness is necessary, remove the pads and measure the remaining friction material – spraying the pad with brake cleaner will help you determine where the friction material ends and the steel backing plate begins

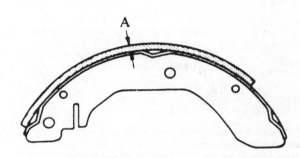

15.14  If the lining is bonded to the brake shoe, measure the lining thickness from the outer surface to the metal shoe, as shown here; if the lining is riveted to the shoe, measure from the lining outer surface to the rivet head

5  After rotation, check and adjust the tire pressures as necessary and be sure to check the lug nut tightness.
6  For further information on the wheels and tires, refer to Chapter 10.

## 15  Brake check

*Refer to illustrations 15.6, 15.9 and 15.14*
**Note:** *For detailed photographs of the brake system, refer to Chapter 9.*
1  In addition to the specified intervals, the brakes should be inspected every time the wheels are removed or whenever a defect is suspected. Any of the following symptoms could indicate a potential brake system defect: The vehicle pulls to one side when the brake pedal is depressed; the brakes make squealing or dragging noises when applied; brake travel is excessive; the pedal pulsates; brake fluid leaks, usually onto the inside of the tire or wheel.
2  The disc brake pads have built-in wear indicators which should make a high pitched squealing or scraping noise when they are worn to the replacement point. When you hear this noise, replace the pads immediately or expensive damage to the rotors can result.
3  Loosen the wheel lug nuts.
4  Raise the vehicle and place it securely on jackstands.
5  Remove the wheels (see Jacking and towing at the front of this book, or your owner's manual, if necessary).

### Disc brakes (front and rear on most models)
6  There are two pads – an outer and an inner – in each caliper. The pads are visible through an inspection hole in each caliper **(see illustration)**.
7  Check the pad thickness by looking at each end of the caliper and through the inspection hole in the caliper body. If the lining material is less than the specified thickness (see this Chapter's Specifications), replace the pads. **Note:** *Keep in mind that the lining material is riveted or bonded to a metal backing plate and the metal portion is not included in this measurement.*
8  If it is difficult to determine the exact thickness of the remaining pad material by the above method, or if you are at all concerned about the condition of the pads, remove the caliper(s), then remove the pads from the calipers for further inspection (see Chapter 9).
9  Once the pads are removed from the calipers, clean them with brake cleaner and remeasure them with a small steel pocket ruler or a vernier caliper **(see illustration)**.
10  Measure the disc rotor thickness with a micrometer to make sure that it still has service life remaining. If any disc is thinner than the specified minimum thickness, replace it (see Chapter 9). Even if the rotor has service life remaining, check its condition. Look for scoring, gouging and burned spots. If these conditions exist, remove the rotor and have it resurfaced (see Chapter 9).
11  Before installing the wheels, check all brake lines and hoses for damage, wear, deformation, cracks, corrosion, leakage, bends and twists, particularly in the vicinity of the rubber hoses at the calipers. Check the clamps for tightness and the connections for leakage. Make sure all hoses and lines are clear of sharp edges, moving parts and the exhaust system. If any of the above conditions are noted, repair, reroute or replace the lines and/or fittings as necessary (see Chapter 9).

### Rear drum brakes (some Integra models)
12  Refer to Chapter 9 and remove the rear brake drums.
13  **Warning:** *Brake dust produced by lining wear and deposited on brake components contains asbestos, which is hazardous to your health. DO NOT blow it out with compressed air and DO NOT inhale it! DO NOT use gasoline or solvents to remove the dust. Brake system cleaner should be used to flush the dust into a drain pan. After the brake components are wiped clean with a damp rag, dispose of the contaminated rag(s) and solvent in a covered and labelled container. Try to use non-asbestos replacement parts whenever possible.*
14  Note the thickness of the lining material on the rear brake shoes **(see illustration)** and look for signs of contamination by brake fluid and grease. If the lining material is within 3/32-inch of the recessed rivets or metal shoes, replace the brake shoes with new ones. The shoes should also be replaced if they are cracked, glazed (shiny lining surfaces) or contaminated with brake fluid or grease. See Chapter 9 for the replacement procedure.

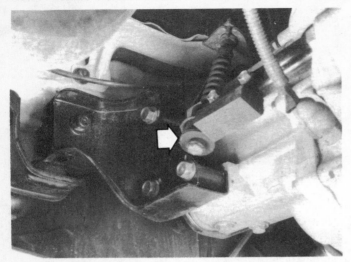

16.2 Move the clutch release lever arm up and down and measure the freeplay (view is from beneath the passenger's side of the vehicle)

16.3 Turn the knurled knob (arrow) (it is easier to reach from the engine compartment) to adjust the clutch release arm freeplay

17.2a Integra air cleaner housing cover screws (arrows)

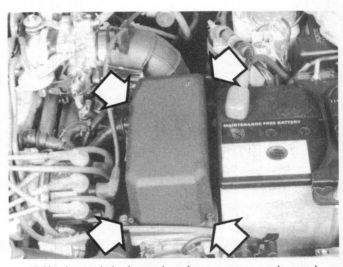

17.2b Legend air cleaner housing cover screws (arrows)

17.5a On Integra models, lift the cover and remove the filter

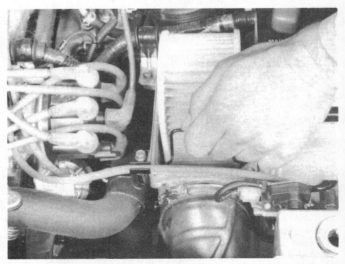

17.5b On Legend models, pull the retaining spring up to release the filter . . .

# Chapter 1  Tune-up and routine maintenance

1 – 25

17.5c  . . . then lift the filter element out of the housing

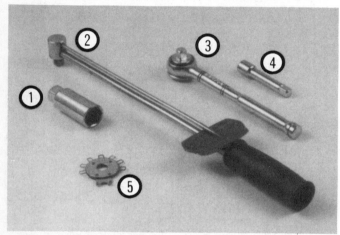

18.1  Tools required for changing spark plugs

1  **Spark plug socket** – This will have special padding inside to protect the spark plug's porcelain insulator
2  **Torque wrench** – Although not mandatory, using this tool is the best way to ensure the plugs are tightened properly
3  **Ratchet** – Standard hand tool to fit the spark plug socket
4  **Extension** – Depending on model and accessories, you may need special extensions and universal joints to reach one or more of the plugs
5  **Spark plug gap gauge** – This gauge for checking the gap comes in a variety of styles. Make sure the gap for your engine is included.

15  Check the shoe return and hold-down springs and the adjusting mechanism to make sure they're installed correctly and in good condition. Deteriorated or distorted springs, if not replaced, could allow the linings to drag and wear prematurely.
16  Check the wheel cylinders for leakage by carefully peeling back the rubber boots. If brake fluid is noted behind the boots, the wheel cylinders must be replaced (see Chapter 9).
17  Check the drums for cracks, score marks, deep scratches and hard spots, which will appear as small discolored areas. If imperfections cannot be removed with emery cloth, the drums must be resurfaced by an automotive machine shop (see Chapter 9 for more detailed information).
18  Refer to Chapter 9 and install the brake drums.
19  Install the wheels and snug the wheel lug nuts finger tight.
20  Remove the jackstands and lower the vehicle.
21  Tighten the wheel lug nuts to the torque listed in this Chapter's Specifications.

### Brake booster check

22  Sit in the driver's seat and perform the following sequence of tests.
23  With the engine stopped, depress the brake pedal several times - the travel distance should not change.
24  With the brake fully depressed, start the engine – the pedal should move down a little when the engine starts.
25  Depress the brake, stop the engine and hold the pedal in for about 30 seconds – the pedal should neither sink nor rise.
26  Restart the engine, run it for about a minute and turn it off. Then firmly depress the brake several times – the pedal travel should decrease with each application.
27  If your brakes do not operate as described above when the preceding tests are performed, the brake booster is either in need of repair or has failed. Refer to Chapter 9 for the removal procedure.

Parking brake
28  Slowly pull up on the parking brake and count the number of clicks you hear until the handle is up as far as it will go. The adjustment is correct if you hear the specified number of clicks (see this Chapter's Specifications). If you hear more or fewer clicks, it's time to adjust the parking brake (see Chapter 9).
29  An alternative method of checking the parking brake is to park the vehicle on a steep hill with the parking brake set and the transmission in Neutral. If the parking brake cannot prevent the vehicle from rolling, it is in need of adjustment (see Chapter 9).

## 16  Clutch release arm freeplay check and adjustment (Integra models only)

*Refer to illustrations 16.2 and 16.3*
1  Symptoms of incorrect clutch release arm freeplay are difficulty in shifting and achieving a smooth takeoff from a standing start. Raise the vehicle and support it securely on jackstands.
2  Move the clutch release arm up and down and measure the freeplay **(see illustration)**.
3  If the freeplay is not as listed in this Chapter's Specifications, turn the knurled knob at the top of the bracket to adjust the freeplay **(see illustration)**. Operate the clutch several times and recheck the freeplay, adjusting as necessary.

## 17  Air filter replacement

*Refer to illustrations 17.2a, 17.2b, 17.5a, 17.5b and 17.5c*
1  At the specified intervals, the air filter should be replaced with a new one.
2  Loosen the cleaner housing cover screws **(see illustrations)**.
3  Lift the cover up.
4  While the air cleaner cover is off, be careful not to drop anything down into the air cleaner assembly.
5  Lift the air filter element out of the housing and wipe out the inside of the air cleaner housing with a clean rag see **(see illustrations)**.
6  Place the new filter in the air cleaner housing. Make sure it seats properly in the bottom of the housing. On Legend models insert the spring as the filter is lowered into place.
7  Install the air cleaner cover and tighten the screws securely.

## 18  Spark plug check and replacement

*Refer to illustrations 18.1, 18.4a, 18.4b, 18.6, 18.8 and 18.10*
1  Spark plug replacement requires a spark plug socket which fits onto a ratchet wrench. This socket is lined with a rubber grommet to protect the porcelain insulator of the spark plug and to hold the plug while you insert it into the spark plug hole. You will also need a wire-type feeler gauge to check and adjust the spark plug gap and a torque wrench to tighten the new plugs to the specified torque **(see illustration)**.
2  If you are replacing the plugs, purchase the new plugs, adjust them to the proper gap and then replace each plug one at a time. **Note:** *When buying new spark plugs, it's essential that you obtain the correct plugs for your*

# Chapter 1 Tune-up and routine maintenance

18.4a Spark plug manufacturers recommend using a wire type gauge when checking the gap – if the wire does not slide between the electrodes with a slight drag, adjustment is required

18.4b To change the gap, bend the *side* electrode only, as indicated by the arrows, and be very careful not to crack or chip the porcelain insulator surrounding the center electrode

18.6 When removing the spark plug wires, pull only on the boot and use a twisting/pulling motion

18.8 Because they are deeply recessed, an extension will be required when removing or installing the spark plugs

specific vehicle. This information can be found on the Vehicle Emissions Control Information (VECI) label located on the underside of the hood or in the owner's manual. If these two sources specify different plugs, purchase the spark plug type specified on the VECI label because that information is provided specifically for your engine.

3   Inspect each of the new plugs for defects. If there are any signs of cracks in the porcelain insulator of a plug, don't use it.

4   Check the electrode gaps of the new plugs. Check the gap by inserting the wire gauge of the proper thickness between the electrodes at the tip of the plug **(see illustration)**. The gap between the electrodes should be identical to that specified on the VECI label. If the gap is incorrect, use the notched adjuster on the feeler gauge body to bend the curved side electrode slightly **(see illustration)**.

5   If the side electrode is not exactly over the center electrode, use the notched adjuster to align them. **Caution:** *If the gap of a new plug must be adjusted, bend only the base of the ground electrode. Do not touch the tip.*

### Removal

6   To prevent the possibility of mixing up spark plug wires, work on one spark plug at a time. Remove the wire and boot from one spark plug. Grasp the boot – not the cable – as shown, give it a half twisting motion and pull straight out **(see illustration)**.

7   If compressed air is available, blow any dirt or foreign material away from the spark plug area before proceeding (a common bicycle pump will also work).

8   Remove the spark plug **(see illustration)**.

9   Whether you are replacing the plugs at this time or intend to reuse the old plugs, compare each old spark plug with those shown in the accompanying photos to determine the overall running condition of the engine.

### Installation

10   It's often difficult to insert spark plugs into their holes without cross-threading them. To avoid this possibility, fit a short piece of 3/16-inch ID rubber hose over the end of the spark plug **(see illustration)**. The flexible hose acts as a universal joint to help align the plug with the plug hole. Should the plug begin to cross-thread, the hose will slip on the spark plug, preventing thread damage. Tighten the plug securely.

11   Attach the plug wire to the new spark plug, again using a twisting motion on the boot until it is firmly seated on the end of the spark plug.

12   Follow the above procedure for the remaining spark plugs, replacing them one at a time to prevent mixing up the spark plug wires.

# Chapter 1 Tune-up and routine maintenance

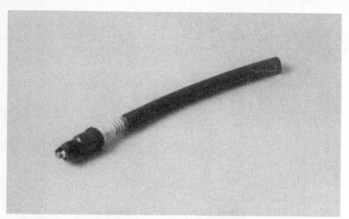

**18.10 A length of 3/16-inch ID rubber hose will save time and prevent damaged threads when installing the spark plugs**

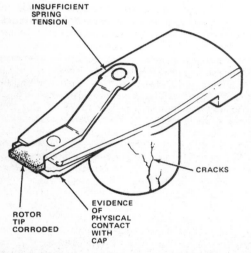

**19.12 The ignition rotor should be checked for wear and corrosion as indicated here (if in doubt about its condition, buy a new one)**

## 19 Spark plug wire, distributor cap and rotor check and replacement

*Refer to illustrations 19.11 and 19.12*

1  The spark plug wires should be checked whenever new spark plugs are installed.
2  Begin this procedure by making a visual check of the spark plug wires while the engine is running. In a darkened garage (make sure there is ventilation) start the engine and observe each plug wire. Be careful not to come into contact with any moving engine parts. If there is a break in the wire, you will see arcing or a small spark at the damaged area. If arcing is noticed, make a note to obtain new wires, then allow the engine to cool and check the distributor cap and rotor.
3  The spark plug wires should be inspected one at a time to prevent mixing up the order, which is essential for proper engine operation. Each original plug wire should be numbered to help identify its location. If the number is illegible, a piece of tape can be marked with the correct number and wrapped around the plug wire.
4  Disconnect the plug wire from the spark plug. A removal tool can be used for this purpose or you can grasp the rubber boot, twist the boot half a turn and pull the boot free. Do not pull on the wire itself.
5  Check inside the boot for corrosion, which will look like a white crusty powder.

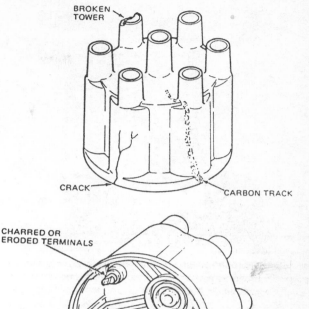

**19.11 Shown here are some of the common defects to look for when inspecting the distributor cap (if in doubt about its condition, install a new one)**

6  Push the wire and boot back onto the end of the spark plug. It should fit tightly onto the end of the plug. If it doesn't, remove the wire and use pliers to carefully crimp the metal connector inside the wire boot until the fit is snug.
7  Using a clean rag, wipe the entire length of the wire to remove built-up dirt and grease. Once the wire is clean, check for burns, cracks and other damage. Do not bend the wire sharply, because the conductor might break.
8  Disconnect the wire from the distributor. Again, pull only on the rubber boot. Check for corrosion and a tight fit. Replace the wire in the distributor.
9  Inspect the remaining spark plug wires, making sure that each one is securely fastened at the distributor and spark plug when the check is complete.
10  If new spark plug wires are required, purchase a set for your specific engine model. Pre-cut wire sets with the boots already installed are available. Remove and replace the wires one at a time to avoid mix-ups in the firing order.
11  Detach the distributor cap by removing the two cap retaining bolts. Look inside it for cracks, carbon tracks and worn, burned or loose contacts **(see illustration)**.
12  Pull the rotor off the distributor shaft (on Legend models remove the retaining screw first) and examine it for cracks and carbon tracks **(see illustration)**. Replace the cap and rotor if any damage or defects are noted.
13  It is common practice to install a new cap and rotor whenever new spark plug wires are installed, but if you wish to continue using the old cap, check the resistance between the spark plug wires and the cap first. If the indicated resistance is more than the specified maximum value (see this Chapter's Specifications), replace the cap and/or wires.
14  When installing a new cap, remove the wires from the old cap one at a time and attach them to the new cap in the exact same location – do not simultaneously remove all the wires from the old cap or firing order mix-ups may occur.

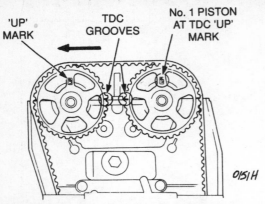

**20.4 The sprockets should be positioned as shown when the number 1 piston is at TDC**

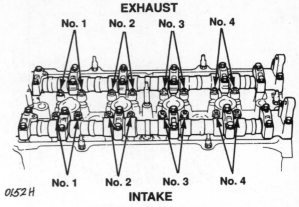

**20.5 Integra valve layout**

## 20 Valve clearance check and adjustment (Integra models only)

*Refer to illustrations 20.4 and 20.5*

1  The valve clearances are checked and adjusted with the engine cold.
2  Remove the air cleaner assembly (see Chapter 4).
3  Remove the camshaft cover (see Chapter 2A).
4  Place the number one piston (closest to the drivebelt end of the engine) at Top Dead Center (TDC) on the compression stroke. This is accomplished by rotating the crankshaft pulley counterclockwise until the UP marks on the backs of the camshaft sprockets are at their uppermost point and the TDC grooves are facing each other **(see illustration)**. The distributor rotor should be pointing toward the number one spark plug wire.
5  With the engine in this position, the number one cylinder valve adjustment can be checked and adjusted **(see illustration)**.
6  Start with the intake valve clearance. Insert a feeler gauge of the specified thickness (see the Specifications) between an intake camshaft lobe and its rocker arm. Withdraw it; you should feel a slight drag. If there's no drag or a heavy drag, loosen the adjuster nut and back off the adjuster screw. Carefully tighten the adjuster screw until you can feel a slight drag on the feeler gauge as you withdraw it.
7  Hold the adjuster screw with a screwdriver (to keep it from turning) and tighten the locknut. Recheck the clearance to make sure it hasn't changed. Repeat the procedure in this Step and the previous Step on the other intake valve, then on the two exhaust valves.
8  Rotate the engine 180-degrees counterclockwise (the camshaft pulley will turn 90-degrees) until the number 3 cylinder is at TDC. With the number 3 cylinder at TDC, the UP marks on the sprockets should be at the exhaust side and the distributor rotor should point at the number 3 spark plug wire. Check and adjust the number 3 cylinder valves.
9  Rotate the engine 180-degrees counterclockwise until the number 4 cylinder is at TDC. With the number 4 cylinder at TDC, the UP marks on the sprockets should be pointed straight down. The distributor rotor should point at the number 4 spark plug wire. Check and adjust the number 4 cylinder valves.
10  Rotate the engine 180-degrees counterclockwise to bring the number 2 cylinder to TDC. The UP marks on the sprockets should be on the intake side. The distributor rotor should point at the number 2 spark plug wire. Check and adjust the number 2 cylinder valves.
11  Install the camshaft cover and the air cleaner assembly.

## 21 Fuel system check

**Warning:** *Certain precautions should be observed when inspecting or servicing the fuel system components. Work in a well ventilated area and do not allow open flames (cigarettes, appliance pilot lights, etc.) near the work area. Mop up spills immediately and do not store fuel-soaked rags where they could ignite. It is a good idea to keep a dry chemical (Class B) fire extinguisher near the work area any time the fuel system is being serviced.*

1  If you smell gasoline while driving or after the vehicle has been sitting in the sun, inspect the fuel system immediately.
2  Remove the gas filler cap and inspect it for damage and corrosion. The gasket should have an unbroken sealing imprint. If the gasket is damaged or corroded, remove it and install a new one.
3  Inspect the fuel feed and return lines for cracks. Make sure all fuel line connections are tight. **Warning:** *It is necessary to relieve the fuel system pressure before servicing fuel system components. The correct procedures for fuel system pressure relief are outlined in Chapter 4.*
4  Since some components of the fuel system – the fuel tank and part of the fuel feed and return lines, for example – are underneath the vehicle, they can be inspected more easily with the vehicle raised on a hoist. If that's not possible, raise the vehicle and secure it on jackstands.
5  With the vehicle raised and safely supported, inspect the gas tank and filler neck for punctures, cracks and other damage. The connection between the filler neck and the tank is particularly critical. Sometimes a rubber filler neck will leak because of loose clamps or deteriorated rubber. These are problems a home mechanic can usually rectify. **Warning:** *Do not, under any circumstances, try to repair a fuel tank (except rubber components). A welding torch or any open flame can easily cause fuel vapors inside the tank to explode.*
6  Carefully check all rubber hoses and metal lines leading away from the fuel tank. Check for loose connections, deteriorated hoses, crimped lines and other damage. Carefully inspect the lines from the tank to the fuel injection system. Repair or replace damaged sections as necessary.

## 22 Manual transaxle lubricant level check

*Refer to illustration 22.1*

1  The manual transaxle does not have a dipstick. To check the fluid level, raise the vehicle and support it securely on jackstands. The fill plug is on the side of the transaxle housing **(see illustration)**. Remove it. If the lubricant level is correct, it should be up to the lower edge of the hole.
2  If the transaxle needs more lubricant (if the level is not up to the hole), use a syringe to add more. Stop filling the transaxle when the lubricant begins to run out the hole.
3  Install the plug and tighten it securely. Drive the vehicle a short distance, then check for leaks.

## 23 Steering and suspension check

*Refer to illustrations 23.8 and 23.9*

**Note:** *For detailed illustrations of the steering and suspension components, refer to Chapter 10.*

# Common spark plug conditions

### NORMAL
**Symptoms:** Brown to grayish-tan color and slight electrode wear. Correct heat range for engine and operating conditions.
**Recommendation:** When new spark plugs are installed, replace with plugs of the same heat range.

### WORN
**Symptoms:** Rounded electrodes with a small amount of deposits on the firing end. Normal color. Causes hard starting in damp or cold weather and poor fuel economy.
**Recommendation:** Plugs have been left in the engine too long. Replace with new plugs of the same heat range. Follow the recommended maintenance schedule.

### CARBON DEPOSITS
**Symptoms:** Dry sooty deposits indicate a rich mixture or weak ignition. Causes misfiring, hard starting and hesitation.
**Recommendation:** Make sure the plug has the correct heat range. Check for a clogged air filter or problem in the fuel system or engine management system. Also check for ignition system problems.

### ASH DEPOSITS
**Symptoms:** Light brown deposits encrusted on the side or center electrodes or both. Derived from oil and/or fuel additives. Excessive amounts may mask the spark, causing misfiring and hesitation during acceleration.
**Recommendation:** If excessive deposits accumulate over a short time or low mileage, install new valve guide seals to prevent seepage of oil into the combustion chambers. Also try changing gasoline brands.

### OIL DEPOSITS
**Symptoms:** Oily coating caused by poor oil control. Oil is leaking past worn valve guides or piston rings into the combustion chamber. Causes hard starting, misfiring and hesitation.
**Recommendation:** Correct the mechanical condition with necessary repairs and install new plugs.

### GAP BRIDGING
**Symptoms:** Combustion deposits lodge between the electrodes. Heavy deposits accumulate and bridge the electrode gap. The plug ceases to fire, resulting in a dead cylinder.
**Recommendation:** Locate the faulty plug and remove the deposits from between the electrodes.

### TOO HOT
**Symptoms:** Blistered, white insulator, eroded electrode and absence of deposits. Results in shortened plug life.
**Recommendation:** Check for the correct plug heat range, over-advanced ignition timing, lean fuel mixture, intake manifold vacuum leaks, sticking valves and insufficient engine cooling.

### PREIGNITION
**Symptoms:** Melted electrodes. Insulators are white, but may be dirty due to misfiring or flying debris in the combustion chamber. Can lead to engine damage.
**Recommendation:** Check for the correct plug heat range, over-advanced ignition timing, lean fuel mixture, insufficient engine cooling and lack of lubrication.

### HIGH SPEED GLAZING
**Symptoms:** Insulator has yellowish, glazed appearance. Indicates that combustion chamber temperatures have risen suddenly during hard acceleration. Normal deposits melt to form a conductive coating. Causes misfiring at high speeds.
**Recommendation:** Install new plugs. Consider using a colder plug if driving habits warrant.

### DETONATION
**Symptoms:** Insulators may be cracked or chipped. Improper gap setting techniques can also result in a fractured insulator tip. Can lead to piston damage.
**Recommendation:** Make sure the fuel anti-knock values meet engine requirements. Use care when setting the gaps on new plugs. Avoid lugging the engine.

### MECHANICAL DAMAGE
**Symptoms:** May be caused by a foreign object in the combustion chamber or the piston striking an incorrect reach (too long) plug. Causes a dead cylinder and could result in piston damage.
**Recommendation:** Repair the mechanical damage. Remove the foreign object from the engine and/or install the correct reach plug.

# Chapter 1 Tune-up and routine maintenance

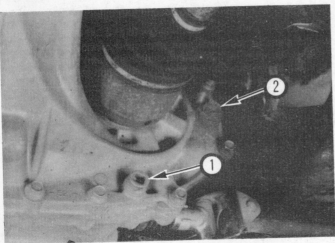

22.1 The manual transaxle drain (1) and fill (2) plugs are located on the side of the transaxle case

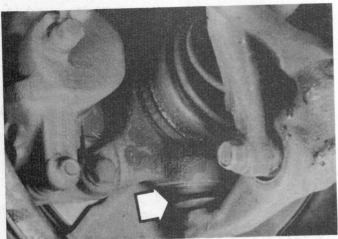

23.9 Push on the balljoint boot (arrow) to check for tears and grease leaks

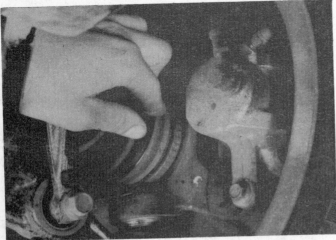

24.2 Flex the driveaxle boots by hand to check for tears, cracks and leaking grease

## With the wheels on the ground

1  With the vehicle stopped and the front wheels pointed straight ahead, rock the steering wheel gently back and forth. If freeplay is excessive, a front wheel bearing, main shaft yoke, intermediate shaft yoke, lower arm

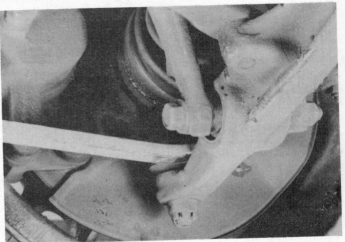

23.8 Pry between the balljoint and lower suspension arm to check for movement indicating balljoint wear

balljoint or steering system joint is worn or the steering gear is out of adjustment, loose on its mounts or broken. Refer to Chapter 10 for the appropriate repair procedure.

2  Other symptoms, such as excessive vehicle body movement over rough roads, swaying (leaning) around corners and binding as the steering wheel is turned, may indicate faulty steering and/or suspension components.

3  Check the shock absorbers by pushing down and releasing the vehicle several times at each corner. If the vehicle does not come back to a level position within one or two bounces, the shocks/struts are worn and must be replaced. When bouncing the vehicle up and down, listen for squeaks and noises from the suspension components. Additional information on suspension components can be found in Chapter 10.

4  Note whether the vehicle looks canted to one side or corner. If it is, try to level it by rocking it down. If this doesn't work, look for bad springs or worn or loose suspension parts..

## Under the vehicle

5  Raise the vehicle with a floor jack and support it securely on jackstands. See Jacking and towing at the front of this book for the proper jacking points.

6  Check the tires for irregular wear patterns (see Section 5) and proper inflation.

7  Inspect the universal joint between the steering shaft and the steering gear housing. Check the steering gear housing for grease leakage or oozing. Make sure that the dust seals and boots are not damaged and that the boot clamps are not loose. Check the steering linkage for looseness or damage. Check the tie-rod ends for excessive play. Look for loose bolts, broken or disconnected parts and deteriorated rubber bushings on all suspension and steering components. While an assistant turns the steering wheel from side to side, check the steering components for free movement, chafing and binding. If the steering components do not seem to be reacting with the movement of the steering wheel, try to determine where the slack is located

8  Check the balljoints for wear by prying between each balljoint and lower suspension arm (**see illustration**) to ensure the balljoint has no play. If any balljoint does have play, replace it. Refer to Chapter 10 for the front balljoint replacement procedure.

9  Inspect the balljoint boots for tears and leaking grease (**see illustration**). Replace the boots with new ones if they are damaged (see Chapter 10).

## 24 Driveaxle boot check

*Refer to illustration 24.2*

1  The driveaxle boots are very important because they prevent dirt, water and foreign material from entering and damaging the constant velocity

# Chapter 1  Tune-up and routine maintenance

**25.2a  Squeeze the PCV hose gently with a pair of pliers – use a rag to protect the hose surface (Integra)**

**25.2b  Checking the PCV valve by squeezing the hose on Legend models**

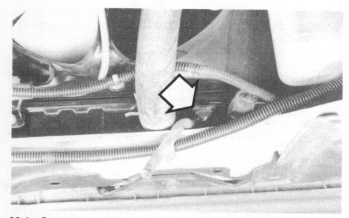

**26.4  On most models you will have to remove a cover for access to the radiator drain fitting located at the bottom of the radiator (arrow)**

(CV) joints. Oil and grease can cause the boot material to deteriorate prematurely, so it's a good idea to wash the boots with soap and water.
2  Inspect the boots for tears and cracks as well as loose clamps **(see illustration)**. If there is any evidence of cracks or leaking grease, they must be replaced as described in Chapter 8.

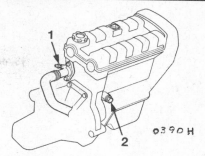

**26.5a  Integra block drain plug and bleeder screw details**
1  Bleeder screw       2  Drain plug

## 25  Positive Crankcase Ventilation (PCV) valve check and replacement

*Refer to illustrations 25.2a and 25.2b*
**Note:** *For a detailed discussion of the PCV system, refer to Chapter 6.*
1  The PCV valve is located in the crankcase breather chamber.

### Check
2  With the engine idling at normal operating temperature, squeeze the PCV hose located at the top of the engine gently shut with a pair of pliers, using a rag to protect the hose surface **(see illustrations)**. Pinch the hose as gently as possible to avoid damaging the hose.
3  If the PCV valve is operating properly, it will make a clicking sound when the hose is pinched shut. If it doesn't, replace the valve.
4  Check the hoses between the intake manifold and breather chamber for plugging, deterioration and other damage. Replace hoses as necessary.

### Replacement
5  Remove the PCV hose from the breather chamber. Remove the two bolts and detach the breather chamber from the valve cover. Withdraw the PCV valve from the valve cover.
6  When purchasing a replacement PCV valve, make sure it's for your particular vehicle and engine size. Compare the old valve with a new one to make sure they're the same.
7  Installation is the reverse of removal. Replace the breather chamber gasket.

## 26  Cooling system servicing (draining, flushing and refilling)

*Refer to illustrations 26.4, 26.5a and 26.5b*
**Warning:** *Antifreeze is a corrosive and poisonous solution, so be careful not to spill any of the coolant mixture on the vehicle's paint or your skin. If this happens, rinse immediately with plenty of clean water. Consult local authorities regarding proper disposal procedures for antifreeze before draining the cooling system. In many areas, reclamation centers have been established to collect used oil and coolant mixtures.*
1  Periodically, the cooling system should be drained, flushed and refilled to replenish the antifreeze mixture and prevent formation of rust and corrosion, which can impair the performance of the cooling system and cause engine damage. When the cooling system is serviced, all hoses and the radiator cap should be checked and replaced if necessary.

### Draining
2  Apply the parking brake and block the wheels. If the vehicle has just been driven, wait several hours to allow the engine to cool down before beginning this procedure.
3  Once the engine is completely cool, remove the radiator cap.
4  Move a large container under the radiator drain fitting to catch the coolant. Attach a 3/8-inch inner diameter hose to the drain fitting to direct the coolant into the container (some models are already equipped with a hose), then open the drain fitting (a pair of pliers may be required to turn it) **(see illustration)**.
5  After the coolant stops flowing out of the radiator, move the container under the engine block drain plug(s) on the engine **(see illustrations)**.

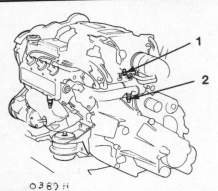

**26.5b Legend block drain plug and bleeder screw details**
1  Bleeder screw
2  Drain plug

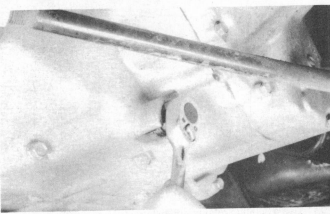

**28.7a Use a 3/8-inch ratchet and extension to remove the automatic transaxle drain plug**

Follow the procedure outlined in the manufacturer's instructions. If the radiator is severely corroded, damaged or leaking, it should be removed (see Chapter 3) and taken to a radiator repair shop.
10  Remove the overflow hose from the coolant recovery reservoir. Drain the reservoir and flush it with clean water, then reconnect the hose.

### Refilling

11  Close and tighten the radiator drain. Install and tighten the block drain plug(s).
12  Place the heater temperature control in the maximum heat position.
13  Loosen the coolant bleeder screw (see illustrations 26.5a and 26.5b).
14  Slowly add new coolant (a 50/50 mixture of water and antifreeze) to the radiator until bubble-free coolant flows from the bleeder screw. Tighten the screw and continue adding coolant to the radiator until it's full. Add coolant to the reservoir until the level is at the upper mark.
15  Leave the radiator cap off and run the engine in a well-ventilated area until the thermostat opens (coolant will begin flowing through the radiator and the upper radiator hose will become hot).
16  Turn the engine off and let it cool. Add more coolant mixture to bring the level back up to the lip on the radiator filler neck.
17  Squeeze the upper radiator hose to expel air, then add more coolant mixture if necessary. Replace the radiator cap.
18  Start the engine, allow it to reach normal operating temperature and check for leaks.

**28.7b Be careful when removing the drain plug because the hot fluid comes out with considerable force**

### 27  Exhaust system check

1  With the engine cold (at least three hours after the vehicle has been driven), check the complete exhaust system from its starting point at the engine to the end of the tailpipe. This should be done on a hoist where unrestricted access is available.
2  Check the pipes and connections for evidence of leaks, severe corrosion or damage. Make sure that all brackets and hangers are in good condition and tight.
3  At the same time, inspect the underside of the body for holes, corrosion, open seams, etc. which may allow exhaust gases to enter the passenger compartment. Seal all body openings with silicone sealer or body putty.
4  Rattles and other noises can often be traced to the exhaust system, especially the mounts and hangers. Try to move the pipes, muffler and catalytic converter. If the components can come in contact with the body or suspension parts, secure the exhaust system with new mounts.
5  Check the running condition of the engine by inspecting inside the end of the tailpipe. The exhaust deposits here are an indication of engine state-of-tune. If the pipe is black and sooty or coated with white deposits, the engine is in need of a tune-up, including a thorough fuel system inspection and adjustment.

**28.9 A funnel will make the job of adding automatic transaxle fluid a lot easier**

Loosen the plug(s) and allow the coolant in the block to drain.
6  While the coolant is draining, check the condition of the radiator hoses, heater hoses and clamps (refer to Section 13 if necessary).
7  Replace any damaged clamps or hoses (refer to Chapter 3 for detailed replacement procedures).

### Flushing

8  Once the system is completely drained, flush the radiator with fresh water from a garden hose until water runs clear at the drain. The flushing action of the water will remove sediments from the radiator but will not remove rust and scale from the engine and cooling tube surfaces.
9  These deposits can be removed by the chemical action of a cleaner.

# Chapter 1  Tune-up and routine maintenance

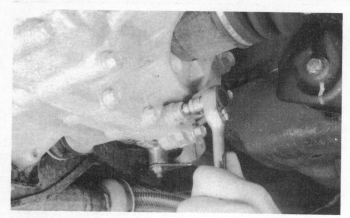

29.1 Use a 3/8-inch ratchet and extension to remove the manual transaxle drain plug

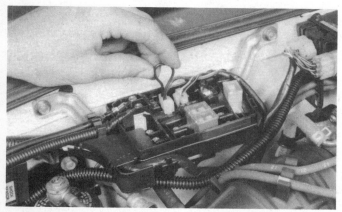

30.2 On 1988 and 1989 Integra models, place a jumper on the main fuse before checking the ignition timing

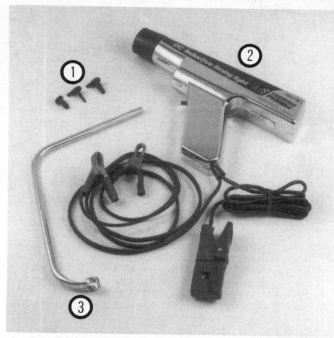

30.1 Tools needed to check and adjust the ignition timing
1 **Vacuum plugs** – Vacuum hoses will, in most cases, have to be disconnected and plugged. Molded plugs in various shapes and sizes are available for this.
2 **Inductive pick-up timing light** – Flashes a bright concentrated beam of light when the number one spark plug fires. Connect the leads according to the instructions supplied with the light.
3 **Distributor wrench** – On some models, the hold-down bolt for the distributor is difficult to reach and turn with conventional wrenches or sockets. A special wrench like this must be used.

## 28  Automatic transaxle fluid change

*Refer to illustrations 28.7a, 28.7b and 28.9*

1  At the specified time intervals, the automatic transaxle fluid should be drained and replaced.
2  Before beginning work, purchase the specified transmission fluid (see *Recommended fluids and lubricants* at the front of this chapter).
3  Other tools necessary for this job include jackstands to support the vehicle in a raised position, a 3/8-inch drive ratchet and extension, a drain pan capable of holding at least eight pints, newspapers and clean rags.
4  The fluid should be drained immediately after the vehicle has been driven. Hot fluid is more effective than cold fluid at removing built-up sediment. **Caution:** *Fluid temperature can exceed 350-degrees F in a hot transaxle. Wear protective gloves.*
5  After the vehicle has been driven to warm up the fluid, raise it and place it on jackstands for access to the transaxle and differential drain plugs.
6  Move the necessary equipment under the vehicle, being careful not to touch any of the hot exhaust components.
7  Place the drain pan under the drain plug in the transaxle and remove the drain plug with the ratchet and, if necessary, extension **(see illustration)**. Be sure the drain pan is in position, as fluid will come out with some force **(see illustration)**. Once the fluid is drained, clean the drain plug and reinstall it securely.
8  Lower the vehicle.
9  With the engine off pull out the dipstick, then add new fluid to the transaxle through the dipstick hole (see *Recommended fluids and lubricants* for the recommended fluid type and capacity) **(see illustration)**. Use a funnel to prevent spills. It is best to add a little fluid at a time, continually checking

the level with the dipstick (see Section 7). Allow the fluid time to drain into the pan.
10  Start the engine and shift the selector into all positions from P through 2, then shift into P and apply the parking brake.
11  Turn off the engine and check the fluid level. Add fluid to bring the level into the hatched area on the dipstick.

## 29  Manual transaxle lubricant change

*Refer to illustration 29.1*

1  Remove the drain plug and drain the fluid **(see illustration)**.
2  Reinstall the drain plug securely.
3  Add new fluid until it begins to run out of the filler hole (see Section 22). See *Recommended lubricants and fluids* for the specified lubricant type.

## 30  Ignition timing check and adjustment

*Refer to illustrations 30.1, 30.2 and 30.11*

**Note:** *It is imperative that the procedures included on the Vehicle Emissions Control Information (VECI) label be followed when adjusting the ignition timing. The label will include all information concerning preliminary steps to be performed before adjusting the timing, as well as the timing specifications.*

1  With the ignition off, locate the VECI label under the hood and read through and perform all preliminary instructions concerning ignition timing. Several special tools will be needed for this procedure **(see illustration)**.

30.11 Drill out the rivets to remove the stay cover from the timing adjuster (1987 and later Legend models)

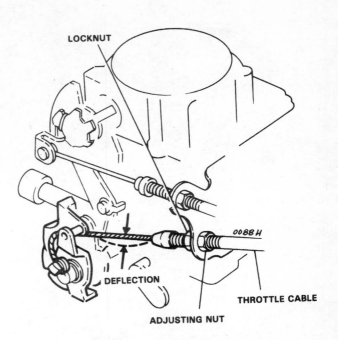

31.3 Throttle cable adjustment details

2  On 1988 and 1989 Integra models, remove the cover from the main fuse block in the engine compartment and connect a jumper between the Br/Bl and Br terminals of the main fuse (see illustration).
3  With the ignition off, hook up an inductive pick-up timing light in accordance with the manufacturer's instructions. Connect the inductive pick-up lead of the timing light to the number one spark plug wire. On Integra models, number one is the one closest to the drivebelt end of the engine. On Legend models, number one is closest to the drivebelt end of the engine, on the rear (firewall side) cylinder bank.
4  Disconnect and plug all vacuum lines connected to the distributor. Locate the timing marks and pointer at the drivebelt end of the engine.
5  With the engine at normal operating temperature, start the engine and point the timing light at the timing marks.
6  The appropriate mark on the front pulley (refer to the VECI label) will appear stationary and be aligned with the pointer if the timing is correct.

### 1986 Legend and all Integra models

7  If an adjustment is required, loosen the three adjusting bolts and rotate the distributor slightly until the timing is correct.
8  Tighten the adjusting bolts and recheck the timing.
9  Turn off the engine and remove the timing light.
10  On 1988 and 1989 models, remove the jumper wire from the main fuse box. Re-connect any vacuum hoses that were disconnected.

### 1987 and later Legend models

11  If adjustment is required, turn the engine off and remove the upper and lower covers of the control box, (located in the engine compartment on the driver's side firewall) for access to the ignition timing adjuster. Using a 3/16-in bit, drill out the rivets, detach the stay cover from the adjuster, then remove the adjuster from the control box (see illustration).
12  Start the engine and turn the adjusting screw on the adjuster until the timing is correct.
13  Turn the engine off, install the stay cover on the adjuster with new rivets, then reinstall the adjuster in the control box. Re-connect any vacuum hoses that were disconnected.

## 31  Throttle linkage inspection

*Refer to illustration 31.3*

1  Inspect the throttle linkage for damage and missing parts and for binding and interference when the accelerator pedal is depressed.
2  Lubricate the various linkage pivot points with engine oil.
3  Push on the throttle cable with your fingers to check the deflection. Compare the measurement with those found in this Chapter's Specifications. If the deflection is incorrect, loosen the locknut and turn the adjusting nut as necessary to adjust the tension (see illustration).
4  Tighten the locknut.

## 32  Idle speed check and adjustment

*Refer to illustrations 32.5a, 32.5b and 32.6*

1  Engine idle speed is the speed at which the engine operates when no accelerator pedal pressure is applied, as when stopped at a traffic light. This speed is critical to the performance of the engine itself, as well as many subsystems.
2  Start the engine and allow it to warm up to normal operating temperature (the cooling fan should come on at least twice).
3  Stop the engine. Hook up a hand-held tachometer in accordance with the manufacturer's instructions.
4  Set the parking brake firmly and block the wheels to prevent the vehicle from rolling. Place the transaxle in Neutral (manual transaxle) or Park (automatic transaxle).

### Integra models

5  Start the engine, unplug the electrical connector from the Electronic Idle Control Valve (EICV) (1987 and earlier models) or Electronic Air Control Valve (EACV) (1988 and later models) (see illustrations).
6  Note the idle speed rpm on the tachometer and compare it to that specified on the VECI label. If the idle speed is too low or too high, adjust it by turning the adjusting screw located on top of the throttle body (see illustration).
7  Turn off the engine and disconnect the tachometer.
8  After adjustment, reconnect the EICV or EACV connector.

### Legend models

9  Start the engine, note the idle speed rpm on the tachometer and compare it to that specified on the VECI label.
10  Check the ECU located under the front passenger's seat and see if the yellow Light Emitting Diode (LED) is lit.
11  If the LED is off, don't adjust the idle adjusting screw on the throttle body (see illustration 32.6).

# Chapter 1  Tune-up and routine maintenance

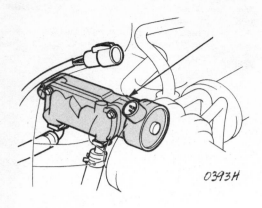

32.5a  On 1987 and earlier models, unplug the connector from the EICV (arrow)

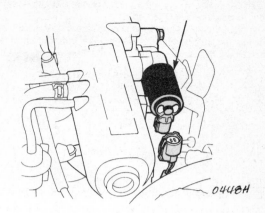

32.5b  On 1988 and later models, disconnect the connector from the EACV (arrow)

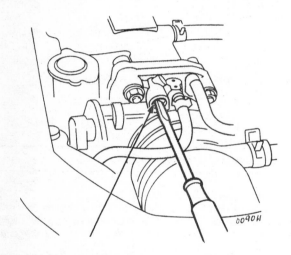

32.6  Adjust the idle speed with the idle speed adjusting screw (arrow) located on the throttle body

12  If the LED is blinking, turn the idle adjusting screw 1/4-turn clockwise.
13  If the LED is lit, turn the idle adjusting screw 1/4-turn counter-clockwise. The LED should go out after approximately 30 seconds. If it doesn't, repeat the operation, turning the screw a further 1/4-turn until the light goes out.

## 33  Evaporative emissions control system check

1  The function of the Fuel Evaporative Emission Control (EVAP) System is to store fuel vapors from the fuel tank in a charcoal canister until they can be routed to the intake manifold where they mix with incoming air before being burned in the cylinder combustion chambers.
2  The most common symptom of a faulty evaporative emissions system is a strong fuel odor in the engine compartment. If a fuel odor is detected, inspect the charcoal canister, located on the firewall in the engine compartment, and the hoses attached to it.
3  The evaporative emissions control system is explained in more detail in Chapter 6.

35.2  The fuel filter is located on the firewall (arrow)

## 34  Exhaust Gas Recirculation (EGR) system check (Legend models only)

1  The EGR valve on these models is located below the distributor, adjacent to the battery. Most of the time when a problem develops in this emissions system, it's due to a stuck or corroded EGR valve.
2  With the engine cold to prevent burns, push on the EGR valve diaphragm. Using moderate pressure, you should be able to press the diaphragm up-and-down within the housing.
3  If the diaphragm doesn't move or moves only with much effort, replace the EGR valve with a new one. If in doubt about the condition of the valve, compare the free movement of your EGR valve with a new valve.
4  Refer to Chapter 6 for more information on the EGR system.

## 35  Fuel filter replacement

*Refer to illustration 35.2*

1  This job should be done with the engine cold (after sitting at least three hours).
2  The fuel filter is located on the firewall in the engine compartment **(see illustration)**. Place an approved gasoline container under the fuel filter.

3  Place a shop towel or rag around the filter and depressurize the fuel system as described in Chapter 4.
4  Remove the banjo and service bolts, disconnect the fittings, remove the clamp and lift the filter from the engine compartment. Use a flare-nut wrench when disconnecting the fuel line fitting at the filter (Legend models). Make sure all the fuel connections and fittings are clean before installation to assure a leak-free seal.
5  Installation is the reverse of removal. Use new banjo and service bolt washers. Tighten the bolts to the torques listed in this Chapter's Specifications. Start the engine and check for leaks.

# Chapter 2 Part A  Integra engine

## Contents

| | |
|---|---|
| Camshaft oil seal – replacement | 9 |
| Camshaft and rocker arms – removal, inspection and installation | 10 |
| Compression check | See Chapter 2C |
| Cylinder head – removal and installation | 12 |
| Drivebelt check, adjustment and replacement | See Chapter 1 |
| Engine mounts – check and replacement | 17 |
| Engine oil and filter change | See Chapter 1 |
| Engine overhaul – general information | See Chapter 2C |
| Engine – removal and installation | See Chapter 2C |
| Exhaust manifold – removal and installation | 6 |
| Flywheel/driveplate – removal and installation | 15 |
| Front crankshaft oil seal – replacement | 8 |
| General information | 1 |
| Intake manifold – removal and installation | 5 |
| Oil pan – removal and installation | 13 |
| Oil pump – removal and installation | 14 |
| Rear crankshaft oil seal – replacement | 16 |
| Repair operations possible with the engine in the vehicle | 2 |
| Spark plug replacement | See Chapter 1 |
| Timing belt – removal, inspection and installation | 7 |
| Top Dead Center (TDC) for number one piston – locating | 3 |
| Valve cover – removal and installation | 4 |
| Valves – servicing | See Chapter 2C |
| Valve springs, retainers and seals – replacement | 11 |
| Water pump – removal and installation | See Chapter 3 |

## Specifications

### General

| | |
|---|---|
| Displacement | 1.6 liters (97 cu. in.) |
| Cylinder numbers (No. 1 at timing belt end) | 1-2-3-4 |
| Firing order | 1-3-4-2 |
| **Camshaft** | |
| Endplay | |
|   Standard | 0.002 to 0.006 inch (0.05 to 0.15 mm) |
|   Service limit | 0.02 inch (0.5 mm) |
| Lobe height | |
|   Intake | 1.2822 inch (32.568 mm) |
|   Exhaust | 1.2735 inch (32.348 mm) |
| Runout | |
|   Standard | 0.001 inch (0.03 mm) |
|   Service limit | 0.002 inch (0.06 mm) |
| Oil clearance | |
|   Standard | 0.002 to 0.004 inch (0.050 to 0.089 mm) |
|   Service limit | 0.006 inch (0.15 mm) |

### Torque specifications

| | Ft-lbs |
|---|---|
| Camshaft bearing cap bolts | 9 |
| Camshaft sprocket bolt | 27 |
| Valve cover nuts | 7 |
| Crankshaft damper bolt | 83 |
| Cylinder head bolts (oiled) | |
|   Step 1 | 22 |
|   Final step | 48 |
| Driveplate bolts (automatic transaxle) | 54 |

**1986 and 1987**

**1988 and 1989**

The blackened terminal shown on the distributor cap indicates the Number One spark plug wire position

**Cylinder location and distributor rotation**

## Chapter 2 Part A Integra engine

**Torque specifications (continued)** — Ft-lbs

| | |
|---|---|
| Engine mount bolts/nuts | |
| Front mount | 33 |
| Left mount | 47 |
| Rear mount | |
| Nuts | 31 |
| Bolts | 54 |
| Exhaust manifold-to-cylinder head nuts | 23 |
| Flywheel bolts (manual transaxle) | 87 |
| Intake manifold-to-cylinder head nuts/bolts | 16 |
| Oil pan bolts | 9 |
| Oil pickup tube nuts | 17 |
| Oil pump-to-block bolts | 9 |
| Oil cooler-to-block center bolt | 54 |
| Rear oil seal housing bolts | 9 |
| Timing belt tensioner bolt | 33 |
| Timing belt cover bolts | 7 |

### 1  General information

This Part of Chapter 2 is devoted to in-vehicle engine repair procedures. All information concerning engine removal and installation and engine block and cylinder head overhaul can be found in Part C of this Chapter.

The following repair procedures are based on the assumption that the engine is installed in the vehicle. If the engine has been removed from the vehicle and mounted on a stand, many of the steps outlined in this Part of Chapter 2 will not apply.

The Specifications included in this Part of Chapter 2 apply only to the procedures contained in this Part. Part C of Chapter 2 contains the Specifications necessary for cylinder head and engine block rebuilding.

### 2  Repair operations possible with the engine in the vehicle

Many major repair operations can be accomplished without removing the engine from the vehicle.

Clean the engine compartment and the exterior of the engine with some type of degreaser before any work is done. It will make the job easier and help keep dirt out of the internal areas of the engine.

Depending on the components involved, it may be helpful to remove the hood to improve access to the engine as repairs are performed (refer to Chapter 11 if necessary). Cover the fenders to prevent damage to the paint. Special pads are available, but an old bedspread or blanket will also work.

If vacuum, exhaust, oil or coolant leaks develop, indicating a need for gasket or seal replacement, the repairs can generally be made with the engine in the vehicle. The intake and exhaust manifold gaskets, valve cover gasket, oil pan gasket, crankshaft oil seals and cylinder head gasket are all accessible with the engine in place.

Exterior engine components, such as the intake and exhaust manifolds, the oil pan, oil pump, water pump, the starter motor, the alternator, the distributor and the fuel system components can be removed for repair with the engine in place.

Since the cylinder head can be removed without pulling the engine, camshafts and valve component servicing can also be accomplished with the engine in the vehicle. Replacement of the timing belt and sprockets is also possible with the engine in the vehicle.

In extreme cases caused by a lack of necessary equipment, repair or replacement of piston rings, pistons, connecting rods and rod bearings is possible with the engine in the vehicle. However, this practice is not recommended because of the cleaning and preparation work that must be done to the components involved.

### 3  Top Dead Center (TDC) for number one piston – locating

*Refer to illustrations 3.6, 3.8 and 3.9*

**Note:** *The following procedure is based on the assumption that the distributor is correctly installed. If you are trying to locate TDC to install the distributor correctly, piston position must be determined by feeling for compression at the number one spark plug hole, then aligning the ignition timing marks as described in step 8.*

1  Top Dead Center (TDC) is the highest point in the cylinder that each piston reaches as it travels up-and-down when the crankshaft turns. Each piston reaches TDC on the compression stroke and again on the exhaust stroke, but TDC generally refers to piston position on the compression stroke.

2  Positioning the piston(s) at TDC is an essential part of many procedures such as camshaft and timing belt/sprocket removal and distributor removal.

3  Before beginning this procedure, be sure to place the transaxle in Neutral and apply the parking brake or block the rear wheels. Also, disable the ignition system by detaching the coil wire from its terminal on the distributor cap and grounding it on the block with a jumper wire. Remove the spark plugs (see Chapter 1).

4  In order to bring any piston to TDC, the crankshaft must be turned using the method outlined here. Turn the crankshaft with a socket and ratchet attached to the bolt threaded into the timing belt end of the crankshaft. Raise the front of the vehicle and support it securely on jackstands. Remove the left (driver's) side front wheel. Working inside the wheel well, put the socket, with an extension attached, through the hole in the wheel well and place it on the crankshaft bolt. Turn it in the normal direction with the ratchet. When looking at the timing-belt end of the engine, normal crankshaft rotation is counter-clockwise.

5  Note the position of the terminal for the number one spark plug wire on the distributor cap. If the terminal isn't marked, follow the plug wire from the number one cylinder spark plug to the cap.

6  Use a felt-tip pen or chalk to make a mark on the distributor body directly adjacent to the terminal **(see illustration)**.

7  Detach the cap from the distributor and set it aside.

8  Turn the crankshaft (see Paragraph 4 above) until the TDC mark on the crankshaft pulley is aligned with the pointer on the timing belt cover **(see illustration)**.

9  Look at the distributor rotor – it should be pointing directly at the mark you made on the distributor body **(see illustration)**. If it is, go to Step 12.

10  If the rotor is 180-degrees off, the number one piston is at TDC on the exhaust stroke. Go to Step 11.

11  To get the piston to TDC on the compression stroke, turn the crankshaft one complete turn (360-degrees) counter-clockwise. The rotor should now be pointing at the mark on the distributor. When the rotor is pointing at the number one spark plug wire terminal in the distributor cap

# Chapter 2 Part A  Integra engine

3.6  Make a mark (arrow) on the distributor body adjacent to the number one spark plug wire terminal

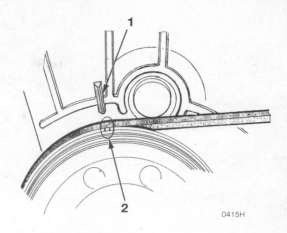

3.8  Align the mark on the crankshaft pulley (2) with the pointer on the timing belt cover (1)

and the timing marks are aligned, the number one piston is at TDC on the compression stroke.
12  After the number one piston has been positioned at TDC on the compression stroke, TDC for any of the remaining pistons can be located by turning the crankshaft and following the firing order. Mark the remaining spark plug wire terminal locations on the distributor body just like you did for the number one terminal, then number the marks to correspond with the cylinder numbers. As you turn the crankshaft, the rotor will also turn. When it's pointing directly at one of the marks on the distributor, the piston for that particular cylinder is at TDC on the compression stroke.

## 4  Valve cover – removal and installation

*Refer to illustration 4.3*

1  Disconnect the negative cable from the battery.
2  Remove the spark plug boots and wires (see Chapter 1).
3  Detach the crankcase ventilation hose **(see illustration)** from the valve cover.
4  Remove the upper timing belt cover.

5  Remove the valve cover retaining nuts, seals and washers. Lift the valve cover from the engine.
6  Thoroughly clean the cover and gasket mating surfaces, removing any traces of old gasket material.
7  Using a new gasket and seals, reinstall the cover and tighten the nuts to the torque listed in this Chapter's Specifications. Be sure to attach the ground cable, if equipped, under the cover nut.
8  Reinstall the remaining parts in the reverse order of removal.
9  Run the engine and check for oil leaks.

## 5  Intake manifold – removal and installation

1  Relieve the fuel pressure (see Chapter 4).
2  Disconnect the negative cable from the battery.
3  Drain the cooling system (see Chapter 1).
4  Remove the throttle body and fuel injectors (see Chapter 4).
5  Label and then disconnect all wiring, control cables and hoses connecting the intake manifold to the vehicle.
6  Remove the manifold-to-cylinder head nuts/bolts and the bolt(s) on the bracket(s) supporting the manifold from underneath.

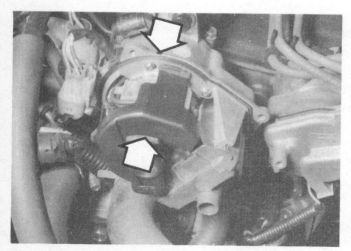

3.9  Align the rotor tip with the mark (arrows)

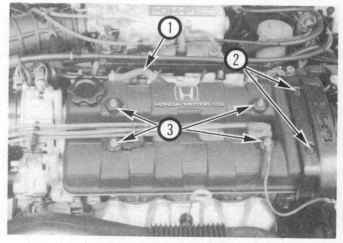

4.3  Valve cover details

1  Crankcase ventilation hose
2  Upper timing belt cover bolts
3  Valve cover mounting nuts

**6.2 Unbolt the intake air duct (arrows)**

**6.3 Remove the heat insulator bolts (arrows)**

7  Lift the manifold off the engine.
8  Thoroughly clean the gasket mating surfaces, removing all traces of old gasket material with a scraper. Be very careful not to damage the delicate aluminum. Wipe the mating surfaces with lacquer thinner or acetone.
9  Using a new gasket, reinstall the manifold. Tighten the nuts/bolts from the center outward in several stages until the torque listed in this Chapter's Specifications is reached.
10  Reinstall the remaining parts in the reverse order of removal.
11  Refill the radiator and bleed the cooling system (see Chapter 1).
12  Run the engine, checking for leaks and proper operation.

## 6  Exhaust manifold – removal and installation

*Refer to illustrations 6.2, 6.3, 6.6 and 6.7*
**Warning:** *The engine must be completely cool before beginning this procedure.*

1  Disconnect the negative cable from the battery.
2  Remove the intake air duct for access **(see illustration)**. Unplug the oxygen sensor wire harness. If you're installing a new manifold, remove the sensor (see Chapter 6).
3  Remove the oil dipstick and heat insulator (shroud) from the manifold **(see illustration)**.
4  Block the rear wheels and set the parking brake. Raise the front of the vehicle and support it securely on jackstands. Remove the front lower splash shield.

5  Apply penetrating oil to the exhaust manifold nuts.
6  Disconnect the exhaust pipe from the exhaust manifold and unbolt the bracket **(see illustration)**.
7  Remove the nuts **(see illustration)** and detach the manifold and gaskets.
8  Use a scraper to remove all traces of old gasket material and carbon deposits from the manifold and cylinder head mating surfaces. Be careful not to damage the delicate aluminum with the scraper. If the gasket was leaking, have the manifold checked for warpage at an automotive machine shop and resurfaced, if necessary.
9  Position a new gasket onto the cylinder head.
10  Install the manifold and thread the mounting nuts into place.
11  Working from the center out, tighten the nuts to the torque listed in this Chapter's Specifications in three equal steps.
12  Reinstall the remaining parts in the reverse order of removal.
13  Run the engine and check for exhaust leaks.

## 7  Timing belt – removal, inspection and installation

*Refer to illustrations 7.8, 7.10, 7.11, 7.14a, 7.14b, 7.15, 7.16a, 7.16b, 7.16c and 7.17*

1  Disconnect the negative cable from the battery.
2  Block the rear wheels and set the parking brake.
3  Loosen the lug nuts on the left front wheel and then raise the front of the vehicle. Support the front of the vehicle securely on jackstands.

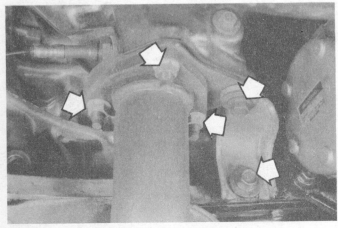

**6.6 Remove the bolts and nuts**

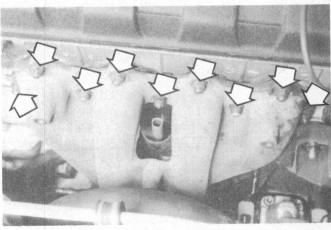

**6.7 Remove the nuts (arrows)**

# Chapter 2 Part A   Integra engine

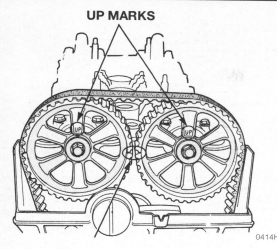

7.8  The marks on the sprockets should align (arrow)

7.10  Remove the upper timing belt cover bolts (arrows)

4   Remove the splash pan and left front wheel for easier access to the end of the crankshaft.
5   Support the engine with a floor jack. Place a block of wood between the jack pad and the oil pan to prevent damage.
6   Unbolt the left engine mount from the engine (see Section 17).
7   Remove the spark plugs and drivebelts (see Chapter 1).
8   Set the engine at Top Dead Center (see Section 3) and then remove the valve cover (see Section 4). The timing marks should align as shown **(see illustration)**.
9   Unbolt the power steering pump without disconnecting the hoses and set it aside (see Chapter 10).
10  Remove the cruise control actuator, (if equipped) and then the upper timing belt cover **(see illustration)**.
11  Loosen the pulley-to-crankshaft bolt **(see illustration)** with a socket and breaker bar. To keep the crankshaft pulley from turning, on manual transaxle models, have an assistant place the transmission in first gear and hold the brakes on. On automatic transaxle models, have an assistant hold a large screwdriver firmly against the ring gear teeth on the flywheel. Recheck the timing marks to ensure the engine is still at TDC, then slip the pulley off the crankshaft.
12  Remove the lower timing belt cover.
13  If you intend to reuse the timing belt, paint match marks on the sprockets and belt and an arrow to indicate direction of rotation.
14  Loosen the timing belt tensioner bolt and slip the belt off. Note the way

7.11  Remove the crankshaft pulley bolt (arrow)

the sprocket outer and inner belt guides are facing for proper reinstallation **(see illustration)**. **Caution:** *Do not rotate the camshaft or crankshaft while the timing belt is removed.* If you are replacing the crankshaft oil seal, slip the sprocket and inner belt guide off the crankshaft **(see illustration)**, noting the way they were installed for later reference.

7.14a  Remove the outer belt guide – note that the curved outer edge faces away from the belt (belt removed for clarity)

7.14b  It's not necessary to remove the crankshaft sprocket unless you intend to replace the oil seal

**7.15 Check the belt tensioner idler pulley for rough rotation and bearing play**

**7.17 After you have detached the camshaft sprocket, remove the Woodruff key (arrow) from the camshaft**

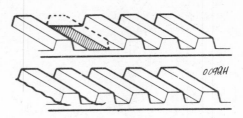

**7.16a Check the timing belt for cracked and missing teeth**

**7.16b If the belt is cracked or worn, check the pulleys for nicks and burrs**

**7.16c Wear on one side of the belt indicates pulley misalignment**

15  Rotate the belt tensioner idler by hand, checking for bearing play and roughness **(see illustration)**. Replace as necessary.
16  Inspect the timing belt for wear (especially on the thrust side of the teeth), cracks, splits, fraying and oil contamination **(see illustrations)**. Replace the belt if it shows signs of deterioration.
17  If you are replacing a camshaft or oil seal, remove the bolt, washer and sprockets. Slip a lever through the camshaft sprocket to hold it from turning. Do not lose the Woodruff key **(see illustration)**.
18  Before installing the timing belt, ensure the camshaft sprockets are installed with the marks aligned as shown in illustration 7.8.
19  If the crankshaft was disturbed, temporarily reinstall the crankshaft bolt and turn the crankshaft until the TDC marks on the crankshaft pulley and camshaft gears are aligned **(see illustrations 3.8 and 7.8)**.
20  Install the timing belt with slight tension between the sprockets on the forward facing side. With the belt tensioner bolt loosened, slowly turn the crankshaft counter-clockwise for a distance of three teeth on the camshaft sprocket.
21  Tighten the belt tensioner bolt to the torque listed in this Chapter's Specifications and reinstall the lower timing belt cover.
22  Carefully turn the crankshaft through two revolutions and recheck the TDC mark and cam sprocket index marks for proper alignment. **Caution:** *If the crankshaft binds or seems to hit something, do not force it, the valves may be hitting the pistons. If this occurs, go back through all steps and check your work carefully.*
23  Reinstall the remaining parts in the reverse order of removal. Be sure to oil the threads on the crankshaft bolt, install the outer belt guide with the concave surface facing out, and tighten the crankshaft pulley bolt to the torque listed in this Chapter's Specifications.
24  Run the engine and check for proper operation.

## 8  Front crankshaft oil seal – replacement

*Refer to illustrations 8.2 and 8.4*

1  Remove the timing belt and crankshaft sprocket (see Section 7).
2  Note how far the seal is seated in the bore, then carefully pry it out of the seal bore with a screwdriver or seal removal tool **(see illustration)**.

**8.2 Use a seal removal tool or screwdriver to pry the crankshaft front oil seal out of the bore**

## Chapter 2 Part A  Integra engine

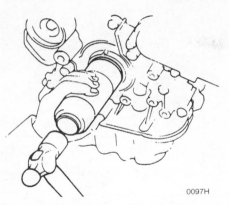

8.4 Lubricate the seal lip and carefully tap the new seal into place with a large socket or piece of pipe and a hammer

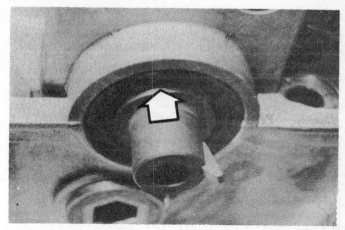

9.2 Insert a small screwdriver between the seal lip and camshaft (arrow) to remove the seal

Don't scratch the housing bore or damage the crankshaft in the process (if the crankshaft is damaged, the new seal will end up leaking).
3   Clean the bore in the housing and coat the outer edge of the new seal with engine oil. Apply moly-base grease to the seal lip. **Note:** *A special factory tool no. 07947-6340000 is available for installing the seals.*
4   Using a special tool or a socket with an outside diameter slightly smaller than the outside diameter of the seal, carefully drive the new seal into place with a hammer **(see illustration)**. Make sure it's installed squarely and driven in to the same depth as the original. If a socket isn't available, a short section of large diameter pipe will also work. Check the seal after installation to make sure the garter spring didn't pop out of place.
5   Reinstall the crankshaft sprocket and timing belt (see Section 7).
6   Run the engine and check for oil leaks at the front seal.

### 9  Camshaft oil seal – replacement

*Refer to illustration 9.2 and 9.4*

1   Remove the timing belt and camshaft sprockets (see Sections 7 and 10). **Note:** *Always replace both camshaft seals, even if only one is leaking.*
2   Note how far the seal is seated in the bore, then carefully pry it out with a small screwdriver **(see illustration)**. Don't scratch the bore or damage the camshaft in the process (if the camshaft is damaged, the new seal will end up leaking).
3   Clean the bore and coat the outer edge of the new seal with engine oil

or multi-purpose grease. Apply moly-base grease to the seal lip.
4   Using a socket with an outside diameter slightly smaller than the outside diameter of the seal, carefully drive the new seals into place with a hammer **(see illustration)**. Make sure they are installed squarely and driven in to the same depth as the original. **Note:** *A special factory tool (no. 07947-SB00100 or equivalent) is available for installing the seal.*
5   Reinstall the camshaft sprockets and timing belt (see Section 7).
6   Run the engine and check for oil leaks at the camshaft seal.

### 10  Camshaft and rocker arms – removal, inspection and installation

*Refer to illustrations 10.4, 10.7, 10.8, 10.9a and 10.9b*

#### Removal

1   Remove the timing belt and camshaft sprockets as described in Section 7.
2   Remove the valve cover (see Section 4).
3   Remove the distributor and crankshaft angle sensor (see Chapter 5).
4   Loosen the valve adjustment locknuts and back off the screws until they no longer hold the valves open. Measure the thrust clearance of both camshafts (endplay) with a dial indicator **(see illustration)**. If the clearance is greater than the maximum listed in this Chapter's Specifications, replace the camshafts and/or the cylinder head.

9.4 If a seal driver isn't available, use a hammer and section of pipe or a large socket to tap the new seal into place

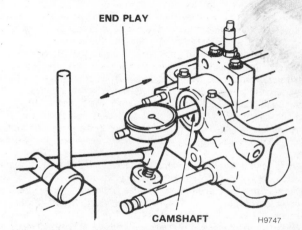

10.4 To check camshaft endplay, mount a dial indicator as shown and move the camshaft forward and backward (arrows)

# 2A–8  Chapter 2 Part A  Integra engine

10.7  Check the camshaft lobes for pitting, wear and score marks – if scoring is excessive, as is the case here, replace the camshaft

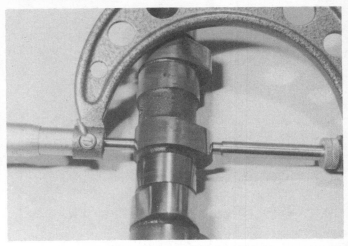

10.8  Measuring camshaft lobe height

5  Loosen the camshaft bearing cap bolts one at a time, working in a diagonal pattern from the center outward in several stages. Lift the caps off, noting the markings. **Caution:** *The bearing caps must always be installed in the same locations, facing the same direction.*

6  Lift the camshafts from the cylinder head and clean thoroughly with solvent. Lift out the rocker arms and store them in a clearly marked rack to ensure reinstallation in the same location.

## Inspection

7  Visually examine the cam lobes **(see illustration)** and bearing journals for score marks, pitting, galling and evidence of overheating (blue, discolored areas). Look for flaking away of the hardened surface layer of each lobe.

8  Using a micrometer, measure the height of each camshaft lobe **(see illustration)** and the diameter of each journal. If the measurements are less than the minimum in this Chapter's Specifications, replace the camshaft.

9  Check the oil clearance for each camshaft journal as follows:
 a) Clean the bearing caps and camshaft journals with lacquer thinner or acetone and a clean cloth.
 b) Carefully lay the camshafts in place in the head. Don't use any lubrication. **Note:** *Do not turn the camshaft during this procedure.*
 c) Lay a strip of Plastigage on each journal **(see illustration)**.
 d) Install the camshaft bearing caps and tighten the bolts two turns at a time to the torque listed in this Chapter's Specifications.
 e) Loosen the cam bearing cap bolts and lift the caps off.
 f) Compare the width of the crushed Plastigage (at it's widest point) to the scale on the Plastigage envelope **(see illustration)**.
 g) If the clearance is greater than allowable in this Chapter's Specifications, replace the worn parts as necessary.
 h) Scrape off the Plastigage with your fingernail or the edge of a credit card – don't nick or scratch the journals or bearing caps.

10  Check the rocker arm contact faces and the bearing cap-to-camshaft surfaces for wear. Also note the condition of the adjustment screws and locknuts. Replace any components which show signs of wear.

## Installation

11  Lubricate the rocker arm components, camshaft lobes, journals and seal lip contact surfaces with engine assembly lube or moly-based grease. Position the rocker arms in their original locations.

12  Fit the oil seals onto the camshafts with the spring side facing inward.

13  Set the camshafts into the cylinder head with the keyways facing up (12 o'clock position).

14  Install the camshaft bearing caps and loosely install the bolts. Apply

10.9a  Lay a strip of Plastigage on each camshaft journal

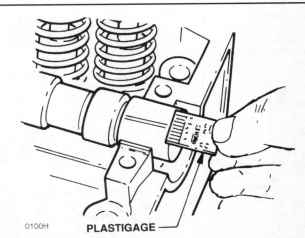

10.9b  Compare the width of the crushed Plastigage to the scale on the envelope to determine the oil clearance

# Chapter 2 Part A  Integra engine

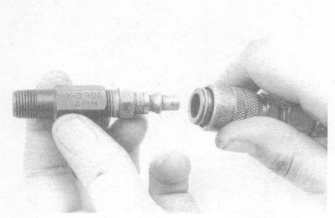

11.4  This is what the air hose adapter that threads into the spark plug hole looks like – they're commonly available from auto parts stores

11.8  Use a spring compressor (1) to release the spring pressure so the valve keepers (2) can be removed

sealant to the head mating surfaces of the number one and number six bearing caps. Apply engine oil to the threads of the four long stud bolts.
15  Gently seat the oil seals with a soft-face hammer.
16  Tighten the bearing caps two turns at a time, working diagonally from the center outward in several stages until the torque listed in this Chapter's Specifications is reached.
17  Reinstall the timing belt and sprockets (see Section 7) and the remaining parts in the reverse order of removal.
18  Adjust the valves (see Chapter 1).

## 11  Valve springs, retainers and seals – replacement

*Refer to illustrations 11.4, 11.8 and 11.16*

**Note:** *Broken valve springs and defective valve stem seals can be replaced without removing the cylinder head. Special tools and a compressed air source are normally required to perform this operation, so read through this Section carefully and rent or buy the tools before beginning the job. If compressed air isn't available, a length of nylon rope can be used to keep the valves from falling into the cylinder during this procedure.*

1  Refer to Section 10 and remove the camshafts and rocker arms.
2  Remove the spark plug from the cylinder which has the defective component. If all of the valve stem seals are being replaced, all of the spark plugs should be removed.
3  Turn the crankshaft until the piston in the affected cylinder is at top dead center on the compression stroke (refer to Section 3 for instructions). If you're replacing all of the valve stem seals, begin with cylinder number one and work on the valves for one cylinder at a time. Move from cylinder-to-cylinder following the firing order sequence (see the Specifications).
4  Thread an adapter into the spark plug hole **(see illustration)** and connect an air hose from a compressed air source to it. Most auto parts stores can supply the air hose adapter. **Note:** *Many cylinder compression gauges utilize a screw-in fitting that may work with your air hose quick-disconnect fitting.*
5  Apply compressed air to the cylinder. **Warning:** *The piston may be forced down by compressed air, causing the crankshaft to turn suddenly. If the wrench used when positioning the number one piston at TDC is still attached to the bolt in the crankshaft nose, it could cause damage or injury when the crankshaft moves.*
6  The valves should be held in place by the air pressure. If the valve faces or seats are in poor condition, leaks may prevent air pressure from retaining the valves – refer to the alternative procedure below.
7  If you don't have access to compressed air, an alternative method can be used. Position the piston at a point just before TDC on the compression stroke, then feed a long piece of nylon rope through the spark plug hole until it fills the combustion chamber.
Be sure to leave the end of the rope hanging out of the engine so it can be removed easily. Use a large ratchet and socket to rotate the crankshaft in the normal direction of rotation (counter-clockwise) until slight resistance is felt.
8  Stuff shop rags into the cylinder head holes above and below the valves to prevent parts and tools from falling into the engine, then use a valve spring compressor to compress the spring. Remove the keepers with small needle-nose pliers or a magnet **(see illustration)**. **Note:** *A couple of different types of tools are available for compressing the valve springs with the head in place. One type grips the lower spring coils and presses on the retainer as the knob is turned, while the other type, shown here, utilizes the camshaft for leverage. Both types work very well, although the lever type is usually less expensive.*
9  Remove the spring retainer and valve spring(s), then remove the guide seal. **Note:** *If air pressure fails to hold the valve in the closed position during this operation, the valve face and/or seat is probably damaged. If so, the cylinder head will have to be removed for additional repair operations.*
10  Wrap a rubber band or tape around the top of the valve stem so the valve won't fall into the combustion chamber, then release the air pressure. **Note:** *If a rope was used instead of air pressure, turn the crankshaft slightly in the direction opposite normal rotation.*
11  Inspect the valve stem for damage. Rotate the valve in the guide and check the end for eccentric movement, which would indicate that the valve is bent.
12  Move the valve up-and-down in the guide and make sure it doesn't bind. If the valve stem binds, either the valve is bent or the guide is damaged. In either case, the head will have to be removed for repair.
13  Reapply air pressure to the cylinder to retain the valve in the closed position, then remove the tape or rubber band from the valve stem. If a rope was used instead of air pressure, rotate the crankshaft in the normal direction of rotation until slight resistance is felt.
14  Lubricate the valve stem with engine oil and install a new guide seal. **Note:** *Intake and exhaust seals are NOT interchangeable. Intake seals have a white spring and exhaust seals have a black spring.*
15  **Note:** *Place the end of the valve springs with closely wound coils or paint marks toward the cylinder head. Install the spring in position over the valve.*
16  Install the valve spring retainer. Compress the valve spring and carefully position the keepers in the groove. Apply a small dab of grease to the

**11.16 Apply a small dab of grease to each keeper before installation to hold it in place on the valve stem until the spring compressor is released**

inside of each keeper to hold it in place if necessary **(see illustration)**.
17  Remove the pressure from the spring tool and make sure the keepers are seated.
18  Disconnect the air hose and remove the adapter from the spark plug hole. If a rope was used in place of air pressure, pull it out of the cylinder.
19  Refer to Section 10 and install the camshafts and rocker arms.
20  Install the spark plug(s) and connect the wire(s).
21  Refer to Section 4 and install the valve cover.
22  Start and run the engine, then check for oil leaks and unusual sounds coming from the valve cover area.

## 12  Cylinder head – removal and installation

**Caution:** *The engine must be completely cool before beginning this procedure.*

### Removal

1  Drain the coolant from the engine block and radiator (see Chapter 1).
2  Drain the engine oil and remove the oil filter (see Chapter 1).
3  Relieve the fuel pressure (see Chapter 4).
4  Disconnect the negative cable from the battery.
5  Remove the intake and exhaust manifolds (see Sections 5 and 6).
**Note:** *Although the cylinder head may be removed with the manifolds and camshafts still in place, the assembly is so cumbersome and heavy that it is not recommended.*
6  Remove the valve cover (see Section 4).
7  Set the engine at Top Dead Center (see Section 3) and then remove the timing belt as described in Section 7.
8  Remove the camshafts and rocker arms (see Section 10).
9  Remove the distributor and the crank angle sensor (see Chapter 5).
10  Check the cylinder head. Label and remove any remaining items, such as coolant fittings, tubes, cables, hoses or wires. At this point the head should be ready for removal.
11  Loosen the cylinder head bolts in 1/3-turn increments until they can be removed by hand.
12  Lift the cylinder head off the engine block. If it's stuck, very carefully pry up on a protrusion beyond the gasket surface. **Caution:** *Do not pry between the block and head.*
13  See Chapter 2, Part C for cylinder head servicing information.

### Installation

*Refer to illustration 12.23*

14  The mating surfaces of the cylinder head and block must be perfectly clean when the head is installed. Pull the oil control jet out and clean out the orifice.

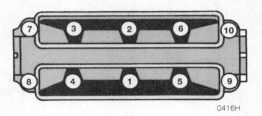

**12.23  Cylinder head bolt tightening sequence – the longer bolt goes in position no. 8**

15  Use a gasket scraper to remove all traces of carbon and old gasket material. Be very careful when scraping on the head, since the aluminum can be easily damaged. Clean the mating surfaces with lacquer thinner or acetone. If there's oil on the mating surfaces when the head is installed, the gasket may not seal correctly and leaks could develop. When working on the block, stuff the cylinders with clean shop rags to keep out debris. Use a vacuum cleaner to remove material that falls into the cylinders.
16  Check the block and head mating surfaces for nicks, deep scratches and other damage. If damage is slight, it can be removed with a file; if it's excessive, machining may be the only alternative.
17  Use a tap of the correct size to chase the threads in the head bolt holes, then clean the holes with compressed air – make sure that nothing remains in the holes. **Warning:** *Wear eye protection.*
18  Mount each bolt in a vise and run a die down the threads to remove corrosion and restore the threads. Dirt, corrosion, sealant and damaged threads will affect torque readings.
19  Install the components that were removed from the head. Be sure the oil control O-ring and jet are in position.
20  Position the new gasket over the dowel pins in the block.
21  Carefully set the head on the block without disturbing the gasket.
22  Before installing the head bolts, apply a small amount of clean engine oil to the threads.
23  Install the bolts in their original locations and tighten them finger tight. Following the recommended sequence, tighten the bolts in two steps to the specified torque **(see illustration)**. **Note:** *the longer bolt goes in the no. 8 position.*
24  The remaining installation steps are the reverse of removal.
25  Check and adjust the valves as necessary (see Chapter 1).
26  Refill and bleed the cooling system, install a new oil filter and add oil to the engine (see Chapter 1).
27  Run the engine and check for leaks. Set the ignition timing (see Chapter 1) and road test the vehicle.

## 13  Oil pan – removal and installation

*Refer to illustration 13.7*

1  Disconnect the negative cable from the battery.
2  Set the parking brake and block the rear wheels.
3  Raise the front of the vehicle and support it securely on jackstands.
4  Remove the splash shields under the engine.
5  Drain the engine oil and remove the oil filter (see Chapter 1). Remove the oil dipstick.
6  Disconnect the exhaust pipe from the exhaust manifold (see Section 6) and remove the clamp behind the engine to allow the pipe to hang down.
7  Attach a hoist and lifting sling to the engine (see Chapter 2, Part C, Section 5) and raise the engine just enough to take the weight off the lower mount. Unbolt the lower engine mount (see Section 17), then remove the crossmember under the oil pan **(see illustration)**.
8  Remove the block-to-transaxle brace, if necessary.

# Chapter 2 Part A Integra engine

13.7 Unbolt the crossmember (arrows) and then remove the bolts around the perimeter of the oil pan

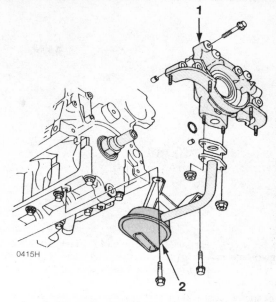

14.1 Oil pump details – exploded view

1  Oil pump  
2  Oil pickup tube and screen assembly

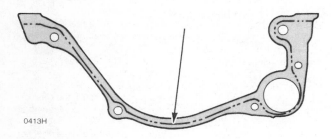

14.6 Apply sealant to the inside of the broken line (arrow)

9  Remove the bolts and detach the oil pan. If it's stuck, pry it loose very carefully with a small screwdriver or putty knife. Don't damage the mating surfaces of the pan and block or oil leaks could develop.
10  Use a scraper to remove all traces of old gasket material and sealant from the block and oil pan. Clean the mating surfaces with lacquer thinner or acetone.
11  Make sure the threaded bolt holes in the block are clean.
12  Check the oil pan flange for distortion, particularly around the bolt holes. If necessary, place the pan on a block of wood and use a hammer to flatten and restore the gasket surface.
13  Inspect the oil pump pick-up tube assembly for cracks and a blocked screen. If the pick-up was removed, install it now, using a new gasket.
14  Apply a 5 mm wide bead of RTV sealant to each oil pan corner (where the half-circle arches meet the flat portions of the gasket surface). **Note:** *The oil pan must be installed within 3 minutes once the sealant has been applied.*
15  Carefully position a new gasket and the oil pan on the engine block and install the bolts. Working from the center out, tighten them to the torque listed in this Chapter's Specifications in three steps.
16  The remainder of installation is the reverse of removal. Be sure to add oil and install a new oil filter.
17  Run the engine and check for oil pressure and leaks.

## 14  Oil pump – removal and installation

*Refer to illustrations 14.1 and 14.6*

1  Remove the oil pan (see Section 13), oil pickup tube and screen **(see illustration)**.
2  Remove the timing belt, crankshaft sprocket and belt guides (see Section 7). **Note:** *Since the oil pan has been removed, the engine should be supported securely from above (see Chapter 2, Part C, Section 5 for hoisting instructions).*
3  Remove any brackets as necessary for access, then remove the pump-to-block mounting bolts **(see illustration 14.1)** and separate the pump from the engine.
4  Use a scraper to remove any traces of old gasket material from the pump and block, then clean the surfaces with lacquer thinner or acetone.
5  Install a new crankshaft oil seal into the pump housing as described in Section 8.
6  Apply a thin film of sealant (Acura no. 06718-550000 OE or equivalent) to the engine side of the pump housing **(see illustration)** and the mounting bolts.
7  Pour an ounce or two of oil into the pump through the pickup tube hole to help prime it. Using a new O-ring, position the pump onto the engine. Use the locating dowels as a guide. **Note:** *Do not allow the sealant to dry – install immediately.*
8  Tighten the bolts to the torque listed in this Chapter's Specifications in three steps. Follow a criss-cross pattern to avoid warping the housing.
9  Using a new gasket, install the oil pickup tube and screen.
10  Reinstall the remaining parts in the reverse order of removal.
11  Add oil, start the engine and check for oil pressure and leaks.
12  Recheck the oil level.

## 15  Flywheel/driveplate – removal and installation

Refer to Chapter 2, Part B, Section 14 for this procedure, but be sure to use the torque specifications in this Part of Chapter 2 for the Integra.

## 16  Rear crankshaft oil seal – replacement

Refer to Chapter 2, Part B, Section 15 for this procedure.

2A-12    Chapter 2 Part A    Integra engine

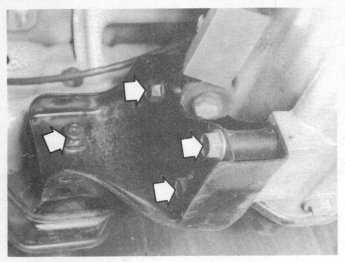

17.10a  Front engine mount bolt locations (arrows)

17.10b  Left engine mount bolt/nut locations (arrows)

### 17  Engine mounts – check and replacement

*Refer to illustrations 17.10a, 17.10b and 17.10c*

1  Engine mounts seldom require attention, but broken or deteriorated mounts should be replaced immediately or the added strain placed on the driveline components may cause damage or wear.

#### Check

2  During the check, the engine must be raised slightly to remove the weight from the mounts.

3  Raise the vehicle and support it securely on jackstands, then position a jack under the engine oil pan. Place a large block of wood between the jack head and the oil pan, then carefully raise the engine just enough to take the weight off the mounts. **Warning:** *DO NOT place any part of your body under the engine when it's supported only by a jack!*

4  Check the mounts to see if the rubber is cracked, hardened or separated from the metal plates. Sometimes the rubber will split right down the center.

5  Check for relative movement between the mount plates and the engine or frame (use a large screwdriver or prybar to attempt to move the mounts). If movement is noted, lower the engine and tighten the mount fasteners.

6  Rubber preservative should be applied to the mounts to slow deterioration.

#### Replacement

7  Disconnect the negative battery cable from the battery, then raise the vehicle and support it securely on jackstands (if not already done).

8  To ensure maximum bushing life and prevent excessive noise and vibration, the vehicle should be level and the engine weight should be on the mounts during the final tightening stage. Ensure that the bushings are not twisted or offset.

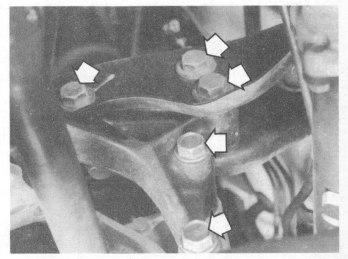

17.10c  Rear engine mount bolt locations (arrows)

9  To remove mounts one at a time, support the engine with a floor jack under the oil pan. Place a block of wood between the jack head and the oil pan to protect it from damage. **Warning:** *DO NOT place any part of your body under the engine when it's supported only by a jack!* **Note:** *Use thread locking compound on the nuts/bolts.*

10  Remove the nuts/bolts and slip the mount out of the vehicle **(see illustrations)**.

11  Install the mount and tighten the fasteners to the torque values listed in this Chapter's Specifications.

# Chapter 2 Part B  Legend engine

## Contents

| | |
|---|---|
| Camshaft oil seals – replacement | 9 |
| Camshafts and valve components – removal, inspection and installation | 10 |
| Compression check | See Chapter 2C |
| Crankshaft front oil seal – replacement | 8 |
| Crankshaft rear oil seal – replacement | 15 |
| Cylinder head covers – removal and installation | 4 |
| Cylinder heads – removal and installation | 11 |
| Drivebelt check, adjustment and replacement | See Chapter 1 |
| Engine mounts and torque rod – check and replacement | 16 |
| Engine oil and filter change | See Chapter 1 |
| Engine overhaul – general information | See Chapter 2C |
| Engine – removal and installation | See Chapter 2C |
| Exhaust manifolds – removal and installation | 6 |
| Flywheel/driveplate – removal and installation | 14 |
| General information | 1 |
| Intake manifold – removal and installation | 5 |
| Oil pan – removal and installation | 12 |
| Oil pump – removal and installation | 13 |
| Repair operations possible with the engine in the vehicle | 2 |
| Spark plug replacement | See Chapter 1 |
| Timing belt and sprockets – removal, inspection and installation | 7 |
| Top Dead Center (TDC) for number one piston – locating | 3 |
| Water pump – removal and installation | See Chapter 3 |

## Specifications

### General

Cylinder numbers (timing belt end-to-transaxle end)
- Rear (firewall) side ............................. 1-2-3
- Front (radiator) side ........................... 4-5-6

Firing order ........................................... 1-4-2-5-3-6
Rocker arm-to-shaft oil clearance service limit ..... 0.003 inch (0.08 mm)

*The blackened terminal shown on the distributor cap indicates the Number One spark plug wire position*

**Cylinder location and distributor rotation**

## Chapter 2 Part B Legend engine

**Camshaft and rocker arms**

Camshaft bearing oil clearance
- Standard: 0.0018 to 0.0032 inch (0.045 to 0.081 mm)
- Service limit: 0.0040 inch (0.10 mm)

Camshaft lobe height
- Intake: 1.5535 inch (39.460 mm)
- Exhaust: 1.5515 inch (39.409 mm)

Camshaft endplay
- Standard: 0.002 to 0.006 inch (0.05 to 0.15 mm)
- Service limit: 0.02 inch (0.5 mm)

Camshaft runout limit (total indicator reading): 0.001 inch (0.03 mm)
Rocker arm-to-shaft oil clearance service limit: 0.003 in (0.08 mm)

**Torque specifications**  Ft-lbs

Camshaft bearing cap bolts
- M6: 9
- M8: 20

Camshaft sprocket bolts: 23

Crankshaft pulley bolt
- 1986 to 1988: 83
- 1989 and later: 117 to 123

Cylinder head bolts
- Step one: 29
- Step two: 56

Cylinder head cover bolts
- Top: 11
- PCV cover: 9
- Side: 9

Driveplate bolts (automatic transaxle models): 54
Engine oil cooler bolts (M8): 16
Exhaust manifold nuts: 25
Flywheel bolts (manual transaxle models): 76
Intake manifold bolts/nuts: 16
Oil pan bolts/nuts: 10
Oil pick-up screen mounting bolts: 9

Oil pump mounting bolts
- M6 bolts: 9
- M8 bolts: 16

Rocker arm guide plate bolts: 9
Rocker arm shaft sealing plug: 33
Timing belt tensioner bolt: 31
Timing belt cover bolts: 7
Rear crankshaft oil seal retainer bolts: 9

### 1 General information

This Part of Chapter 2 is devoted to in-vehicle repair procedures for the V6 engine. All information concerning engine removal and installation and engine block and cylinder head overhaul can be found in Part C of this Chapter.

The following repair procedures are based on the assumption that the engine is installed in the vehicle. If the engine has been removed from the vehicle and mounted on a stand, many of the steps outlined in this Part of Chapter 2 will not apply.

The Specifications included in this Part of Chapter 2 apply only to the procedures contained in this Part. Part C of Chapter 2 contains the Specifications necessary for cylinder head and engine block rebuilding.

### 2 Repair operations possible with the engine in the vehicle

Many major repair operations can be accomplished without removing the engine from the vehicle.

Clean the engine compartment and the exterior of the engine with some type of degreaser before any work is done. It will make the job easier and help keep dirt out of the internal areas of the engine.

Depending on the components involved, it may be helpful to remove the hood to improve access to the engine as repairs are performed (refer to Chapter 11 if necessary). Cover the fenders to prevent damage to the paint. Special pads are available, but an old bedspread or blanket will also work.

If vacuum, exhaust, oil or coolant leaks develop, indicating a need for gasket or seal replacement, the repairs can generally be made with the engine in the vehicle. The intake and exhaust manifold gaskets, oil pan gasket, crankshaft oil seals and cylinder head gaskets are all accessible with the engine in place.

Exterior engine components, such as the intake and exhaust manifolds, the oil pan, the oil pump, the water pump, the starter motor, the alternator, the distributor and the fuel system components can be removed for repair with the engine in place.

Since the cylinder heads can be removed without pulling the engine, valve component servicing can also be accomplished with the engine in the vehicle. Replacement of the camshafts, timing belt and sprockets is also possible with the engine in the vehicle.

In extreme cases caused by a lack of necessary equipment, repair or replacement of piston rings, pistons, connecting rods and rod bearings is possible with the engine in the vehicle. However, this practice is not recommended because of the cleaning and preparation work that must be done to the components involved.

### 3 Top Dead Center (TDC) for number one piston – locating

*Refer to illustrations 3.6 and 3.8*

**Note:** *The following procedure is based on the assumption that the distributor is correctly installed. If you are trying to locate TDC to install the distributor correctly, piston position must be determined by feeling for*

# Chapter 2 Part B  Legend engine

**3.6  Mark the distributor below the number one terminal (arrow)**

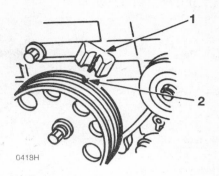

**3.8  Align the TDC mark with the pointer**

1  Pointer on timing cover     2  TDC mark

compression at the number one spark plug hole, then aligning the ignition timing marks as described in step 8.

1  Top Dead Center (TDC) is the highest point in the cylinder that each piston reaches as it travels up-and-down when the crankshaft turns. Each piston reaches TDC on the compression stroke and again on the exhaust stroke, but TDC generally refers to piston position on the compression stroke.

2  Positioning the piston(s) at TDC is an essential part of several procedures such as camshaft and timing belt/sprocket removal and distributor removal.

3  Before beginning this procedure, be sure to place the transaxle in Neutral and apply the parking brake or block the rear wheels. Also, disable the ignition system by detaching the coil wire from its terminal on the distributor cap and grounding it on the block with a jumper wire. Remove the spark plugs (see Chapter 1).

4  In order to bring any piston to TDC, the crankshaft must be turned using one of the methods outlined below. When looking at the timing belt end of the engine, normal crankshaft rotation is counter-clockwise.
   a) The preferred method is to remove the lower splash shield on the driver's side and turn the crankshaft with a socket and ratchet attached to the bolt threaded into the front of the crankshaft.
   b) A remote starter switch, which may save some time, can also be used. Follow the instructions included with the switch. Once the piston is close to TDC, use a socket and ratchet as described in the previous paragraph.
   c) If an assistant is available to turn the ignition switch to the Start position in short bursts, you can get the piston close to TDC without a remote starter switch. Make sure your assistant is out of the vehicle, away from the ignition switch, then use a socket and ratchet as described in Paragraph a) to complete the procedure.

5  Note the position of the terminal for the number one spark plug wire on the distributor cap. If the plug wire isn't marked, follow the plug wire from the number one cylinder spark plug to the cap.

6  Use a felt-tip pen or chalk to make a mark on the distributor body directly under the terminal **(see illustration)**.

7  Detach the cap from the distributor and set it aside (see Chapter 1 if necessary).

8  Turn the crankshaft (see Paragraph 4 above) until the TDC notch in the crankshaft pulley is aligned with the pointer on the timing belt cover **(see illustration)**.

9  Look at the distributor rotor – it should be pointing directly at the mark you made on the distributor body.

10  If the rotor is 180-degrees off, the number one piston is at TDC on the exhaust stroke.

11  To get the piston to TDC on the compression stroke if the rotor is 180-degrees off, turn the crankshaft one complete turn (360-degrees) clockwise. The rotor should now be pointing at the mark on the distributor.

**4.2  Rear top cylinder head cover details**

1  Spark plug wire bracket     3  Oil temperature sensor
2  Breather hose

When the rotor is pointing at the number one spark plug wire terminal in the distributor cap and the ignition timing marks are aligned, the number one piston is at TDC on the compression stroke.

12  After the number one piston has been positioned at TDC on the compression stroke, TDC for any of the remaining pistons can be located by turning the crankshaft and following the firing order. Mark the remaining spark plug wire terminal locations on the distributor body just like you did for the number one terminal, then number the marks to correspond with the cylinder numbers. As you turn the crankshaft, the rotor will also turn. The crankshaft must be turned 120-degrees to move from one cylinder to the next one in the firing order. When it's pointing directly at one of the marks on the distributor, the piston for that particular cylinder is at TDC on the compression stroke.

## 4  Cylinder head covers – removal and installation

*Refer to illustrations 4.2, 4.4 and 4.6*

### Top covers

1  Disconnect the negative cable from the battery.
2  Remove the spark plug connectors, wires and brackets from the spark plugs and set them aside **(see illustration)**. Unplug the oil temperature sensor located on top of the rear cover.
3  Detach the breather hose from the cover fitting.

# 2B–4  Chapter 2 Part B  Legend engine

4.4  Remove the retaining bolts (arrows) – front top cylinder head cover shown, rear cover similar

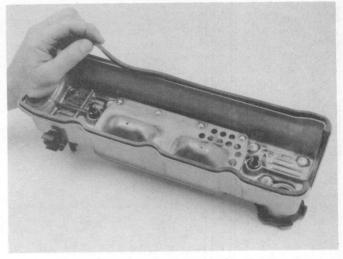

4.6  Position a new gasket in the groove

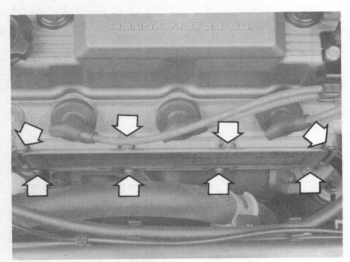

4.9  Remove the bolts (arrows)

4.11  Be sure both surfaces are clean

4   Remove the retaining bolts **(see illustration)**, then detach the top cover. If the cover is stuck to the head, bump the end with a block of wood and a hammer to jar it loose. **Caution:** *Don't pry at the cover-to-head joint or damage to the sealing surfaces may occur, leading to oil leaks after the cover is reinstalled.*
5   Remove the old gasket and seal washers and clean the mating surfaces of the cylinder head and cover. Remove the small PCV cover from the front cylinder head cover to clean and inspect the PCV valve (see Chapter 1) and replace the seals.
6   Position a new gasket in the groove **(see illustration)** and install new seal washers on the bolts.
7   Install the cover and tighten the bolts to the torque listed in this Chapter's Specifications in three equal steps.
8   Reinstall the remaining parts, run the engine and check for oil leaks.

## Side covers

*Refer to illustrations 4.9 and 4.11*

9   Remove the bolts around the perimeter of the cover **(see illustration)**. On the front cover, detach and set aside the ground cable. On both the front and rear covers, detach the wiring harness brackets.
10  Detach the cover from the engine. If the cover is stuck to the head, bump the end with a block of wood and a hammer to jar it loose.

**Caution:** *Don't pry at the cover-to-head joint or damage to the sealing surfaces may occur, leading to oil leaks after the cover is reinstalled.*
11  Remove the old gasket and clean the mating surfaces of the cylinder head **(see illustration)** and cover.
12  Position a new gasket in the groove.
13  Install the cover and tighten the bolts to the torque listed in this Chapter's Specifications in two equal steps.
14  Reinstall the remaining parts, run the engine and check for oil leaks.

## 5   Intake manifold – removal and installation

*Refer to illustrations 5.3, 5.8a, 5.8b, 5.10 and 5.11*

1   Relieve the fuel pressure (see Chapter 4) and then disconnect the negative cable from the battery.
2   Drain the coolant into a clean container (see Chapter 1).
3   Clearly label, then detach all wires, hoses and brackets attached to the intake manifold **(see illustration)**.
4   Remove the alternator and power steering pump (see Chapters 5 and 10).
5   Detach the spark plug wires from the spark plugs and wire holders (leave them connected at the distributor).
6   Disconnect the accelerator cable from the throttle body and remove the fuel injectors, lines and hoses (see Chapter 4).

# Chapter 2 Part B  Legend engine

5.3  Label and then remove the wiring and hoses

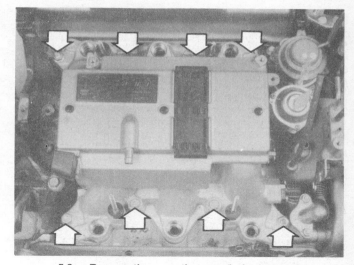

5.8a  Remove the mounting nuts/bolts (arrows)

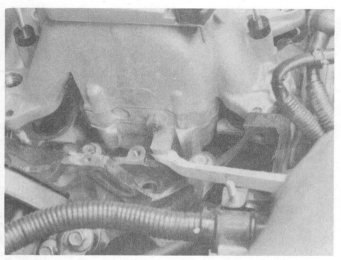

5.8b  Pry against external tabs such as this

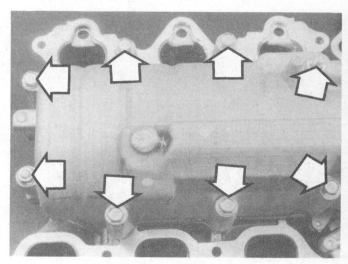

5.10  Remove the bolts/nuts (arrows)

manifold from the engine. If it's stuck, don't pry between the gasket mating surfaces or damage may result **(see illustration)**.

9  Carefully use a scraper to remove all traces of old gasket material and sealant from the manifold and cylinder heads, then clean the mating surfaces with lacquer thinner or acetone.

10  If necessary, remove the bolts/nuts **(see illustration)** and separate the manifold and bypass valve body. Reassemble with new gaskets and tighten securely.

11  Install new gaskets **(see illustration)**, then position the manifold on the engine. Make sure the gaskets haven't shifted and install the nuts/bolts.

12  Tighten the nuts/bolts, in three equal steps, to the torque listed in this Chapter's Specifications. Work from the center out towards the ends to avoid warping the manifold.

13  Install the remaining parts in the reverse order of removal.

14  Refill the cooling system. Run the engine and check for fuel, vacuum and coolant leaks.

5.11  Slip new gaskets over the studs as shown

7  Detach the air suction pipe from the front exhaust manifold.
8  Remove the mounting nuts/bolts **(see illustration)**, then detach the

6  Exhaust manifolds – removal and installation

*Refer to illustrations 6.4a, 6.4b, 6.6a, 6.6b and 6.8*
**Note:** *The engine must be completely cool when this procedure is done.*

1  Disconnect the negative cable from the battery.

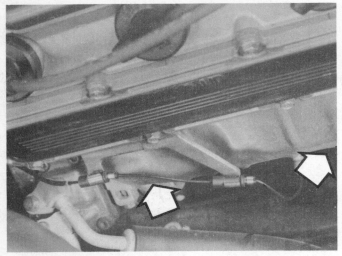

6.4a  Remove the shroud bolts (arrows point to bolts out of view)

6.4b  Also remove this shroud nut (arrow)

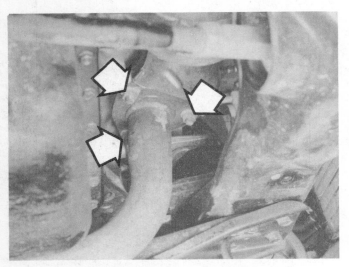

6.6a  Front exhaust pipe nut locations (arrows)

6.6b  Rear exhaust pipe nut locations (arrows)

2  Spray penetrating oil on the exhaust manifold fasteners and allow it to soak in.
3  Unplug the oxygen sensor from the wiring harness.
4  Unbolt the heat insulator shroud from the manifold(s) being removed **(see illustrations)**.
5  Block the rear wheels to prevent the vehicle from rolling. Set the parking brake and place the transaxle in Park (automatic) or in first gear (manual). Raise the front of the vehicle and support it securely on jackstands. Remove the lower splash guard.
6  Disconnect the exhaust pipes from the manifolds **(see illustrations)** and detach the pipes from the vehicle.
7  Remove the air suction pipe from the front manifold (see Chapter 6).
8  Remove the nuts retaining the manifold to the cylinder head **(see illustration)** and slip it off the mounting studs.
9  Carefully inspect the manifold and fasteners for cracks and damage.
10  Use a scraper to remove any traces of old gasket material and carbon deposits from the manifold and cylinder head mating surfaces. If the gasket was leaking, have the manifold checked for warpage at an automotive machine shop and resurfaced if necessary.
11  Position a new gasket over the cylinder head studs.
12  Install the manifold and thread the mounting nuts into place.
13  Working from the center out, tighten the nuts to the torque listed in this Chapter's Specifications in three equal steps.
14  Reinstall the remaining parts in the reverse order of removal. Use new

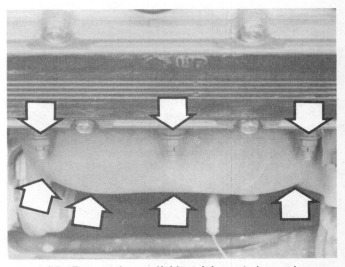

6.8  Remove the manifold retaining nuts (arrows)

gaskets when connecting the exhaust pipes and air suction pipe.
15  Run the engine and check for exhaust leaks.

# Chapter 2 Part B  Legend engine

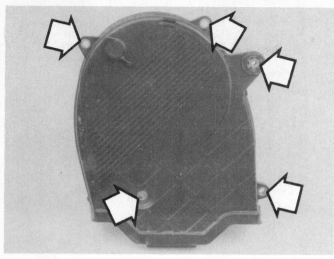

7.9a  Front timing belt cover bolt locations (arrows)

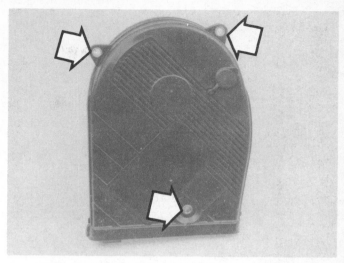

7.9b  Rear timing belt cover bolt locations (arrows)

7.11  Mark the timing belt with the direction of rotation

7.12a  Front camshaft timing marks (arrows)

## 7  Timing belt and sprockets – removal, inspection and installation

### Removal

*Refer to illustrations 7.9a, 7.9b, 7.11, 7.12a, 7.12b, 7.13, 7.14, 7.15a, 7.15b, 7.16 and 7.17*

1  Disconnect the negative cable from the battery.
2  Loosen the lug nuts on the right front wheel, but don't remove them yet.
3  Place the manual transaxle in Neutral or the automatic in Park.
4  Apply the parking brake and block the rear wheels. Raise the front of the vehicle and support it securely on jackstands. Remove the right front wheel.
5  Remove the right front inner fender splash guard.
6  Remove the drivebelts (see Chapter 1).
7  Remove the alternator and power steering pump (see Chapters 5 and 10).
8  Support the engine by placing a floor jack under the oil pan and remove the upper engine mount (see Section 16). Place a block of wood on the jack pad to protect the oil pan.
9  Remove the upper timing belt covers **(see illustrations)**.
10  Remove the spark plugs to make it easier to turn the crankshaft (see Chapter 1). Position the number one piston at TDC (see Section 3).

7.12b  Rear camshaft timing marks (arrows)

11  If you intend to reuse the belt, mark the belt to indicate the direction of rotation **(see illustration)**.
12  Make sure the timing marks are properly aligned **(see illustrations)**.

7.13  Remove the crankshaft pulley bolt (arrow)

7.14  Lower timing belt cover bolt locations (arrows)

7.15a  Crankshaft sprocket timing marks (engine removed for clarity)

7.15b  Loosen the timing belt tensioner bolt (arrow)

7.16  Hold the camshaft from turning with a heavy screwdriver while you loosen the bolts, then pull the sprocket off

13  Remove the crankshaft pulley (see illustration). Wrap a rag and then a chain wrench around the pulley on automatic transaxle models or have an assistant place the transaxle in gear and hold the brakes on manual transaxle models. Note: *When the crankshaft pulley bolt is loosened, the position of the timing marks on the crankshaft pulley and the camshafts may be disturbed. Check and align them again. Temporarily reinstall the crankshaft pulley bolt to turn the crankshaft.*

14  Remove the lower timing belt cover (see illustration).

15  Slip the timing belt guide off the crankshaft sprocket, noting how it's installed. Also note the alignment of the crankshaft sprocket timing marks (see illustration). Loosen the timing belt tensioner bolt and remove the timing belt (see illustration).

16  The camshaft sprockets can be removed at this point, if they are damaged or to replace the oil seals (see illustration). On the rear sprocket, remove the lower bolt last. Remove the keys from the shafts so they don't fall out and get lost.

17  If it's worn or damaged, or if you're replacing the front crankshaft oil seal, the crankshaft sprocket can now be removed. If it won't come off by hand, lever it off (see illustration). Also remove the inner timing belt guide, noting how it's installed.

### Inspection

18  Refer to Chapter 2, Part A, Section 7 for the timing belt inspection procedure.

# Chapter 2 Part B  Legend engine

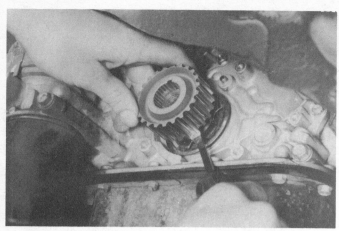

**7.17 Carefully work the crankshaft sprocket off without damaging anything**

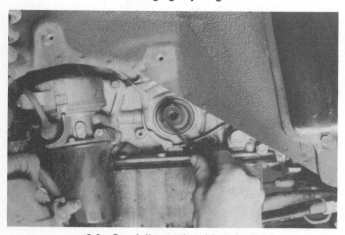

**8.2 Carefully pry the old seal out**

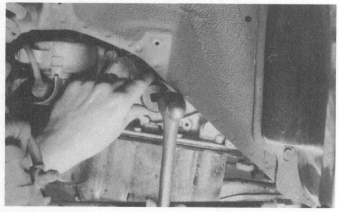

**8.4 Lubricate the seal lip and tap the new crankshaft seal into place with a large socket or piece of pipe and a hammer**

## Installation

*Refer to illustration 7.24*

19  Remove all dirt and oil from the timing belt area.
20  If any of the timing belt sprockets were removed, install them now with their keys and tighten the bolts to the torque listed in this Chapter's Specifications.
21  Temporarily advance the crankshaft about 15-degrees.
22  If removed, install the inner timing belt guide over the crankshaft sprocket with the chamfered edge facing away from the belt. Also install the crankshaft sprocket.

**7.24 Install the outer timing belt guide (arrow) with the chamfered side out**

23  Recheck the position of the timing marks (see illustrations 7.12a, 7.12b and 7.15a). Install the timing belt, starting at the crankshaft sprocket, then front camshaft sprocket, water pump, belt tensioner and rear camshaft sprocket. For ease of installation, temporarily advance the rear camshaft sprocket about one-half tooth. If you're reusing the original belt, the arrow you made in Step 11 should point in the normal direction of rotation.
24  Install the outer timing belt guide over the crankshaft sprocket with the chamfered edge facing away from the belt **(see illustration)**.
25  Loosen the tensioner adjustment bolt, allow the tensioner to move into position and retighten the bolt.
26  Turn the crankshaft slowly six revolutions clockwise using a socket and breaker bar on the crankshaft pulley bolt to seat the belt and return to TDC. **Caution:** *If anything hits, back up and recheck the belt timing. Do not force the crankshaft to turn or engine damage will occur!*
27  Install the lower timing belt cover.
28  Slip the drivebelt pulley onto the crankshaft, aligning the pulley keyway with the crankshaft key. Install the bolt and tighten it to the specified torque. Use the method described above to keep the crankshaft from turning.
29  Turn the crankshaft clockwise nine teeth on the CAMSHAFT sprocket. The blue mark on the CRANKSHAFT pulley should align with the pointer on the belt cover.
30  Loosen the tensioner bolt, allow the tensioner to move into position and tighten the bolt to the torque listed in this Chapter's Specifications.
31  Recheck the timing marks (see illustrations 7.12a and 7.12b). **Caution:** *If the timing marks are not aligned exactly as shown, repeat the timing belt installation procedure. DO NOT start the engine until you're absolutely certain that the timing belt is installed correctly. Serious and costly engine damage could occur if the belt is installed wrong.*
32  Reinstall the remaining parts in the reverse order of removal.

## 8  Crankshaft front oil seal – replacement

*Refer to illustrations 8.2 and 8.4*

1  Remove the timing belt and crankshaft sprocket (see Section 7). **Note:** *Leave the engine mount in place to support the engine.*
2  Carefully pry the seal out of the engine with a screwdriver or seal removal tool **(see illustration)**. If you use a screwdriver, don't scratch the housing bore or damage the crankshaft (if the crankshaft is damaged, the new seal will end up leaking).
3  Clean the oil seal bore and coat the outer edge of the new seal with a small amount of engine oil to ease installation. Apply moly-base grease to the seal lip.
4  Using Acura tool no. 07GAD-PH70200 or a socket with an outside diameter slightly smaller than the outside diameter of the seal, carefully drive the new seal into place with a hammer **(see illustration)**. Make sure it's installed squarely and driven in to the same depth as the original. If a

# 2B-10  Chapter 2 Part B  Legend engine

9.2  Remove the oil seal from its bore (arrow)

9.4  If a seal driver is not available, use a hammer and a section of pipe or a large socket to tap the new seal into place

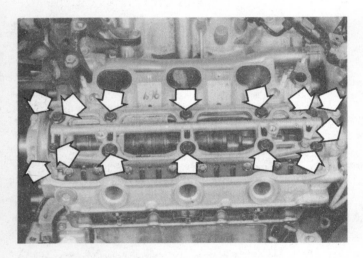

10.3  Remove the bolts (arrows)

10.4  Carefully lift off the bearing cap assembly

socket isn't available, a short section of large diameter pipe will also work. Check the seal after installation to make sure the garter spring didn't pop out of place.

5   Reinstall the crankshaft sprocket and timing belt (see Section 7).
6   Run the engine and check for oil leaks at the front seal.

## 9  Camshaft oil seals – replacement

*Refer to illustrations 9.2 and 9.4*

**Note:** *Both camshaft oil seals should be replaced at the same time, even if only one is leaking.*

1   Remove the timing belt and camshaft sprockets (see Section 7).
2   Note how far each seal is seated in the bore, then carefully pry it out with a straight screwdriver **(see illustration)**. Wrap the screwdriver tip with tape – don't scratch the bore or damage the camshaft (if the camshaft is damaged, the new seal will end up leaking).
3   Clean the bore and coat the outer edge of the new seal with engine oil or multi-purpose grease. Apply moly-base grease to the seal lip.
4   Using a socket with an outside diameter slightly smaller than the outside diameter of the seal, carefully drive the new seal into place with a hammer **(see illustration)**. Make sure it's installed squarely and driven in to the same depth as the original. If a socket of the correct size isn't available, a short section of pipe will also work. After the seal is installed, make sure the garter spring did not pop loose.

5   Reinstall the camshaft sprocket and timing belt (see Section 7).
6   Run the engine and check for oil leaks at the camshaft seals.

## 10  Camshafts and valve components – removal, inspection and installation

### Removal

*Refer to illustrations 10.3, 10.4, 10.5, 10.7, 10.9a, 10.9b, 10.9c, 10.9d, 10.10a and 10.10b*

**Note:** *To remove all the rocker arm components, it is necessary to remove the rocker arm shafts, which is difficult with the cylinder head installed in the vehicle. Therefore, if the rocker arm shafts must be removed, we recommend first removing the cylinder head from the vehicle (see the next Section).*

1   Remove the cylinder head covers (see Section 4) and the timing belt, sprockets and inner covers (see Section 7).
2   Remove the distributor (see Chapter 5).
3   Working in a sequence around the bolt pattern, loosen the camshaft bearing cap bolts **(see illustration)** in 1/3-turn increments until they can be removed by hand.
4   Remove the bearing cap oil pipe and cap assemblies **(see illustration)**. Remove the rear camshaft sealing plug. Note the position of the sprocket locating dowel pin for reassembly and gently lift out the camshaft. Be sure to keep it level during removal.

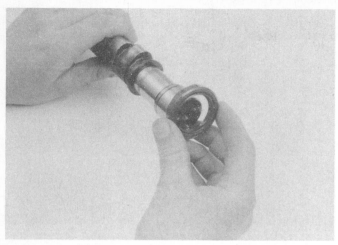

10.5  Slip the oil seal off the end of the camshaft

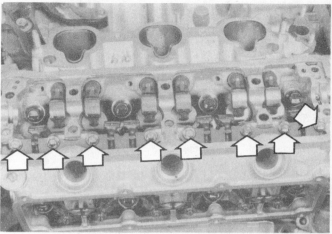

10.7  Remove the rocker arm guide plate bolts (arrows), then remove the rocker arms from the top of the cylinder head

10.9a  Unscrew the end caps . . .

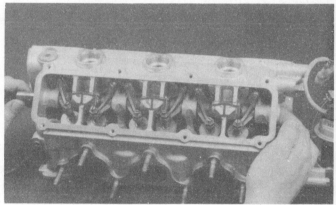

10.9b  . . . and guide the rocker shaft out

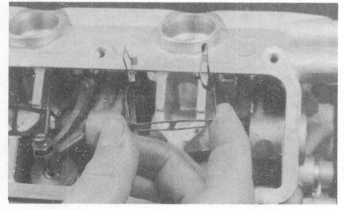

10.9c  Catch the wave washers (spring clips) as the shaft is removed

10.9d  The rocker arms will be released when the wave washers are removed

5  Remove the oil seal by slipping it off the end of the camshaft (see illustration).
6  Mark the bearing caps and oil pipes to indicate which cylinder head they came from.
7  Remove the rocker arm guide plates, then lift out the rocker arms (see illustration). Be sure to keep them separate so they can be reinstalled in their original locations. Label egg cartons to store and organize them.
8  Working in the lower cylinder head cover openings, loosen the exhaust rocker arm lock nuts and back off the adjustment screws. Lift out the pushrods. Push them through holes in a cardboard box and label their positions so they can be reinstalled in their original locations.
9  Remove the end caps and slide the exhaust rocker arm shaft out (see illustrations). If the shaft is stuck, install a long 12 x 1.25 mm bolt and use it as a handle. Remove the spring clips and lift the exhaust rocker arms out. Store them in order so they can be returned to their original locations.
**Note:** *Rocker shaft removal is difficult with the engine in the vehicle. Several components must be removed for clearance and rear cylinder head access is limited. Therefore, we recommend removing the cylinder head*

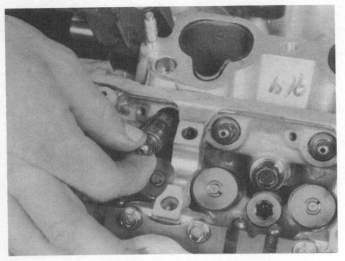

**10.10a** Pull the lash adjusters out

**10.10b** Store the lash adjusters in an organized manner

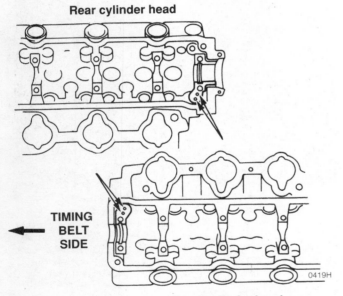

**10.15** Pour oil into the holes marked with arrows

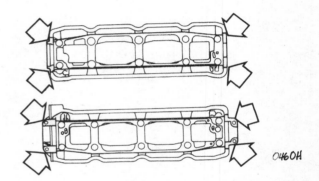

**10.20** Apply sealant at these locations (arrows)

*(see the next Section) when the rocker arm shaft must be removed.*
10  Remove the lash adjusters by pulling them out **(see illustration)**. Store them in order so they may be reinstalled in the same position **(see illustration)**.

### Inspection

11  Roll the pushrods on a piece of glass to check for bent ones. Replace as necessary.
12  Visually inspect the rocker arms. Slip the exhaust rockers over the rocker shaft and check for excess clearance and/or shaft wear. Check all the rockers for wear and damage, especially where the rockers contact the rocker shaft, pushrods and camshaft.
13  Check the lash adjusters for wear and damage, especially where the rockers contact them.
14  Refer to Chapter 2, Part A, Section 11 for camshaft checking procedures. Be sure to use the Specifications in this Part of Chapter 2 for the V6 engine.

### Installation

*Refer to illustrations 10.15, 10.20 and 10.22*

15  Pour engine oil into the lash adjuster holes in the cylinder heads, then insert the lash adjusters into the heads in the same locations they were originally in. Do not rotate the lash adjusters during installation. If you suspect any of the lash adjusters are faulty, take them to a dealer for air bleeding and inspection using tool no. 07GAJ-PH70100 or equivalent. Pour oil into the oil filler holes **(see illustration)**.
16  Apply moly-base grease or engine assembly lube to the contact surfaces of all moving parts. Install the exhaust rocker arms, wave washers and shafts into their original locations in the heads.
17  Install the upper rocker arms and pushrods in their original positions.
18  Lubricate the lips with engine oil, then install a new camshaft oil seal by slipping it over the end of the camshaft. Be sure the spring side faces in. Install the rear camshaft sealing plug.
19  Set the camshaft in place in the cylinder head, being sure the seal is positioned correctly. **Note:** *The front camshaft has a groove for driving the distributor.* Locate the front camshaft dowel pin at the one o'clock position and the rear camshaft dowel pin at the 11 o'clock position.
20  Apply a thin coat of non-hardening sealant to the outer edges of the bearing cap-to-cylinder head mating surfaces **(see illustration)**.
21  Install the bearing cap and oil pipe assemblies.
22  Tighten the bearing cap bolts in 1/3-turn increments until the torque listed in this Chapter's Specifications is reached. Follow the factory recommended sequence **(see illustration)**.
23  Apply moly-base grease or engine assembly lube to the camshaft lobes, bearing journals and gear thrust faces.
24  Install the other camshaft in the same manner.

# Chapter 2 Part B  Legend engine

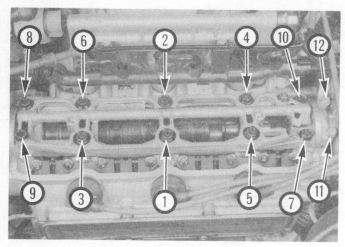

10.22  Bolt tightening sequence

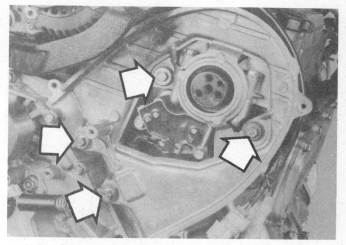

11.8a  Front inner timing belt cover bolt locations (arrows)

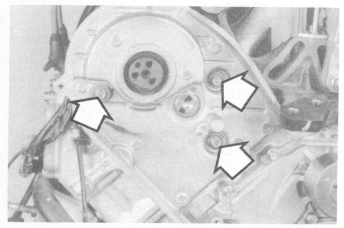

11.8b  Rear inner timing belt cover bolt locations (arrows)

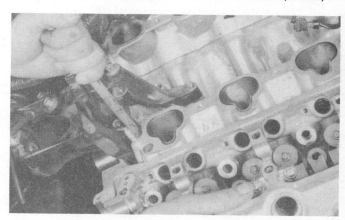

11.12  Pry up carefully on a casting protrusion

28  Reinstall the remaining components in the reverse order of removal.
29  Remove the spark plugs and crank the engine. Check for compression at each cylinder. If any cylinder lacks compression, it may be necessary to disassemble the head and check for faulty lash adjusters.
30  Run the engine, then check for leaks and proper operation.

## 11  Cylinder heads – removal and installation

**Caution:** *Allow the engine to cool completely before beginning this procedure.*

### Removal

*Refer to illustrations 11.8a, 11.8b, 11.10 and 11.12*

1  Relieve the fuel pressure (see Chapter 4).
2  Disconnect the negative cable from the battery.
3  Drain the cooling system, including both block drains (see Chapter 1).
4  Remove the alternator and distributor (see Chapter 5).
5  Remove the intake manifold (see Section 5).
6  Remove the exhaust manifold (see Section 6).
7  Detach the timing belt and camshaft sprockets (see Section 7).
8  Remove the inner timing belt cover **(see illustrations)**.
9  Remove the camshaft(s) from the head(s) you intend to remove (see Section 10).
10  Detach the coolant temperature sensor and the coolant passage assembly **(see illustration)**.
11  Using a socket, loosen the cylinder head bolts in 1/4-turn increments until they can be removed by hand.
12  Lift the cylinder head off the engine block. If the head is stuck, pry against an external casting protrusion **(see illustration)**. **Caution:** *Don't*

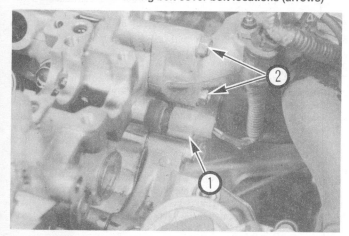

11.10  Remove these parts attached to the cylinder heads

1  *Coolant temperature sensor (EFI)*
2  *Coolant passage assembly bolts (two on each side)*

25  Reinstall the timing belt, covers and sprockets (see Section 7).
26  With the engine at Top Dead Center for number one cylinder (see Section 3), adjust the valves for cylinders number one, two and four. Loosen the locknut, tighten the adjusting screw until the tip contacts the valve, then turn it 1.5 turns more and tighten the locknut securely.
27  Turn the crankshaft one full turn (360-degrees). Adjust the valves for cylinders three, five and six as described in the previous Step.

**11.15 Use a scraper to remove all traces of old gasket material**

**11.19 Fit the new gasket over the oil control jet and locating dowels (arrows)**

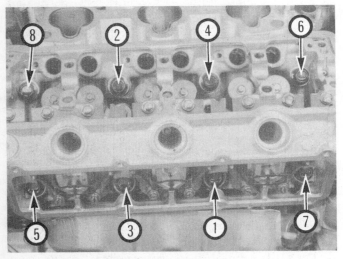

**11.22 Cylinder head bolt tightening sequence**

pry between the head and block. The gasket surfaces may be damaged and leaks could result.
13  Repeat Steps 6, 8, 11 and 12 for the other head.

### Installation

*Refer to illustrations 11.15, 11.19 and 11.22*

14  The mating surfaces of the cylinder heads and block must be perfectly clean when the heads are installed.
15  Use a gasket scraper to remove all traces of carbon and old gasket material **(see illustration)**. Be careful not to gouge the delicate aluminum. Clean the mating surfaces with lacquer thinner or acetone. If there's oil on the mating surfaces when the head is installed, the gasket may not seal correctly and leaks could develop. When working on the block, stuff the cylinders with clean shop rags to keep out debris. Use a vacuum cleaner to remove material that falls into the cylinders.
16  Check the block and head mating surfaces for nicks, deep scratches and other damage. If damage is slight, it can be removed with a file; if it's excessive, machining may be the only alternative.
17  Use a tap of the correct size to chase the threads in the head bolt holes, then clean the holes with compressed air – make sure that nothing remains in the holes. **Warning:** *Wear eye protection when using compressed air!*

18  Mount each bolt in a vise and run a die down the threads to remove corrosion and restore the threads. Dirt, corrosion, sealant and damaged threads will affect torque readings.
19  Position the new gaskets over the locating dowels in the block **(see illustration)**. Be sure to install the oil control jets with new O-rings.
20  Carefully set the head on the block without disturbing the gasket.
21  Before installing the head bolts, apply a small amount of clean engine oil to the threads.
22  Install the bolts and special washers and tighten them finger tight. Following the recommended sequence **(see illustration)**, tighten the bolts to the torque listed in this Chapter's Specifications in two steps.
23  Repeat the entire procedure to install the other cylinder head, if necessary.
24  The remaining installation steps are the reverse of removal.
25  Refill the cooling system, change the oil and filter (see Chapter 1), run the engine and check for leaks.

## 12  Oil pan – removal and installation

### Removal

*Refer to illustrations 12.6a, 12.6b, 12.9a and 12.9b*

1  Disconnect the negative cable from the battery.
2  Block the rear wheels and set the parking brake. Raise the front of the vehicle and support it securely on jackstands.
3  Remove the lower splash shields.
4  Drain the engine oil and remove the oil filter (see Chapter 1).
5  Remove the center engine mount (see Section 16).
6  Unbolt the longitudinal crossmember **(see illustrations)**.
7  Remove the exhaust pipe from the manifolds (see Section 6).
8  Detach the lower bellhousing cover, if equipped.
9  Remove the bolts **(see illustration)** and lower the oil pan. The bolts adjacent to the driveaxle can be removed with a 1/4-inch drive socket, extension and ratchet. If the pan is stuck, break it loose with a soft-face hammer **(see illustration)**. Don't damage the mating surfaces of the pan and block or oil leaks could develop.

### Installation

10  Use a scraper to remove all traces of old sealant from the block and oil pan. Be careful not to gouge the delicate aluminum block. Clean the mating surfaces with lacquer thinner or acetone.
11  Make sure the threaded bolt holes in the block are clean.
12  Check the oil pan flange for distortion, particularly around the bolt holes. If necessary, place the pan on a block of wood and use a hammer to flatten and restore the gasket surface.

# Chapter 2 Part B  Legend engine

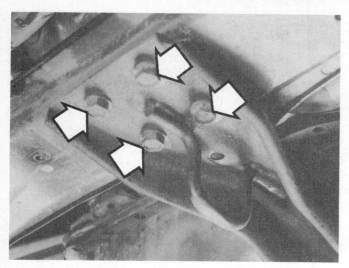

12.6a  Remove the bolts (arrows) . . .

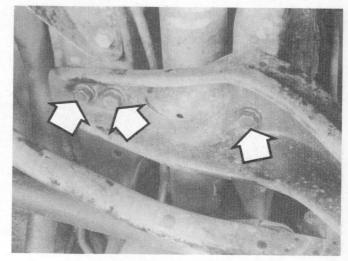

12.6b  . . . from the ends of the crossmember

12.9a  Remove the bolts from around the perimeter of the oil pan

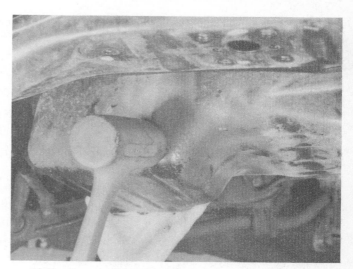

12.9b  Use a soft-face hammer to break the oil pan loose

13  Inspect the oil pump pick-up screen assembly for damage and a blocked strainer. If the pick-up and/or baffle was removed, install it now. Use a new O-ring on the pick-up. Tighten the fasteners to the torque listed in this Chapter's Specifications.
14  Position a new gasket on the oil pan.
15  Carefully position the oil pan on the engine block and install the bolts. Working from the center out, tighten them to the torque listed in this Chapter's Specifications in three steps.
16  The remainder of installation is the reverse of removal. Be sure to add oil and install a new oil filter.
17  Run the engine and check for oil pressure and leaks.

## 13  Oil pump – removal and installation

### Removal

*Refer to illustrations 13.3, 13.4, 13.5 and 13.6*

1  Remove the oil cooler (see Chapter 3).
2  Remove the timing belt and crankshaft sprocket (see Section 7). **Note:** *Leave the engine mount on the timing belt end of the engine connected to support the engine. It's not necessary to completely remove the belt.*

13.3  Oil pickup screen bolt locations (arrows)

3  Remove the oil pan (see Section 12) and oil pickup screen **(see illustration)**. On LS models, remove the oil level sensor.

## 2B–16  Chapter 2 Part B  Legend engine

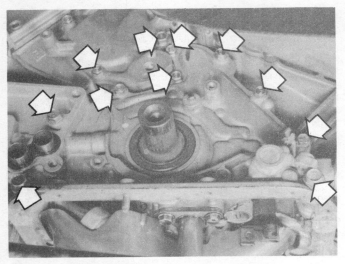

13.4  Oil pump bolt locations (arrows)

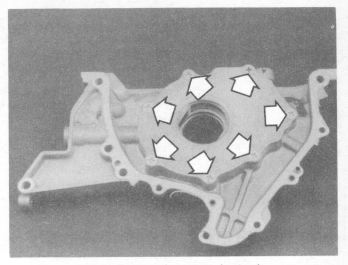

13.5  Remove the screws (arrows)

13.6  Inspect the condition of the rotors and the inside of the cover

14.3  Remove the bolts (arrows) with a 12-point socket

4  Remove the bolts and detach the oil pump from the engine (see illustration). You may have to pry carefully between the main bearing cap and the pump body with a screwdriver.
5  Use a large Phillips screwdriver to remove the screws holding the body cover to the rear of the oil pump (see illustration).
6  Lift the cover off and inspect the pump rotors (see illustration). If any wear or damage is evident, replace the pump.
7  Use a scraper to remove any traces of old sealant from the pump body and engine block, being careful not to damage the delicate aluminum.

### Installation

8  Replace the old crankshaft oil seal (see Section 8). Apply moly-base grease to the seal lip.
9  Pack the pump cavity with petroleum jelly and install the cover. Tighten the screws securely following a criss-cross pattern.
10  Use acetone or lacquer thinner and a clean rag to remove all traces of oil from the gasket surfaces.
11  Apply a bead of sealant (Honda no. 08718-5500000 OE or equivalent) to the oil pump flange and M8 x 1.25 x 45 mm bolt. Avoid using an excessive amount of sealant, especially around oil passages and bolt holes. Parts must be assembled within five minutes of sealant application, otherwise the material must be removed and reapplied. Use new O-rings where needed.

12  If removed, position the oil pass pipe for installation. Be sure to use a new seal.
13  Engage the flat surfaces on the oil pump drive rotor with the matching flats on the crankshaft and slide the pump into place.
14  Install the pump mounting bolts in their original locations and tighten them to the torque listed in this Chapter's Specifications in a criss-cross pattern.
15  Using a new O-ring, install the oil pick-up screen and tighten the fasteners to the torque listed in this Chapter's Specifications.
16  Reinstall the remaining parts in the reverse order of removal.
17  Add oil, start the engine and check for oil leaks and pressure.
18  Recheck the engine oil level.

### 14  Flywheel/driveplate – removal and installation

*Refer to illustrations 14.3 and 14.8*

1  Raise the vehicle and support it securely on jackstands, then refer to Chapter 7 and remove the transaxle. If it's leaking, now would be a very good time to replace the front pump seal/O-ring (automatic transaxle only).

## Chapter 2 Part B  Legend engine

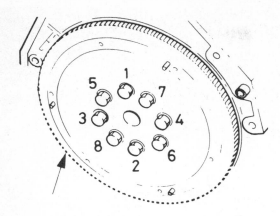

**14.8  Flywheel/driveplate bolt tightening sequence**

**15.2  Carefully pry the rear main seal out – don't damage the surface of the crankshaft or the new seal will leak**

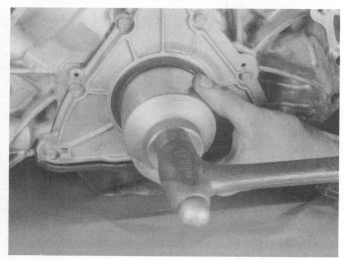

**15.3  Drive the new seal in squarely**

2  On manual transaxle equipped vehicles, remove the pressure plate and clutch disc (see Chapter 8). Now is a good time to check/replace the clutch components and pilot bearing.
3  Remove the bolts that secure the flywheel/driveplate to the crank-

shaft **(see illustration)**. Hold the crankshaft from turning with Acura tool no. 07924-PD20003 or 07924-PD20002. If these are unavailable, wedge a large screwdriver into the ring gear to jam the flywheel.
4  Remove the flywheel/driveplate from the crankshaft. Since it's fairly heavy, be sure to support it while removing the last bolt.
5  Clean the flywheel to remove grease and oil. Inspect the surface for cracks, rivet grooves, burned areas and score marks. Light scoring can be removed with emery cloth. Check for cracked and broken ring gear teeth. Lay the flywheel on a flat surface and use a straightedge to check for warpage.
6  Clean and inspect the mating surfaces of the flywheel/driveplate and the crankshaft. If the crankshaft oil seal is leaking, replace it before reinstalling the flywheel/driveplate.
7  Position the flywheel/driveplate against the crankshaft. Note that offset bolt holes ensure correct installation. Be sure to install the spacer washer with the driveplate.
8  Hold the crankshaft from turning as described above. Working in several stages, following the tightening sequence **(see illustration)** tighten the bolts to the torque listed in this Chapter's Specifications.
9  The remainder of installation is the reverse of the removal procedure.

## 15  Crankshaft rear oil seal – replacement

*Refer to illustrations 15.2 and 15.3*

1  The transaxle must be removed from the vehicle for this procedure and the flywheel/driveplate must be separated from the engine. Refer to Chapter 7 and Section 14 as necessary.
2  The seal can be replaced without dropping the oil pan or removing the seal retainer. However, the lip of the seal is quite stiff and it's possible to cock the seal in the retainer bore or damage it during installation. If you want to take the chance, pry out the old seal with a screwdriver **(see illustration)**.
3  Apply moly-base grease to the crankshaft seal journal and the lip of the new seal and carefully drive the new seal into place **(see illustration)**. Install the seal with the spring side in. Use a socket, section of pipe or Acura special tool no. 07749-0010000. The lip is stiff so carefully work it onto the seal journal of the crankshaft. Don't rush it or you may damage the seal.
4  The crankshaft seal replacement method described in Chapter 2C is recommended but requires removal of the oil pan and the seal retainer.
5  The remaining steps are the reverse of removal.

## 16  Engine mounts and torque rod – check and replacement

*Refer to illustrations 16.11, 16.13, 16.15, 16.19 and 16.21*

1  Engine mounts seldom require attention, but broken or deteriorated mounts should be replaced immediately or the added strain placed on the driveline components may cause damage or wear.

### Check

2  During the check, the engine must be raised slightly to remove the weight from the mounts.
3  Raise the vehicle and support it securely on jackstands, then position a jack under the engine oil pan. Place a large block of wood between the jack head and the oil pan, then carefully raise the engine just enough to take the weight off the mounts. **Warning:** *DO NOT place any part of your body under the engine when it's supported only by a jack!*
4  Check the mounts to see if the rubber is cracked, hardened or separated from the metal plates. Sometimes the rubber will split right down the center.
5  Check for relative movement between the mount plates and the engine or frame (use a large screwdriver or prybar to attempt to move the mounts). If movement is noted, lower the engine and tighten the mount fasteners.
6  Rubber preservative should be applied to the mounts to slow deterioration.

# 2B–18   Chapter 2 Part B   Legend engine

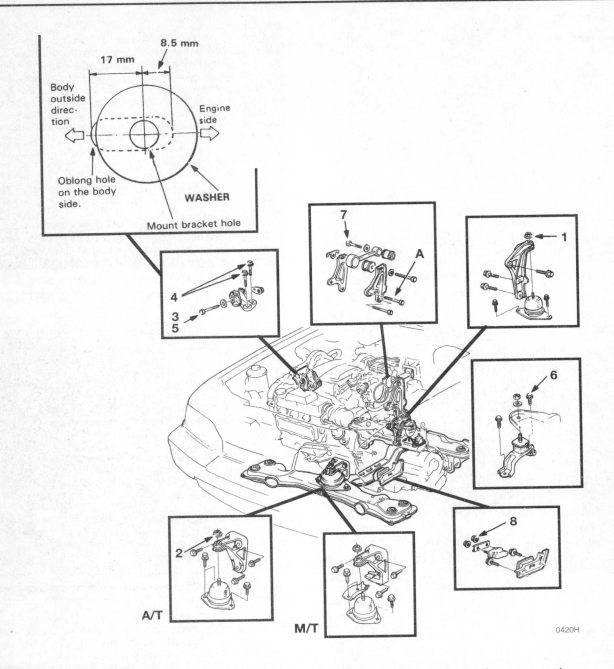

16.11  Engine mounts and torque rod – exploded view – when tightening all the mounts, use the sequence shown

1. Tighten to 28 ft-lbs
2. Tighten to 28 ft-lbs
3. Tighten temporarily
4. Tighten to 40 ft-lbs
5. Position the upper mount as shown here and then tighten to 28 ft-lbs
6. Tighten to 18 ft-lbs (automatic transaxle only)
7. Avoid bushing distortion, tighten to 54 ft-lbs
8. Center the rubber insulator and tighten to 40 ft-lbs
A. If these bolts are loosened, replace the bolts, then tighten to 28 ft-lbs

7   Disconnect the negative battery cable from the battery, then set the parking brake, block the rear wheels, raise the front of the vehicle and support it securely on jackstands (if not already done).

## Replacement

8   The engine mounting system is composed of weight-bearing mounts and a torque rod. The upper mount is attached to the engine near the tim-

# Chapter 2 Part B  Legend engine

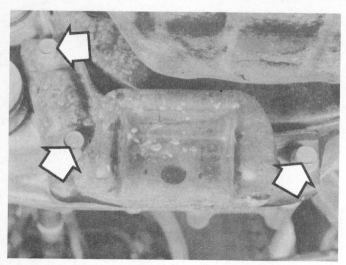

16.13  Lower mount retaining bolts (arrows)

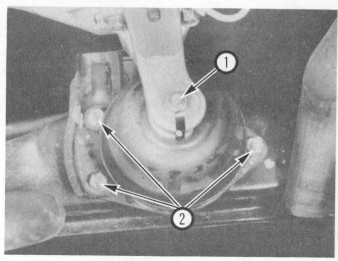

16.15  Typical front and rear mount
1  Engine mount-to-bracket nut   2  Mount retaining bolts

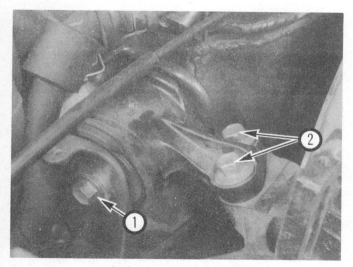

16.19  Upper mount details
1  Through bolt   2  Bracket retaining bolts

16.21  Remove the torque rod through bolt (arrow)

ing belt. The lower mount is sandwiched between the center fore-and-aft crossmember and the engine below the bellhousing area.

9  The front and rear mounts, which bear the majority of the engine/transaxle weight, are bolted to the respective crossmembers. They are located low between the engine and firewall and between the engine and radiator.

10  The torque rod reduces twisting forces on the engine mounts. Automatic transaxle models also have an upper transaxle mount.

11  To ensure maximum bushing life and prevent excessive noise and vibration, the vehicle should be level and the engine weight should be on the mounts during the final tightening stage. Ensure that the bushings are not twisted or offset. If you intend to replace more than one mount, or when you are installing the engine, tighten the mounts in the order shown (**see illustration**).

12  To remove mounts one at a time, support the engine with a floor jack under the oil pan. Place a block of wood between the jack head and the oil pan to protect it from damage. **Warning:** *DO NOT place any part of your body under the engine when it's supported only by a jack!*
**Note:** *Use thread locking compound on the nuts/bolts.*

## Lower mount

13  Remove the nuts/bolts (**see illustration**) and slip the lower mount out of the vehicle.

14  Install the mount and tighten the fasteners securely.

## Front and rear mounts

15  Remove the engine mount-to-bracket nut and raise the engine until the stud protruding from the mount comes out of the hole in the bracket (**see illustration**).

16  Unbolt the mount from the bracket and lift it from the vehicle.

17  Install the mount on the crossmember, lower the engine until it rests on the mount and install the nut. Be sure all fasteners are tightened securely.

## Upper mount

18  The upper mount carries little weight and seldom fails in service.

19  Whenever the engine is removed the mounts should be inspected and replaced as necessary. To change the mount, remove the through bolt (**see illustration**) and pull the mount out of its bracket.

20  Installation is the reverse of removal. Tighten the bolts securely.

**Torque rod**

21 Unbolt the torque rod from the engine bracket (see illustration).
22 Unbolt the rod from the frame bracket and lift it from the vehicle.

23 Installation is the reverse of removal. Be sure to tighten the bolts securely.

# Chapter 2 Part C
# General engine overhaul procedures

## Contents

| | |
|---|---|
| Crankshaft – inspection | 18 |
| Crankshaft – installation and main bearing oil clearance check | 22 |
| Crankshaft – removal | 13 |
| Cylinder head – cleaning and inspection | 9 |
| Cylinder head – disassembly | 8 |
| Cylinder head – reassembly | 11 |
| Compression check | 3 |
| Cylinder honing | 16 |
| Engine block – cleaning | 14 |
| Engine block – inspection | 15 |
| Engine overhaul – disassembly sequence | 7 |
| Engine overhaul – general information | 2 |
| Engine overhaul – reassembly sequence | 20 |
| Engine rebuilding alternatives | 6 |
| Engine – removal and installation | 5 |
| Engine removal – methods and precautions | 4 |
| General information | 1 |
| Initial start-up and break-in after overhaul | 25 |
| Main and connecting rod bearings – inspection and bearing selection | 19 |
| Pistons/connecting rods – inspection | 17 |
| Pistons/connecting rods – installation and rod bearing oil clearance check | 24 |
| Pistons/connecting rods – removal | 12 |
| Piston rings – installation | 21 |
| Rear main oil seal installation | 23 |
| Valves – servicing | 10 |

## Specifications

### Four-cylinder engine

#### General
| | |
|---|---|
| Displacement | 122 cu in (2.0 liters) |
| Cylinder compression pressure | |
|   Standard | 192 psi |
|   Minimum | 135 psi |
|   Maximum variation between cylinders | 28 psi |
| Oil pressure (engine warm) | |
|   At 3000 rpm | 60 to 78 psi |
|   At idle | 21 psi minimum |
| Cylinder head warpage service limit | 0.002 in (0.05 mm) |

## Four-cylinder engine (continued)

### Valves and related components

Minimum valve margin width
- Intake
  - Standard .......................................... 0.041 to 0.053 in (1.05 to 1.35 mm)
  - Service limit ..................................... 0.039 in (1.0 mm)
- Exhaust
  - Standard .......................................... 0.065 to 0.077 in (1.65 to 1.95 mm)
  - Service limit ..................................... 0.057 in (1.45 mm)

Intake valve
- Stem diameter
  - Standard .......................................... 0.2591 to 0.2594 in (6.58 to 6.59 mm)
  - Service limit ..................................... 0.258 in (6.55 mm)
- Valve stem-to-guide clearance
  - Standard .......................................... 0.002 to 0.004 in (0.004 to 0.010 mm)
  - Service limit ..................................... 0.006 in (0.16 mm)
- Length ............................................... 4.141 to 4.153 in (105.18 to 105.48 mm)

Exhaust valve
- Stem diameter
  - Standard .......................................... 0.2579 to 0.2583 in (6.55 to 6.56 mm)
  - Service limit ..................................... 0.257 in (6.52 mm)
- Valve stem-to-guide clearance
  - Standard .......................................... 0.004 to 0.006 in (0.010 to 0.016 mm)
  - Service limit ..................................... 0.009 in (0.22 mm)
- Length ............................................... 4.113 to 4.125 in (104.47 to 104.77 mm)

Valve spring
- Out-of-square limit ................................. 0.063 in (1.6 mm)
- Free length
  - Intake
    - Standard ........................................ 1.80 in (45.6 mm)
    - Service limit ................................... 1.76 in (44.6 mm)
  - Exhaust
    - Standard ........................................ 1.82 in (46.3 mm)
    - Service limit ................................... 1.78 in (45.3 mm)
- Installed height
  - Intake
    - Standard ........................................ 1.268 in (32.195 mm)
    - Service limit ................................... 1.299 in (32.985 mm)
  - Exhaust
    - Standard ........................................ 1.268 in (32.195 mm)
    - Service limit ................................... 1.299 in (32.985 mm)

### Crankshaft and connecting rods

Connecting rod journal
- Diameter ............................................ 1.7707 to 1.7717 in (44.976 to 45.000 mm)
- Taper and out-of-round
  - Standard .......................................... 0.0002 in (0.005 mm)
  - Service limit ..................................... 0.0004 in (0.010 mm)
- Rod bearing oil clearance
  - Standard .......................................... 0.0008 to 0.0015 in (0.020 to 0.038 mm)
  - Service limit ..................................... 0.002 in (0.05 mm)

Connecting rod side clearance (endplay)
- Standard ............................................ 0.006 to 0.012 in (0.15 to 0.30 mm)
- Service limit ....................................... 0.016 in (0.40 mm)

Main bearing journal
- Diameter ............................................ 2.1644 to 2.1654 in (54.976 to 55.000 mm)
- Taper and out-of-round
  - Standard .......................................... 0.0002 in (0.005 mm)
  - Service limit ..................................... 0.0004 in (0.010
- Runout
  - Standard .......................................... 0.0012 in (0.03 mm)
  - Service limit ..................................... 0.002 in (0.06 mm)
- Bearing oil clearance
  - No. 1,2,4 and 5 journals
    - Standard ........................................ 0.0009 to 0.0017 in (0.024 to 0.042 mm)
    - Service limit ................................... 0.002 in (0.05 mm)
  - No. 3 journals
    - Standard ........................................ 0.0012 to 0.0019 in (0.030 to 0.048 mm)
    - Service limit ................................... 0.002 in (0.05 mm)

# Chapter 2 Part C  General engine overhaul procedures

Crankshaft endplay
  Standard ............................................. 0.004 to 0.0014 in (0.10 to 0.35 mm)
  Service limit ......................................... 0.018 in (0.45 mm)

## Cylinder block
Cylinder block deck warpage
  Standard ............................................. 0.003 in (0.07 mm)
  Service limit ......................................... 0.004 in (0.10 mm)
Cylinder bore
  Diameter
    Standard ......................................... 2.9528 to 2.9535 in (75.00 to 75.02 mm)
    Service limit ..................................... 2.956 in (75.07 mm)
  Taper
    Standard ......................................... 0.0003 to 0.0005 in (0.007 to 0.012 mm)
    Service limit ..................................... 0.002 in (0.05 mm)

## Pistons and rings
Piston diameter ......................................... 2.9520 to 2.9524 in (74.98 to 74.99 mm)
  **Note:** *Measured 5/8-inch (16 mm) from bottom of skirt*
Piston-to-bore clearance
  Standard ............................................. 0.0004 to 0.0024 in (0.01 to 0.06 mm)
  Service limit ......................................... 0.003 in (0.07 mm)
Piston ring end gap
  Top ring
    Standard ......................................... 0.006 to 0.014 in (0.15 to 0.35 mm)
    Service limit ..................................... 0.02 in (0.6 mm)
  Middle ring
    Standard ......................................... 0.012 to 0.018 in (0.30 to 0.45 mm)
    Service limit ..................................... 0.02 in (0.6 mm)
  Oil ring
    Standard ......................................... 0.008 to 0.028 in (0.20 to 0.70 mm)
    Service limit ..................................... 0.03 in (0.8 mm)
Piston ring side clearance
  Top ring
    Standard ......................................... 0.0012 to 0.0024 in (0.030 to 0.060 mm)
    Service limit ..................................... 0.005 in (0.13 mm)
  Middle ring
    Standard ......................................... 0.0012 to 0.0022 in (0.030 to 0.055 mm)
    Service limit ..................................... 0.005 in (0.13 mm)

## Torque specifications*
Main bearing cap bolts ................................ **Ft-lbs (unless otherwise indicated)**
Main bearing cap bolts ................................ 46
Connecting rod cap nuts ............................. 23
Rear main oil seal retainer bolts ..................... 108 in-lbs
*\* Note: Refer to Part A for additional torque specifications.*


## V6 engine
### General
Displacement ........................................... 2.675 liters
Cylinder compression pressure at 200 rpm with wide open throttle
  Standard ............................................. 171 psi
  Minimum ............................................. 142 psi
  Maximum variation between cylinders ........... 28 psi
Oil pressure (coolant temperature 178 degrees F.)
  At 3000 rpm ........................................ 71 to 82 psi
  At idle ................................................. 20 psi minimum
Cylinder head warpage limit ........................... 0.002 in (0.05 mm)

### Valves and related components
Minimum valve margin width
  Intake
    Standard ......................................... 0.053 to 0.065 in (1.35 to 1.65 mm)
    Service limit ..................................... 0.045 in (1.15 mm)
  Exhaust
    Standard ......................................... 0.065 to 0.077 in (1.65 to 1.95 mm)
    Service limit ..................................... 0.057 in (1.45 mm)

## V6 engine (continued)

### Valves and related components (continued)

Intake valve
- Stem diameter .................................................... 0.2591 to 0.2594 in (6.58 to 6.59 mm)
- Valve stem-to-guide clearance
  - Standard ...................................................... 0.001 to 0.002 in (0.02 to 0.05 mm)
  - Service limit .................................................. 0.003 in (0.08 mm)
- Length ........................................................... 4.106 to 4.118 in (104.30 to 104.60 mm)

Exhaust valve
- Stem diameter
  - Standard ...................................................... 0.2579 to 0.2583 in (6.55 to 6.56 mm)
  - Service limit .................................................. 0.257 in (6.52 mm)
- Valve stem-to-guide clearance
  - Standard ...................................................... 0.002 to 0.003 in (0.05 to 0.08 mm)
  - Service limit .................................................. 0.004 in (0.11 mm)
- Length ........................................................... 4.189 to 4.201 in (106.40 to 106.70 mm)

Valve guide inside diameter (all)
- Standard ........................................................ 0.260 to 0.261 in (6.61 to 6.63 mm)
- Service limit .................................................... 0.262 in (6.65 mm)

Valve spring
- Free length
  - Intake
    - Standard .................................................... 2.12 in (53.90 mm)
    - Service limit ................................................ 2.08 in (52.90 mm)
  - Exhaust – Inner
    - Standard .................................................... 1.77 in (44.95 mm)
    - Service limit ................................................ 1.73 in (43.95 mm)
  - Exhaust – Outer
    - Standard .................................................... 1.95 in (49.55 mm)
    - Service limit ................................................ 1.91 in (48.55 mm)
- Out-of-square limit
  - Intake ......................................................... 0.073 in (1.86 mm)
  - Exhaust – Inner ................................................ 0.062 in (1.57 mm)
  - Exhaust – Outer ................................................ 0.068 in (1.73 mm)

### Crankshaft and connecting rods

Connecting rod journal
- Diameter ........................................................ 2.0463 to 2.0472 in (51.976 to 52.000 mm)
- Taper and out-of-round
  - Standard ...................................................... 0.0002 in (0.005 mm)
  - Service limit .................................................. 0.0004 in (0.010 mm)
- Bearing oil clearance
  - Standard ...................................................... 0.0010 to 0.002 in (0.026 to 0.050 mm)
  - Service limit .................................................. 0.002 in (0.05 mm)

Connecting rod side clearance (endplay)
- Standard ........................................................ 0.006 to 0.012 in (0.15 to 0.30 mm)
- Service limit .................................................... 0.016 in (0.40 mm)

Main bearing
- Journal diameter ................................................ 2.5187 to 2.5197 in (63.976 to 64.000 mm)
- Journal taper and out-of-round
  - Standard ...................................................... 0.0002 in (0.005 mm)
  - Service limit .................................................. 0.0004 in (0.010 mm)
- Oil clearance
  - Standard ...................................................... 0.0009 to 0.0019 in (0.024 to 0.048 mm)
  - Service limit .................................................. 0.002 in (0.05 mm)

Crankshaft endplay
- Standard ........................................................ 0.004 to 0.014 in (0.10 to 0.35 mm)
- Service limit .................................................... 0.018 in (0.45 mm)

### Cylinder block

Cylinder block warpage
- Standard ........................................................ 0.003 in (0.07 mm)
- Service limit .................................................... 0.004 in (0.10 mm)

Cylinder bore
- Diameter
  - Standard ...................................................... 3.4252 to 3.4260 in (87.00 to 87.02 mm)
  - Service limit .................................................. 3.4279 in (87.07 mm)
- Taper and out-of-round limit .................................... 0.002 in (0.05 mm)

# Chapter 2 Part C  General engine overhaul procedures

**Pistons and rings**
*Note: measure piston 23/32-inch (18 mm) above bottom of skirt*
Piston diameter "A"
  Standard .................................................. 3.4244 to 3.4250 in (86.981 to 86.994 mm)
  Service limit ............................................. 3.4240 in (86.97 mm)
Piston diameter "B"
  Standard .................................................. 3.4240 to 3.4246 in (86.971 to 86.984 mm)
  Service limit ............................................. 3.4236 in (86.96 mm)
Piston-to-bore clearance
  Standard .................................................. 0.0018 to 0.0026 in (0.045 to 0.065 mm)
  Service limit ............................................. 0.0033 in (0.085 mm)
Piston ring end gap
  No. 1 (top) compression ring
    Standard ................................................ 0.008 to 0.014 in (0.20 to 0.35 mm)
    Service limit ........................................... 0.02 in (0.6 mm)
  No. 2 (middle) compression ring
    Standard ................................................ 0.014 to 0.019 in (0.35 to 0.50 mm)
    Service limit ........................................... 0.03 in (0.75 mm)
  Oil ring
    Standard ................................................ 0.008 to 0.028 in (0.20 to 0.70 mm)
    Service limit ........................................... 0.03 in (0.8 mm)
Piston ring side clearance
  No. 1 (top) and no. 2 (middle) compression rings
    Standard ................................................ 0.0006 to 0.0018 in (0.015 to 0.045 mm)
    Service limit ........................................... 0.005 in (0.13 mm)

**Torque specifications***  **Ft-lbs (unless otherwise indicated)**
Main bearing bolts
  Cap bridge bolt .......................................... 41
  Side bolt (M10 x 1.25) ................................ 36
  Cap bolt (M9 x 1.25) .................................. 29
Connecting rod cap nuts ................................ 33
Rear main oil seal retainer bolts .................... 108 in-lbs

*\* Note: Refer to Part B for additional torque specifications.*

## 1  General information

Included in this portion of Chapter 2 are the general overhaul procedures for the cylinder head(s) and internal engine components.

The information ranges from advice concerning preparation for an overhaul and the purchase of replacement parts to detailed, step-by-step procedures covering removal and installation of internal engine components and the inspection of parts.

The following Sections have been written based on the assumption that the engine has been removed from the vehicle. For information concerning in-vehicle engine repair, as well as removal and installation of the external components necessary for the overhaul, see Part A or B of this Chapter and Section 7 of this Part.

The Specifications included in this Part are only those necessary for the inspection and overhaul procedures which follow. Refer to Parts A and B for additional Specifications.

## 2  Engine overhaul – general information

*Refer to illustrations 2.4a, 2.4b and 2.4c*

It's not always easy to determine when, or if, an engine should be completely overhauled, as a number of factors must be considered.

High mileage is not necessarily an indication that an overhaul is needed, while low mileage doesn't preclude the need for an overhaul. Frequency of servicing is probably the most important consideration. An engine that's had regular and frequent oil and filter changes, as well as other required maintenance, will most likely give many thousands of miles of reliable service. Conversely, a neglected engine may require an overhaul very early in its life.

Excessive oil consumption is an indication that piston rings, valve seals and/or valve guides are in need of attention. Make sure that oil leaks aren't responsible before deciding that the rings and/or guides are bad. Perform a cylinder compression check to determine the extent of the work required (see Section 3).

Check the oil pressure with a gauge installed in place of the oil pressure sending unit **(see illustrations)** and compare it to that listed in this Chapter's specifications. If it's extremely low, the bearings and/or oil pump are probably worn out.

**2.4a  Temporarily install a gauge in place of the oil pressure sending unit**

## 2C–6  Chapter 2 Part C  General engine overhaul procedures

**2.4b  On Legend models, the oil pressure sending unit (arrow) is located on the oil cooler, above the oil filter**

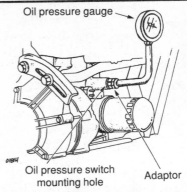

**2.4c  On Integra models, the gauge is connected above the oil filter, as shown here**

**3.6  A compression gauge with a threaded fitting for the spark plug hole is preferred over the type that requires hand pressure to maintain the seal – be sure to open the throttle as far as possible during the compression check**

Loss of power, rough running, knocking or metallic engine noises, excessive valvetrain noise and high fuel consumption rates may also point to the need for an overhaul, especially if they're all present at the same time. If a complete tune-up doesn't remedy the situation, major mechanical work is the only solution.

An engine overhaul involves restoring the internal parts to the specifications of a new engine. During an overhaul, the piston rings are replaced and the cylinder walls are reconditioned (rebored and/or honed). If a rebore is done by an automotive machine shop, new oversize pistons will also be installed. The main bearings, connecting rod bearings and camshaft bearings are generally replaced with new ones and, if necessary, the crankshaft may be reground to restore the journals. Generally, the valves are serviced as well, since they're usually in less-than-perfect condition at this point. While the engine is being overhauled, other components, such as the distributor, starter and alternator, can be rebuilt as well. The end result should be a like-new engine that will give many trouble free miles. **Note:** *Critical cooling system components such as the hoses, drivebelts, thermostat and water pump MUST be replaced with new parts when an engine is overhauled. The radiator should be checked carefully to ensure that it isn't clogged or leaking (see Chapter 3). Also, we don't recommend overhauling the oil pump – always install a new one when an engine is rebuilt.*

Before beginning the engine overhaul, read through the entire procedure to familiarize yourself with the scope and requirements of the job. Overhauling an engine isn't difficult, but it is time consuming. Plan on the vehicle being tied up for a minimum of two weeks, especially if parts must be taken to an automotive machine shop for repair or reconditioning. Check on availability of parts and make sure that any necessary special tools and equipment are obtained in advance. Most work can be done with typical hand tools, although a number of precision measuring tools are required for inspecting parts to determine if they must be replaced. Often an automotive machine shop will handle the inspection of parts and offer advice concerning reconditioning and replacement. **Note:** *Always wait until the engine has been completely disassembled and all components, especially the engine block, have been inspected before deciding what service and repair operations must be performed by an automotive machine shop.* Since the block's condition will be the major factor to consider when determining whether to overhaul the original engine or buy a rebuilt one, never purchase parts or have machine work done on other components until the block has been thoroughly inspected. As a general rule, time is the primary cost of an overhaul, so it doesn't pay to install worn or substandard parts.

As a final note, to ensure maximum life and minimum trouble from a rebuilt engine, everything must be assembled with care in a spotlessly clean environment.

### 3  Compression check

*Refer to illustration 3.6*

1   A compression check will tell you what mechanical condition the upper end (pistons, rings, valves, head gasket[s]) of your engine is in. Specifically, it can tell you if the compression is down due to leakage caused by worn piston rings, defective valves and seats or a blown head gasket. **Note:** *The engine must be at normal operating temperature and the battery must be fully charged for this check.*

2   Begin by cleaning the area around the spark plugs before you remove them (compressed air should be used, if available, otherwise a small brush or even a bicycle tire pump will work). The idea is to prevent dirt from getting into the cylinders as the compression check is being done.

3   Remove all of the spark plugs from the engine (see Chapter 1).

4   Block the throttle wide open.

5   Detach the coil wire from the center of the distributor cap and ground it on the engine block. Use a jumper wire with alligator clips on each end to ensure a good ground. The fuel pump circuit should also be disabled (see Chapter 4).

6   Install the compression gauge in the spark plug hole **(see illustration)**.

7   Crank the engine over at least seven compression strokes and watch the gauge. The compression should build up quickly in a healthy engine. Low compression on the first stroke, followed by gradually increasing pressure on successive strokes, indicates worn piston rings. A low compression reading on the first stroke, which doesn't build up during successive strokes, indicates leaking valves or a blown head gasket (a cracked

# Chapter 2 Part C  General engine overhaul procedures

5.6a  Unplug the electrical connectors (arrows)

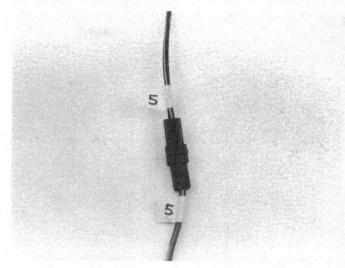

5.6b  Label each wire before unplugging the connector

head could also be the cause). Deposits on the undersides of the valve heads can also cause low compression. Record the highest gauge reading obtained.
8   Repeat the procedure for the remaining cylinders and compare the results to this Chapter's Specifications.
9   Add some engine oil (about three squirts from a plunger-type oil can) to each cylinder, through the spark plug hole, and repeat the test.
10  If the compression increases after the oil is added, the piston rings are definitely worn. If the compression doesn't increase significantly, the leakage is occurring at the valves or head gasket. Leakage past the valves may be caused by burned valve seats and/or faces or warped, cracked or bent valves.
11  If two adjacent cylinders have equally low compression, there's a strong possibility that the head gasket between them is blown. The appearance of coolant in the combustion chambers or the crankcase would verify this condition.
12  If one cylinder is slightly lower than the others, and the engine has a slightly rough idle, a worn lobe on the camshaft could be the cause.
13  If the compression is unusually high, the combustion chambers are probably coated with carbon deposits. If that's the case, the cylinder head(s) should be removed and decarbonized.
14  If compression is way down or varies greatly between cylinders, it would be a good idea to have a leak-down test performed by an automotive repair shop. This test will pinpoint exactly where the leakage is occurring and how severe it is.

### 4  Engine removal – methods and precautions

If you've decided that an engine must be removed for overhaul or major repair work, several preliminary steps should be taken.

Locating a suitable place to work is extremely important. Adequate work space, along with storage space for the vehicle, will be needed. If a shop or garage isn't available, at the very least a flat, level, clean work surface made of concrete or asphalt is required.

Cleaning the engine compartment and engine before beginning the removal procedure will help keep tools clean and organized.

An engine hoist or A-frame will also be necessary. Make sure the equipment is rated in excess of the combined weight of the engine and transaxle. Safety is of primary importance, considering the potential hazards involved in lifting the engine out of the vehicle.

If the engine is being removed by a novice, a helper should be available. Advice and aid from someone more experienced would also be helpful. There are many instances when one person cannot simultaneously perform all of the operations required when lifting the engine out of the vehicle.

Plan the operation ahead of time. Arrange for or obtain all of the tools and equipment you'll need prior to beginning the job. Some of the equipment necessary to perform engine removal and installation safely and with relative ease are (in addition to an engine hoist) a heavy duty floor jack, complete sets of wrenches and sockets as described in the front of this manual, wooden blocks and plenty of rags and cleaning solvent for mopping up spilled oil, coolant and gasoline. If the hoist must be rented, make sure that you arrange for it in advance and perform all of the operations possible without it beforehand. This will save you money and time.

Plan for the vehicle to be out of use for quite a while. A machine shop will be required to perform some of the work which the do-it-yourselfer can't accomplish without special equipment. These shops often have a busy schedule, so it would be a good idea to consult them before removing the engine in order to accurately estimate the amount of time required to rebuild or repair components that may need work.

Always be extremely careful when removing and installing the engine. Serious injury can result from careless actions. Plan ahead, take your time and a job of this nature, although major, can be accomplished successfully.

### 5  Engine – removal and installation

*Refer to illustrations 5.6a, 5.6b, 5.14, 5.17a, 5.17b, 5.17c, 5.17d and 5.18*
**Note:** *Read through the entire Section before beginning this procedure. The engine and transaxle are removed as a unit and then separated outside the vehicle.*

#### Removal

1   Relieve the fuel system pressure. Remove the air cleaner assembly and ducts (see Chapter 4).
2   Disconnect the negative cable from the battery and remove the battery (and battery tray on Integras).
3   Place protective covers on the fenders and cowl and remove the hood (see Chapter 11). **Note:** *On Legends, the hood may be propped open by repositioning the stay instead of removing it, if desired.*
4   On Integras, remove the alternator, distributor and spark plug wires (see Chapter 5).
5   Block the rear wheels and set the parking brake. Raise the vehicle and support it securely on jackstands. Remove the oil filter, drain the cooling system, transaxle and engine oil and remove the drivebelts (see Chapter 1). On Legends, remove the oil cooler (see Chapter 3).
6   Clearly label and disconnect all vacuum lines, coolant and emissions hoses, electrical connectors **(see illustration)**, ground straps and fuel lines. Masking tape and/or a touch-up paint applicator work well for marking items **(see illustration)**. Take instant photos or sketch the locations of

## 2C-8    Chapter 2 Part C  General engine overhaul procedures

5.14  Attach a lifting sling and take up the slack

5.17a  Lift the engine/transaxle clear of its mounts (Legend shown, Integra similar)

5.17b  Turn the engine as necessary to clear obstructions

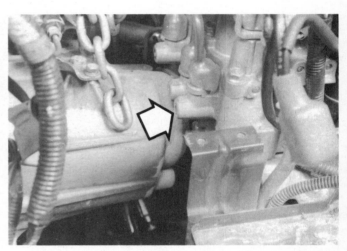

5.17c  If the engine or transaxle gets caught on something (arrow), stop and free it before proceeding

components and brackets as necessary.
7  Remove the cooling fan(s) and radiator (see Chapter 3).
8  Release any residual pressure in the tank by removing the gas cap, then undo the fuel lines connecting the engine to the chassis (see Chapter 4). Plug or cap all open fittings.
9  Disconnect the throttle linkage (and TV linkage and speed control cable, when equipped) from the engine (see Chapter 4).
10  Unbolt the power steering pump. If clearance allows, tie the pump aside without disconnecting the hoses. If necessary, remove the pump (see Chapter 10).
11  On air conditioned vehicles, unbolt the compressor and set it aside. Don't disconnect the refrigerant hoses.
12  Detach the exhaust pipe(s) from the manifold(s) (see Chapter 4).
13  Remove the driveaxles (see Chapter 8) and refer to Chapter 7 for information on how to disconnect the wire harness, shift linkage and speedometer cable from the transaxle prior to removal.
14  Attach a lifting sling to the brackets on the engine. Position a hoist and connect the sling to it. Take up the slack until there is slight tension on the hoist (see illustration).
15  Recheck to be sure nothing except the mounts are still connecting the engine/transaxle to the vehicle. Disconnect anything still remaining.
16  Support the transaxle with a floor jack. Place a block of wood on the jack head to prevent damage to the transaxle. Remove the nuts/bolts from the engine and transaxle mounts (see the appropriate engine mount procedure in Chapter 2A or 2B). **Warning:** *DO NOT place any part of your body under the engine/transaxle when it's supported only by a hoist or other lifting device.*
17  Slowly lift the engine/transaxle out of the vehicle **(see illustrations)**. It may be necessary to pry the mounts away from the frame brackets.
18  Move the engine/transaxle away from the vehicle and carefully lower the hoist until the transaxle is supported on the floor or a sturdy workbench **(see illustration)**.
19  Remove the engine block-to-transaxle brace.
20  On automatic transaxle equipped models, remove the torque converter-to-driveplate fasteners (see Chapter 7) and push the converter back slightly into the bellhousing.
21  Remove the engine-to-transaxle bolts and separate the engine from the transaxle. The torque converter should remain in the transaxle.
22  Place the engine on the floor or remove the flywheel/driveplate and mount the engine on an engine stand.

### Installation

23  Check the engine/transaxle mounts. If they're worn or damaged, replace them.
24  On manual transaxle equipped models, inspect the clutch components (see Chapter 8) and on automatic models inspect the converter seal and bushing.

**5.17d  Lift the engine/transaxle clear of the vehicle**

**5.18  Lower the engine/transaxle onto a work surface, then remove the transaxle and mount the engine on a stand**

25  On manual transaxle vehicles, apply a dab of high temperature grease to the pilot bearing.
26  On automatic transaxle equipped models, apply a dab of grease to the nose of the converter and to the seal lips.
27  Carefully guide the transaxle into place, following the procedure outlined in Chapter 7. **Caution:** *Do Not use the bolts to force the engine and transaxle into alignment. It may crack or damage major components.*
28  Install the engine-to-transaxle bolts and tighten them securely.
29  Attach the hoist to the engine and carefully lower the engine/transaxle assembly into the engine compartment.
30  Install the mount bolts and tighten them securely.
31  Reinstall the remaining components and fasteners in the reverse order of removal.
32  Add coolant, oil, power steering and transmission fluids and an oil filter as needed (see Chapter 1).
33  Run the engine and check for proper operation and leaks. Shut off the engine and recheck the fluid levels.

## 6  Engine rebuilding alternatives

The do-it-yourselfer is faced with a number of options when performing an engine overhaul. The decision to replace the engine block, piston/connecting rod assemblies and crankshaft depends on a number of factors, with the number one consideration being the condition of the block. Other considerations are cost, access to machine shop facilities, parts availability, time required to complete the project and the extent of prior mechanical experience on the part of the do-it-yourselfer.

Some of the rebuilding alternatives include:

**Individual parts** – If the inspection procedures reveal that the engine block and most engine components are in reusable condition, purchasing individual parts may be the most economical alternative. The block, crankshaft and piston/connecting rod assemblies should all be inspected carefully. Even if the block shows little wear, the cylinder bores should be surface honed.

**Short block** – A short block consists of an engine block with a crankshaft and piston/connecting rod assemblies already installed. All new bearings are incorporated and all clearances will be correct. The existing camshaft, valvetrain components, cylinder head(s) and external parts can be bolted to the short block with little or no machine shop work necessary.

**Long block** – A long block consists of a short block plus an oil pump, oil pan, cylinder head(s), cylinder head covers or valve cover(s), camshaft and valvetrain components, timing sprockets, belt and timing cover. All components are installed with new bearings, seals and gaskets incorporated throughout. The installation of manifolds and external parts is all that's necessary.

Give careful thought to which alternative is best for you and discuss the situation with local automotive machine shops, auto parts dealers and experienced rebuilders before ordering or purchasing replacement parts.

## 7  Engine overhaul – disassembly sequence

*Refer to illustrations 7.5a, 7.5b, and 7.5c*

1  It's much easier to disassemble and work on the engine if it's mounted on a portable engine stand. A stand can often be rented quite cheaply from an equipment rental yard. Before the engine is mounted on a stand, the flywheel/driveplate and rear oil seal retainer should be removed from the engine.
2  If a stand isn't available, it's possible to disassemble the engine with it blocked up on the floor. Be extra careful not to tip or drop the engine when working without a stand.
3  If you're going to obtain a rebuilt engine, all external components must come off first, to be transferred to the replacement engine, just as they will if you're doing a complete engine overhaul yourself. These include:

  Alternator and brackets
  Emissions control components
  Distributor, spark plug wires and spark plugs
  Thermostat and housing cover
  Water pump
  Fuel injection components
  Intake and exhaust manifolds
  Oil filter
  Engine mounts
  Clutch and flywheel/driveplate
  Engine rear plate

**Note:** *When removing the external components from the engine, pay close attention to details that may be helpful or important during installation. Note the installed position of gaskets, seals, spacers, pins, brackets, washers, bolts and other small items.*

4  If you're obtaining a short block, which consists of the engine block, crankshaft, pistons and connecting rods all assembled, then the cylinder head(s), oil pan and oil pump will have to be removed as well. See *Engine rebuilding alternatives* for additional information regarding the different possibilities to be considered.
5  If you're planning a complete overhaul, the engine must be disassembled and the internal components removed in the following order **(see**

7.5a  Legend engine – radiator side

7.5b  Legend engine – timing belt end

7.5c  Legend engine – firewall side

8.2  A small plastic bag, with an appropriate label, can be used to store the valve train components so they can be kept together and reinstalled in the original location

illustrations).
  Cylinder head/valve cover(s)
  Intake and exhaust manifolds
  Timing belt covers
  Timing belt and sprockets
  Cylinder head(s)
  Oil pan
  Oil pump
  Piston/connecting rod assemblies
  Crankshaft rear oil seal retainer
  Crankshaft and main bearings

6  Before beginning the disassembly and overhaul procedures, make sure the following items are available. Also, refer to *Engine overhaul – reassembly sequence* for a list of tools and materials needed for engine reassembly.
  Common hand tools
  Small cardboard boxes or plastic bags for storing parts
  Gasket scraper
  Ridge reamer
  Micrometers
  Telescoping gauges

  Dial indicator set
  Valve spring compressor
  Cylinder surfacing hone
  Piston ring groove cleaning tool
  Electric drill motor
  Tap and die set
  Wire brushes
  Oil gallery brushes
  Cleaning solvent

## 8  Cylinder head – disassembly

*Refer to illustrations 8.2 and 8.3*

**Note:** *New and rebuilt cylinder heads are commonly available for most engines at dealerships and auto parts stores. Due to the fact that some specialized tools are necessary for the disassembly and inspection procedures, and replacement parts may not be readily available, it may be more practical and economical for the home mechanic to purchase re-*

# Chapter 2 Part C  General engine overhaul procedures

8.3  Use a valve spring compressor to compress the spring, then remove the keepers from the valve stem

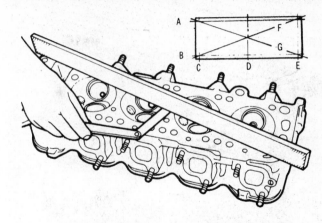

9.12  Check the cylinder head gasket surface for warpage by trying to slip a feeler gauge under the straightedge (see the Specifications for the maximum warpage allowed and use a feeler gauge of that thickness)

*placement head(s) rather than taking the time to disassemble, inspect and recondition the original(s).*

1  Cylinder head disassembly involves removal of the intake and exhaust valves and related components. It's assumed that the rocker arms and camshaft(s) have already been removed (see Part A or B as needed).
2  Before the valves are removed, arrange to label and store them, along with their related components, so they can be kept separate and reinstalled in the same valve guides they are removed from **(see illustration)**.
3  Compress the springs on the first valve with a spring compressor and remove the keepers **(see illustration)**. Carefully release the valve spring compressor and remove the retainer, the spring and the spring seat (if used).
4  Pull the valve out of the head, then remove the oil seal from the guide. If the valve binds in the guide (won't pull through), push it back into the head and deburr the area around the keeper groove with a fine file or whetstone.
5  Repeat the procedure for the remaining valves. Remember to keep all the parts for each valve together so they can be reinstalled in the same locations.
6  Once the valves and related components have been removed and stored in an organized manner, the head should be thoroughly cleaned and inspected. If a complete engine overhaul is being done, finish the engine disassembly procedures before beginning the cylinder head cleaning and inspection process.

## 9  Cylinder head – cleaning and inspection

*Refer to illustrations 9.12, 9.14, 9.15, 9.16, 9.17 and 9.18*

1  Thorough cleaning of the cylinder head(s) and related valvetrain components, followed by a detailed inspection, will enable you to decide how much valve service work must be done during the engine overhaul. **Note:** *If the engine was severely overheated, the cylinder head is probably warped (see Step 12).*

### Cleaning

2  Scrape all traces of old gasket material and sealing compound off the head gasket, intake manifold and exhaust manifold sealing surfaces. Be very careful not to gouge the cylinder head. Special gasket removal solvents that soften gaskets and make removal much easier are available at auto parts stores.

3  Remove all built up scale from the coolant passages.
4  Run a stiff wire brush through the various holes to remove deposits that may have formed in them.
5  Run an appropriate size tap into each of the threaded holes to remove corrosion and thread sealant that may be present. If compressed air is available, use it to clear the holes of debris produced by this operation. **Warning:** *Wear eye protection when using compressed air!*
6  Clean the exhaust and intake manifold stud threads with a wire brush.
7  Clean the cylinder head with solvent and dry it thoroughly. Compressed air will speed the drying process and ensure that all holes and recessed areas are clean. **Note:** *Decarbonizing chemicals are available and may prove very useful when cleaning cylinder heads and valvetrain components. They are very caustic and should be used with caution. Be sure to follow the instructions on the container.*
8  Clean the rocker arms with solvent and dry them thoroughly (don't mix them up during the cleaning process). Compressed air will speed the drying process and can be used to clean out the oil passages.
9  Clean all the valve springs, spring seats, keepers and retainers with solvent and dry them thoroughly. Do the components from one valve at a time to avoid mixing up the parts.
10  Scrape off any heavy deposits that may have formed on the valves, then use a motorized wire brush to remove deposits from the valve heads and stems. Again, make sure the valves don't get mixed up.

### Inspection

**Note:** *Be sure to perform all of the following inspection procedures before concluding that machine shop work is required. Make a list of the items that need attention. The inspection procedures for the lifters and rocker arms, as well as the camshafts, can be found in Part A.*

#### Cylinder head

11  Inspect the head very carefully for cracks, evidence of coolant leakage and other damage. If cracks are found, check with an automotive machine shop concerning repair. If repair isn't possible, a new cylinder head should be obtained.
12  Using a straightedge and feeler gauge, check the head gasket mating surface for warpage **(see illustration)**. If the warpage exceeds the specified limit, it can be resurfaced at an automotive machine shop. **Note:** *If the V6 engine heads are resurfaced, the intake manifold flanges will also require machining.*
13  Examine the valve seats in each of the combustion chambers. If they're pitted, cracked or burned, the head will require valve service that's beyond the scope of the home mechanic.

# Chapter 2 Part C  General engine overhaul procedures

9.14  A dial indicator can be used to measure valve stem-to-guide clearance (move the valve stem back and forth as shown)

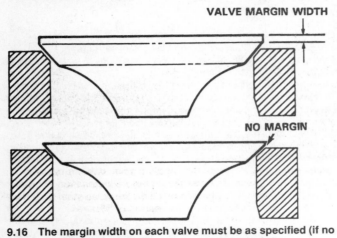

9.16  The margin width on each valve must be as specified (if no margin exists, the valve cannot be reused)

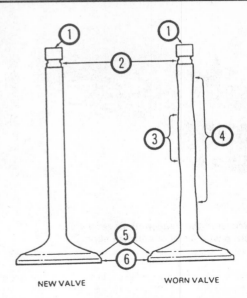

9.15  Check for valve wear at the points shown here

1  Valve tip
2  Keeper groove
3  Stem (least worn area)
4  Stem (most worn area)
5  Valve face
6  Margin

9.17  Check each valve spring for squareness

14  Check the valve stem-to-guide clearance with a small hole gauge and micrometer. Also check the valve stem deflection crosswise (parallel to the rocker arm) with a dial indicator attached securely to the head **(see illustration)**. The valve must be in the guide and approximately 1/16-inch off the seat. The total valve stem movement indicated by the gauge needle must be noted. If it exceeds the stem-to-guide clearance limit listed in this Chapter's specifications, the valve guides should be replaced. After this is done, if there's still some doubt regarding the condition of the valve guides, they should be checked by an automotive machine shop (the cost should be minimal).

### Valves

15  Carefully inspect each valve face for uneven wear **(see illustration)**, deformation, cracks, pits and burned areas. Check the valve stem for scuffing and galling and the neck for cracks. Rotate the valve and check for any obvious indication that it's bent. Look for pits and excessive wear on the end of the stem. The presence of any of these conditions indicates the need for valve service by an automotive machine shop.

16  Measure the margin width on each valve **(see illustration)**. Any valve with a margin narrower than that listed in this Chapter's specifications will have to be replaced with a new one.

### Valve components

17  Check each valve spring for wear (on the ends) and pits. Stand each spring on a flat surface and check it for squareness **(see illustration)**. If any of the springs are distorted or sagged, replace all of them with new parts.

18  Measure the free length of each valve spring with a dial or vernier caliper **(see illustration)**. The tension of springs decreases with age and usage. We recommend replacing the valve springs during an overhaul.

19  Check the spring retainers and keepers for obvious wear and cracks. Any questionable parts should be replaced with new ones, as extensive damage will occur if they fail during engine operation.

20  Any damaged or excessively worn parts must be replaced with new ones.

21  If the inspection process indicates that the valve components are in generally poor condition and worn beyond the limits specified, which is usually the case in an engine that's being overhauled, reassemble the valves in the cylinder head and refer to Section 10 for valve servicing recommendations.

# Chapter 2 Part C  General engine overhaul procedures

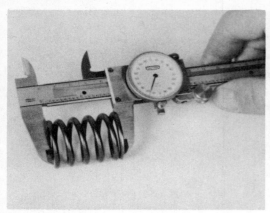

9.18  Measure the free length of each valve spring with a dial or vernier caliper

11.3a  Drop the spring seats over the valve guides

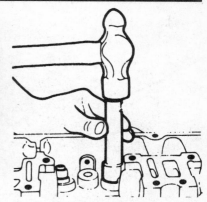

11.3b  Gently tap the valve seals into place with a seal installation tool or a deep socket and hammer

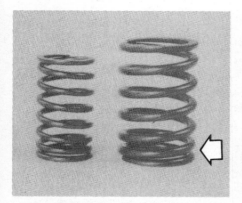

11.5a  Make sure each outer valve spring (right) is installed with the narrow pitch end (arrow) against the cylinder head

11.5b  Install the valve retainer over the springs and valve stem

11.6  Apply a small dab of grease to each keeper before installation to hold them in place on the valve stem until the spring is released

## 10  Valves – servicing

1  Because of the complex nature of the job and the special tools and equipment needed, servicing of the valves, the valve seats and the valve guides, commonly known as a valve job, should be done by a professional.
2  The home mechanic can remove and disassemble the head, do the initial cleaning and inspection, then reassemble and deliver it to a dealer service department or an automotive machine shop for the actual service work. Doing the inspection will enable you to see what condition the head and valvetrain components are in and will ensure that you know what work and new parts are required when dealing with an automotive machine shop.
3  The dealer service department, or automotive machine shop, will remove the valves and springs, recondition or replace the valves and valve seats, recondition the valve guides, check and replace the valve springs, spring retainers and keepers (as necessary), replace the valve seals with new ones, reassemble the valve components and make sure the installed spring height is correct. The cylinder head gasket surface should also be resurfaced if it's warped. If you're working on a Legend model and one of the heads is warped, have both of them resurfaced. Also, the intake manifold mating surfaces should be machined to match the cylinder heads (since the heads will sit slightly lower on the engine).
4  After the valve job has been performed by a professional, the head will be in like new condition. When the head is returned, be sure to clean it again before installation on the engine to remove any metal particles and abrasive grit that may still be present from the valve service or head resurfacing operations. Use compressed air, if available, to blow out all the oil holes and passages.

## 11  Cylinder head – reassembly

*Refer to illustrations 11.3a, 11.3b, 11.5a, 11.5b, 11.6 and 11.8*

1  Regardless of whether or not the head was sent to an automotive repair shop for valve servicing, make sure it's clean before beginning reassembly.
2  If the head was sent out for valve servicing, the valves and related components will already be in place. Begin the reassembly procedure with Step 8.
3  Slip the spring seats over the guides (**see illustration**), then install new seals on each of the valve guides. **Note:** *Intake and exhaust valves require different seals – DO NOT mix them up! Exhaust seals have black springs and intake seals have white or silver springs.* Gently tap each valve seal into place with Acura tool no. 07GAD-PH70100 until it's seated on the guide (**see illustration**). **Caution:** *Don't hammer on the valve seals once they're seated or you may damage them. Don't twist or cock the seals during installation or they won't seat properly on the valve stems.*
4  Beginning at one end of the head, lubricate and install the first valve. Apply moly-base grease or clean engine oil to the valve stem.
5  Set the valve spring and retainer in place (**see illustrations**).
6  Compress the springs with a valve spring compressor and carefully install the keepers in the upper groove, then slowly release the compressor and make sure the keepers seat properly. Apply a small dab of grease to each keeper to hold it in place if necessary (**see illustration**).
7  Repeat the procedure for the remaining valves. Be sure to return the components to their original locations – don't mix them up!

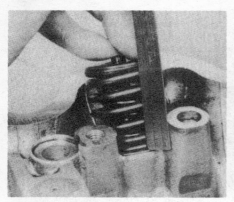

11.8 Double-check the height with the valve springs installed (do this for each valve)

12.1 A ridge reamer is required to remove the ridge from the top of each cylinder – do this before removing the pistons!

12.3 Check the connecting rod side clearance with a feeler gauge as shown here (Legend shown, Integra similar)

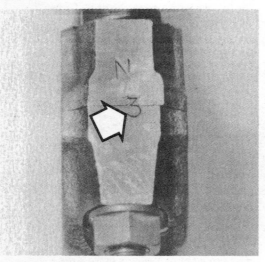

12.4a DO NOT confuse the stamped numbers on the parting surface, such as this 3 (arrow) with cylinder numbers – this number indicates big-end bore size

12.4b Before they're removed, the connecting rods and caps should be marked with a center punch to indicate in which cylinder they're installed

8  Check the valve spring installed height with a dial or vernier caliper **(see illustration)**.

## 12  Pistons/connecting rods – removal

*Refer to illustrations 12.1, 12.3, 12.4a, 12.4b and 12.6*

**Note:** *Prior to removing the piston/connecting rod assemblies, remove the cylinder head(s), the oil pan, oil pump pick-up tube and bearing cap bridge by referring to the appropriate Sections in Chapter 2A or 2B.*

1  Use your fingernail to feel if a ridge has formed at the upper limit of ring travel (about 1/4-inch down from the top of each cylinder). If carbon deposits or cylinder wear have produced ridges, they must be completely removed with a special tool **(see illustration)**. Follow the manufacturer's instructions provided with the tool. Failure to remove the ridges before attempting to remove the piston/connecting rod assemblies may result in piston breakage.

2  After the cylinder ridges (if any) have been removed, turn the engine upside-down so the crankshaft is facing up.

3  Before the connecting rods are removed, check the side clearance (endplay) with feeler gauges. Slide them between the first connecting rod and the crankshaft throw until the play is removed **(see illustration)**. The endplay is equal to the thickness of the feeler gauge(s). If the endplay exceeds the service limit, new connecting rods will be required. If new rods (or a new crankshaft) are installed, the endplay may fall under the specified minimum (if it does, the rods will have to be machined to restore it – consult an automotive machine shop for advice if necessary). Repeat the procedure for the remaining connecting rods.

4  The existing numbers on the connecting rods indicate the rod bore size, not the position in the engine **(see illustration)**. Use a small center punch to make the appropriate number of indentations on each rod and cap (1, 2, 3, etc., depending on the engine type and cylinder they're associated with) **(see illustration)**.

5  Loosen each of the connecting rod cap nuts 1/2-turn at a time until they can be removed by hand. Remove the number one connecting rod cap and bearing insert. Don't drop the bearing insert out of the cap.

6  Slip a short length of plastic or rubber hose over each connecting rod cap bolt to protect the crankshaft journal and cylinder wall as the piston is removed **(see illustration)**.

7  Remove the bearing insert and push the connecting rod/piston assembly out through the top of the engine. Use a wooden hammer handle to push on the upper bearing surface in the connecting rod. If resistance is felt, double-check to make sure that all of the ridge was removed from the cylinder.

8  Repeat the procedure for the remaining cylinders.

# Chapter 2 Part C  General engine overhaul procedures

12.6  To prevent damage to the crankshaft journals and cylinder walls, slip sections of hose over the rod bolts before removing the pistons

13.1  Position the dial indicator as shown and move the crankshaft back and forth with a screwdriver

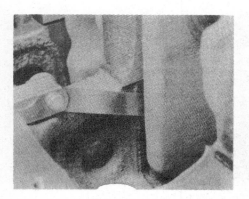

13.3  The endplay can also be checked with a feeler gauge at the thrust bearing journal

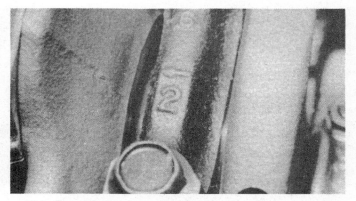

13.4  The main bearing caps should have numbers and arrows – this number indicates it is the second cap from the timing belt end and the arrows point toward the timing belt end

9  After removal, reassemble the connecting rod caps and bearing inserts in their respective connecting rods and install the cap nuts finger tight. Leaving the old bearing inserts in place until reassembly will help prevent the connecting rod bearing surfaces from being accidentally nicked or gouged.

10  Don't separate the pistons from the connecting rods (see Section 17 for additional information).

## 13  Crankshaft – removal

*Refer to illustrations 13.1, 13.3 and 13.4*

**Note:** *The crankshaft can be removed only after the engine has been removed from the vehicle. It's assumed that the flywheel or driveplate, timing belt, oil pan, oil pick-up tube, oil pump and piston/connecting rod assemblies have already been removed. The rear main oil seal retainer must be unbolted and separated from the block before proceeding with crankshaft removal.*

1  Before the crankshaft is removed, check the endplay. Mount a dial indicator with the stem in line with the crankshaft and just touching one of the crank throws **(see illustration)**.

2  Push the crankshaft all the way to the rear and zero the dial indicator. Next, pry the crankshaft to the front as far as possible and check the reading on the dial indicator. The distance that it moves is the endplay. If it's greater than specified, check the crankshaft thrust surfaces for wear. If no wear is evident, new thrust washers should correct the endplay.

3  If a dial indicator isn't available, feeler gauges can be used. Gently pry or push the crankshaft all the way to the front of the engine. Slip feeler gauges between the crankshaft and the front face of the thrust main bearing to determine the clearance **(see illustration)**. The thrust bearing on four-cylinder engines is number four, while on the V6 engine it's number three.

4  Check the main bearing caps to see if they're marked to indicate their locations. They should be numbered consecutively from the front of the engine to the rear **(see illustration)**. If they aren't, mark them with number stamping dies or a center punch. Main bearing caps generally have a cast-in arrow, which points to the front of the engine. Loosen the main bearing cap bolts 1/4-turn at a time each, working around the engine until they can be removed by hand.

5  Gently tap the caps with a soft-face hammer, then separate them from the engine block. If necessary, use the bolts as levers to remove the caps. Try not to drop the bearing inserts if they come out with the caps.

6  Carefully lift the crankshaft out of the engine. It may be a good idea to have an assistant available, since the crankshaft is quite heavy. With the bearing inserts in place in the engine block, return the caps to their respective locations on the engine block, install the main bearing cap bridge and tighten the bolts finger tight.

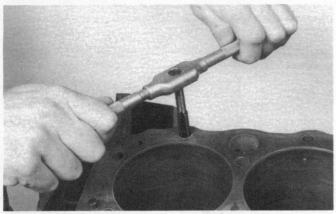

**14.7 All bolt holes in the block – particularly the main bearing cap and head bolt holes – should be cleaned and restored with a tap (be sure to remove debris from the holes after this is done)**

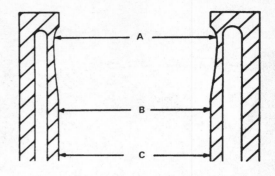

**15.4a Measure the diameter of each cylinder just under the wear ridge (A), at the center (B) and at the bottom (C)**

## 14 Engine block – cleaning

*Refer to illustration 14.7*

1  Using a gasket scraper, remove all traces of gasket material from the engine block. Be very careful not to nick or gouge the gasket sealing surfaces.
2  Remove the main bearing caps and bridge and separate the bearing inserts from the caps and the engine block. Tag the bearings, indicating which cylinder they were removed from. The oil groove identifies the upper bearings.
3  Remove all of the threaded oil gallery plugs from the block. The plugs are usually very tight – they may have to be drilled out and the holes re-tapped. Use new plugs when the engine is reassembled.
4  If the block is extremely dirty it should be taken to an automotive machine shop to be steam cleaned or tanked.
5  After the block is returned, clean all oil holes and oil galleries one more time. Brushes specifically designed for this purpose are available at most auto parts stores. Flush the passages with warm water until the water runs clear, dry the block thoroughly and wipe all machined surfaces with a light, rust preventive oil. If you have access to compressed air, use it to speed the drying process and to blow out all the oil holes and galleries. **Warning:** *Wear eye protection when using compressed air!*
6  If the block isn't extremely dirty or sludged up, you can do an adequate cleaning job with hot soapy water and a stiff brush. Take plenty of time and do a thorough job. Regardless of the cleaning method used, be sure to clean all oil holes and galleries very thoroughly, dry the block completely and coat all machined surfaces with light oil.
7  The threaded holes in the block must be clean to ensure accurate torque readings during reassembly. Run the proper size tap into each of the holes to remove rust, corrosion, thread sealant or sludge and restore damaged threads **(see illustration)**. If possible, use compressed air to clear the holes of debris produced by this operation. Now is a good time to clean the threads on the head bolts and the main bearing cap bolts as well.
8  Reinstall the main bearing caps and tighten the bolts finger tight.
9  Apply non-hardening sealant (such as Permatex no. 2 or Teflon pipe sealant) to the new oil gallery plugs and thread them into the holes in the block. Make sure they're tightened securely.
10  If the engine isn't going to be reassembled right away, cover it with a large plastic trash bag to keep it clean.

## 15 Engine block – inspection

*Refer to illustrations 15.4a, 15.4b, 15.4c, 15.12a and 15.12b*

1  Before the block is inspected, it should be cleaned as described in Section 14.

**15.4b The ability to "feel" when the telescoping gauge is at the correct point will be developed over time, so work slowly and repeat the check until you're satisfied the bore measurement is accurate**

2  Visually check the block for cracks, rust and corrosion. Look for stripped threads in the threaded holes. It's also a good idea to have the block checked for hidden cracks by an automotive machine shop that has the special equipment to do this type of work. If defects are found, have the block repaired, if possible, or replaced.
3  Check the cylinder bores for scuffing and scoring.
4  Measure the diameter of each cylinder at the top (just under the ridge area), center and bottom of the cylinder bore, parallel to the crankshaft axis **(see illustrations)**.
5  Next, measure each cylinder's diameter at the same three locations across the crankshaft axis. Compare the results to those listed in this Chapter's Specifications.
6  If the required precision measuring tools aren't available, the piston-to-cylinder clearances can be obtained, though not quite as accurately, using feeler gauge stock. Feeler gauge stock comes in 12-inch lengths and various thicknesses and is generally available at auto parts stores.
7  To check the clearance, select a feeler gauge and slip it into the cylinder along with the matching piston. The piston must be positioned exactly as it normally would be. The feeler gauge must be between the piston and cylinder on one of the thrust faces (90-degrees to the piston pin bore).
8  The piston should slip through the cylinder (with the feeler gauge in place) with moderate pressure.
9  If it falls through or slides through easily, the clearance is excessive and a new piston will be required. If the piston binds at the lower end of the

# Chapter 2 Part C  General engine overhaul procedures

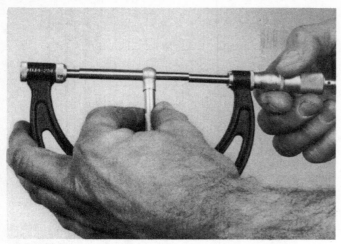

15.4c  The gauge is then measured with a micrometer to determine the bore size

15.12a  Check the block deck for distortion with a precision straightedge and feeler gauges

cylinder and is loose toward the top, the cylinder is tapered. If tight spots are encountered as the piston/feeler gauge is rotated in the cylinder, the cylinder is out-of-round.
10  Repeat the procedure for the remaining pistons and cylinders.
11  If the cylinder walls are badly scuffed or scored, or if they're out-of-round or tapered beyond the limits given in this Chapter's Specifications, have the engine block rebored and honed at an automotive machine shop. If a rebore is done, oversize pistons and rings will be required.
12  Using a precision straightedge and feeler gauge, check the block deck (the surface that mates with the cylinder head[s]) for distortion (**see illustrations**). If it's distorted beyond the specified limit, it can be resurfaced by an automotive machine shop.
13  If the cylinders are in reasonably good condition and not worn to the outside of the limits, and if the piston-to-cylinder clearances can be maintained properly, then they don't have to be rebored. Honing is all that's necessary (see Section 16).

## 16  Cylinder honing

*Refer to illustrations 16.3a and 16.3b*

1  Prior to engine reassembly, the cylinder bores must be honed so the new piston rings will seat correctly and provide the best possible combustion chamber seal. **Note:** *If you don't have the tools or don't want to tackle the honing operation, most automotive machine shops will do it for a reasonable fee.*
2  Before honing the cylinders, install the main bearing caps or cap assembly (without bearing inserts) and tighten the bolts to the torque listed in this Chapter's Specifications.
3  Two types of cylinder hones are commonly available – the flex hone or "bottle brush" type and the more traditional surfacing hone with spring-loaded stones. Both will do the job, but for the less experienced mechanic the "bottle brush" hone will probably be easier to use. You'll also need some kerosene or honing oil, rags and an electric drill motor. Proceed as follows:
   a) Mount the hone in the drill motor, compress the stones and slip it into the first cylinder (**see illustration**). Be sure to wear safety goggles or a face shield!
   b) Lubricate the cylinder with plenty of honing oil, turn on the drill and move the hone up-and-down in the cylinder at a pace that will produce a fine crosshatch pattern on the cylinder walls. Ideally, the crosshatch lines should intersect at approximately a 60-degree angle (**see illustration**). Be sure to use plenty of lubricant and don't take off any more material than is absolutely necessary to produce the desired finish. **Note:** *Piston ring manufacturers may specify a*

15.12b  Lay the straightedge across the block, diagonally and from end-to-end when making the check

16.3a  A "bottle brush" hone will produce better results if you've never honed cylinders before

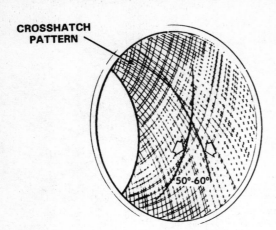

**16.3b  The cylinder hone should leave a smooth, crosshatch pattern with the lines intersecting at approximately a 60-degree angle**

**17.4a  The piston ring grooves can be cleaned with a special tool, as shown here, . . .**

smaller crosshatch angle than the traditional 60-degrees – read and follow any instructions included with the new rings.

c) Don't withdraw the hone from the cylinder while it's running. Instead, shut off the drill and continue moving the hone up-and-down in the cylinder until it comes to a complete stop, then compress the stones and withdraw the hone. If you're using a "bottle brush" type hone, stop the drill motor, then turn the chuck in the normal direction of rotation while withdrawing the hone from the cylinder.

d) Wipe the oil out of the cylinder and repeat the procedure for the remaining cylinders.

4   After the honing job is complete, chamfer the top edges of the cylinder bores with a small file so the rings won't catch when the pistons are installed. Be very careful not to nick the cylinder walls with the end of the file.

5   The entire engine block must be washed again very thoroughly with warm, soapy water to remove all traces of the abrasive grit produced during the honing operation. **Note:** *The bores can be considered clean when a lint-free white cloth – dampened with clean engine oil – used to wipe them out doesn't pick up any more honing residue, which will show up as gray areas on the cloth.* Be sure to run a brush through all oil holes and galleries and flush them with running water.

6   After rinsing, dry the block and apply a coat of light rust preventive oil to all machined surfaces. Wrap the block in a plastic trash bag to keep it clean and set it aside until reassembly.

## 17  Pistons/connecting rods – inspection

*Refer to illustrations 17.4a, 17.4b, 17.10, 17.11 and 17.12*

1   Before the inspection process can be carried out, the piston/connecting rod assemblies must be cleaned and the original piston rings removed from the pistons. **Note:** *Always use new piston rings when the engine is reassembled.*

2   Using a piston ring tool, carefully remove the rings from the pistons. Be careful not to nick or gouge the pistons in the process.

3   Scrape all traces of carbon from the top of the piston. A hand-held wire brush or a piece of fine emery cloth can be used once the majority of the deposits have been scraped away. Do not, under any circumstances, use a wire brush mounted in a drill motor to remove deposits from the pistons. The piston material is soft and may be eroded away by the wire brush.

4   Use a piston ring groove cleaning tool to remove carbon deposits from the ring grooves. If a tool isn't available, a piece broken off the old ring will do the job. Be very careful to remove only the carbon deposits – don't remove any metal and do not nick or scratch the sides of the ring grooves **(see illustrations)**.

**17.4b  . . . or a section of a broken ring**

5   Once the deposits have been removed, clean the piston/rod assemblies with solvent and dry them with compressed air (if available). Make sure the oil return holes in the back sides of the ring grooves and the oil hole in the lower end of each rod are clear. **Warning:** *Wear eye protection when using compressed air.*

6   If the pistons and cylinder walls aren't damaged or worn excessively, and if the engine block is not rebored, new pistons won't be necessary. Normal piston wear appears as even vertical wear on the piston thrust surfaces and slight looseness of the top ring in its groove. New piston rings, however, should always be used when an engine is rebuilt.

7   Carefully inspect each piston for cracks around the skirt, at the pin bosses and at the ring lands.

8   Look for scoring and scuffing on the thrust faces of the skirt, holes in the piston crown and burned areas at the edge of the crown. If the skirt is scored or scuffed, the engine may have been suffering from overheating and/or abnormal combustion, which caused excessively high operating temperatures. The cooling and lubrication systems should be checked thoroughly. A hole in the piston crown is an indication that abnormal combustion (preignition) was occurring. Burned areas at the edge of the piston crown are usually evidence of spark knock (detonation). If any of the above problems exist, the causes must be corrected or the damage will occur again. The causes may include intake air leaks, incorrect fuel/air mixture, incorrect ignition timing and EGR system malfunctions.

# Chapter 2 Part C  General engine overhaul procedures

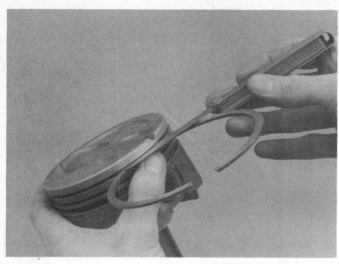

17.10  Check the ring side clearance with a feeler gauge at several points around the groove

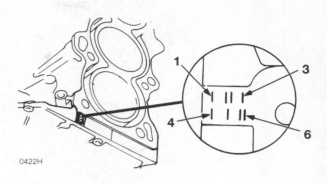

17.11  Match the marks with the letters on the pistons (I equals A, II equals B) – the marks on the engine read from left to right, 1, 2 and 3 on top, 4, 5, and 6 on the bottom

17.12  Measure the piston diameter at a 90-degree angle to the piston pin and in line with it

18.1  The oil holes should be chamfered so sharp edges don't gouge or scratch the new bearings

9  Corrosion of the piston, in the form of small pits, indicates that coolant is leaking into the combustion chamber and/or the crankcase. Again, the cause must be corrected or the problem may persist in the rebuilt engine.

10  Measure the piston ring side clearance by laying a new piston ring in each ring groove and slipping a feeler gauge in beside it **(see illustration)**. Check the clearance at three or four locations around each groove. Be sure to use the correct ring for each groove – they are different. If the side clearance is greater than specified, new pistons will have to be used.

11  Check the piston-to-bore clearance by measuring the bore (see Section 15) and the piston diameter. Make sure the pistons and bores are correctly matched. **Note:** *On Legends, there are two standard size pistons (marked A and B on the piston). Additionally, the engine block is marked* **(see illustration)** *to indicate which size piston was originally fitted to each bore.*

12  Measure the piston across the skirt, at a 90-degree angle to the piston pin, the distance listed in this Chapter's Specifications from the bottom edge of the piston skirt **(see illustration)**. Subtract the piston diameter from the bore diameter to obtain the clearance. If it's greater than specified, the block will have to be rebored and new pistons and rings installed.

13  Check the piston pin-to-rod clearance by twisting the piston and rod in opposite directions. Any noticeable play indicates excessive wear, which must be corrected. The piston/connecting rod assemblies should be taken to an automotive machine shop to have the pistons and rods resized and new pins installed.

14  If the pistons must be removed from the connecting rods for any reason, they should be taken to an automotive machine shop. While they are there have the connecting rods checked for bend and twist, since automotive machine shops have special equipment for this purpose. **Note:** *Unless new pistons and/or connecting rods must be installed, do not disassemble the pistons and connecting rods.*

15  Check the connecting rods for cracks and other damage. Temporarily remove the rod caps, lift out the old bearing inserts, wipe the rod and cap bearing surfaces clean and inspect them for nicks, gouges and scratches. After checking the rods, replace the old bearings, slip the caps into place and tighten the nuts finger tight. **Note:** *If the engine is being rebuilt because of a connecting rod knock, be sure to install new rods.*

## 18  Crankshaft – inspection

*Refer to illustration 18.1, 18.3, 18.4 and 18.6*

1  Remove all burrs from the crankshaft oil holes with a stone, file **(see illustration)** or scraper.

**18.3 Rubbing a penny lengthwise on each journal will reveal its condition – if copper rubs off and is embedded in the crankshaft, the journals should be reground**

**18.4 Use a wire or stiff plastic bristle brush to clean the oil passages in the crankshaft**

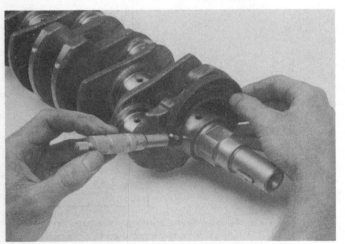

**18.6 Measure the diameter of each crankshaft journal at several points to detect taper and out-of-round conditions**

2  Check the main and connecting rod bearing journals for uneven wear, scoring, pits and cracks.
3  Rub a penny across each journal several times **(see illustration)**. If a journal picks up copper from the penny, it's too rough and must be reground.
4  Clean the crankshaft with solvent and dry it with compressed air (if available). Be sure to clean the oil holes with a stiff brush **(see illustration)** and flush them with solvent.
5  Check the rest of the crankshaft for cracks and other damage. It should be magnafluxed to reveal hidden cracks – an automotive machine shop will handle the procedure.
6  Using a micrometer, measure the diameter of the main and connecting rod journals and compare the results to this Chapter's Specifications **(see illustration)**. By measuring the diameter at a number of points around each journal's circumference, you'll be able to determine whether or not the journal is out-of-round. Take the measurement at each end of the journal, near the crank throws, to determine if the journal is tapered. Crankshaft runout should be checked also, but large V-blocks and a dial indicator are needed to do it correctly. If you don't have the equipment, have a machine shop check the runout.
7  If the crankshaft journals are damaged, tapered, out-of-round or worn beyond the limits given in this Chapter's Specifications, have the crankshaft reground by an automotive machine shop. Be sure to use the correct size bearing inserts if the crankshaft is reconditioned.
8  Check the oil seal journals at each end of the crankshaft for wear and damage. If the seal has worn a groove in the journal, or if it's nicked or scratched, the new seal may leak when the engine is reassembled. In some cases, an automotive machine shop may be able to repair the journal by pressing on a thin sleeve. If repair isn't feasible, a new or different crankshaft should be installed.
9  Refer to Section 19 and examine the main and rod bearing inserts.

## 19  Main and connecting rod bearings – inspection and bearing selection

### Inspection

*Refer to illustration 19.1*

1  Even though the main and connecting rod bearings should be replaced with new ones during the engine overhaul, the old bearings should be retained for close examination, as they may reveal valuable information about the condition of the engine **(see illustration)**.
2  Bearing failure occurs because of lack of lubrication, the presence of dirt or other foreign particles, overloading the engine and corrosion. Regardless of the cause of bearing failure, it must be corrected before the engine is reassembled to prevent it from happening again.
3  When examining the bearings, remove them from the engine block, the main bearing caps, the connecting rods and the rod caps and lay them out on a clean surface in the same general position as their location in the engine. This will enable you to match any bearing problems with the corresponding crankshaft journal.
4  Dirt and other foreign particles get into the engine in a variety of ways. It may be left in the engine during assembly, or it may pass through filters or the PCV system. It may get into the oil, and from there into the bearings. Metal chips from machining operations and normal engine wear are often present. Abrasives are sometimes left in engine components after reconditioning, especially when parts are not thoroughly cleaned using the proper cleaning methods. Whatever the source, these foreign objects often end up embedded in the soft bearing material and are easily recognized. Large particles will not embed in the bearing and will score or gouge the bearing and journal. The best prevention for this cause of bearing failure is to clean all parts thoroughly and keep everything spotlessly clean during engine assembly. Frequent and regular engine oil and filter changes are also recommended.
5  Lack of lubrication (or lubrication breakdown) has a number of interrelated causes. Excessive heat (which thins the oil), overloading (which squeezes the oil from the bearing face) and oil leakage or throw off (from

# Chapter 2 Part C  General engine overhaul procedures

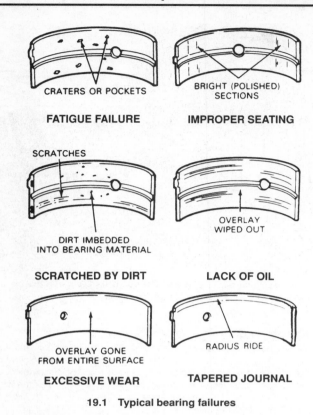

**19.1 Typical bearing failures**

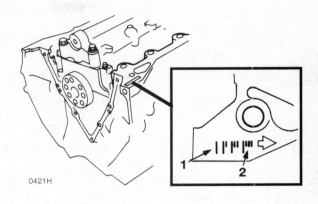

**19.10a** On the Legend, the codes are stamped on the block adjacent to the oil pan surface

excessive bearing clearances, worn oil pump or high engine speeds) all contribute to lubrication breakdown. Blocked oil passages, which usually are the result of misaligned oil holes in a bearing shell, will also oil starve a bearing and destroy it. When lack of lubrication is the cause of bearing failure, the bearing material is wiped or extruded from the steel backing of the bearing. Temperatures may increase to the point where the steel backing turns blue from overheating.

6  Driving habits can have a definite effect on bearing life. Full throttle, low speed operation (lugging the engine) puts very high loads on bearings, which tends to squeeze out the oil film. These loads cause the bearings to flex, which produces fine cracks in the bearing face (fatigue failure). Eventually the bearing material will loosen in pieces and tear away from the steel backing. Short trip driving leads to corrosion of bearings because insufficient engine heat is produced to drive off the condensed water and corrosive gases. These products collect in the engine oil, forming acid and sludge. As the oil is carried to the engine bearings, the acid attacks and corrodes the bearing material.

7  Incorrect bearing installation during engine assembly will lead to bearing failure as well. Tight fitting bearings leave insufficient bearing oil clearance and will result in oil starvation. Dirt or foreign particles trapped behind a bearing insert result in high spots on the bearing which lead to failure.

## Selection

*Refer to illustrations 19.10a, 19.10b, 19.11a, 19.11b, 19.12a, 19.12b, 19.14, 19.15a, 19.15b, 19.16a and 19.16b*

8  If the original bearings are worn or damaged, or if the oil clearances are incorrect (see Section 22 or 24), the following procedures should be used to select the correct new bearings for engine reassembly. However, if the crankshaft has been reground, new undersize bearings must be installed – the following procedure should not be used if undersize bearings are required! The automotive machine shop that reconditions the crankshaft will provide or help you select the correct size bearings. Regardless of how the bearing sizes are determined, use the oil clearance, measured with Plastigage, as a guide to ensure the bearings are the right size.

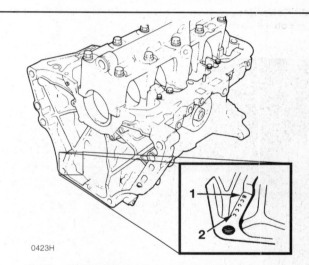

**19.10b** On the Integra, the codes are stamped on the transaxle mounting surface

### Main bearings

9  If you need to use a STANDARD size main bearing, install one that has the same color code as the original bearing.

10  If the color code on the original main bearing has been obscured, locate the codes stamped into the block for the corresponding cap location **(see illustrations)**.

11  Locate the main journal grade numbers on the crankshaft as well **(see illustrations)**.

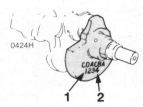

**19.11a** On the Legend, the main journal grade numbers are stamped on the timing belt end of the crankshaft

# 2C–22    Chapter 2 Part C  General engine overhaul procedures

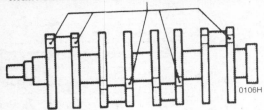

**19.11b  On the Integra, the main bearing journal grade numbers are stamped adjacent to their respective journals**

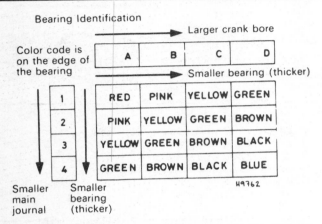

**19.12b  Find the correct main bearing color code for the Integra by using the letter on the block and the Arabic number on the crankshaft – example: C3 would be brown**

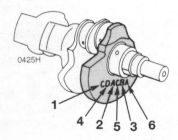

**19.15a  On Legends, the code letters for the connecting rod bearing journals are stamped on the timing belt end of the crankshaft**

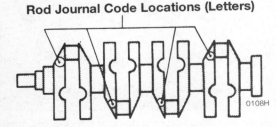

**19.15b  On Integras, the code letters for the connecting rod bearing journals are stamped adjacent to their respective journals**

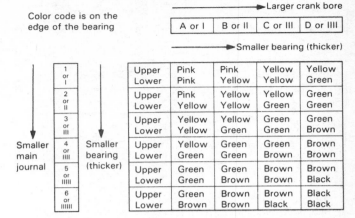

**19.12a  Find the correct main bearing color code for the Legend by using the Roman numeral or letter on the block and the number on the crankshaft – example: C3 would be green (note that some call for different upper and lower bearing codes)**

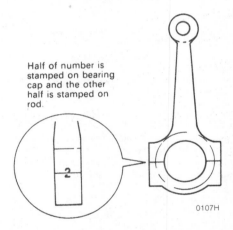

**19.14  The number stamped on the parting surface of the connecting rod and cap indicates the big-end bearing bore size – it does NOT indicate the cylinder number it came from**

12  Use the accompanying chart to determine the correct bearings for each journal (**see illustrations**).

### Connecting rod bearings

13  If you need to use a STANDARD size rod bearing, install one that has the same color code as the original.
14  If the color code has been obscured, locate the number stamped on each connecting rod cap (**see illustration**). This code indicates the connecting rod big-end bearing bore size.
15  Locate the letters stamped on the crankshaft (**see illustrations**). These denote the size of their respective connecting rod journals.
16  Use the accompanying chart (**see illustrations**) to determine the correct bearings for each journal.

### All bearings

17  Remember, the oil clearance is the final judge when selecting new bearing sizes. If you have any questions or are unsure which bearings to use, get help from your dealer parts or service department.

# Chapter 2 Part C  General engine overhaul procedures

## BEARING IDENTIFICATION

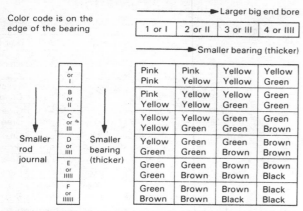

**19.16a** Find the correct connecting rod bearing color code for Legends by using the letter on each crankshaft throw and the number on the respective connecting rod – example: D4 would be brown (note that some call for different upper and lower bearing codes)

### Bearing Identification

| | 1 | 2 | 3 | 4 |
|---|---|---|---|---|
| A | RED | PINK | YELLOW | GREEN |
| B | PINK | YELLOW | GREEN | BROWN |
| C | YELLOW | GREEN | BROWN | BLACK |
| D | GREEN | BROWN | BLACK | BLUE |

Color code is on the edge of the bearing → Larger rod big end → Smaller bearing (thicker) ↓ Smaller rod journal ↓ Smaller bearing (thicker)

**19.16b** Find the correct connecting rod bearing color code for Integras by using the letter on each crankshaft throw and the number on the respective connecting rod – example: D4 would be blue

---

## 20  Engine overhaul – reassembly sequence

1  Before beginning engine reassembly, make sure you have all the necessary new parts, gaskets and seals as well as the following items on hand:
   Common hand tools
   A 1/2-inch drive torque wrench
   Piston ring installation tool
   Piston ring compressor
   Short lengths of rubber or plastic hose to fit over connecting rod bolts
   Plastigage
   Feeler gauges
   A fine-tooth file
   New engine oil
   Engine assembly lube or moly-base grease
   Gasket sealer
   Thread locking compound

2  In order to save time and avoid problems, engine reassembly must be done in the following general order:

### Four-cylinder engine
Piston rings
Crankshaft and main bearings
Piston/connecting rod assemblies
Rear main (crankshaft) oil seal
Cylinder head (see Part A)
Camshafts and rockers (see Part A)
Valve cover (see Part A)
Oil pump (see Part A)
Oil pick-up (see Part A)
Oil pan (see Part A)
Timing belt and sprockets (see Part A)
Timing belt cover (see Part A)
Intake and exhaust manifolds (see Part A)
Flywheel/driveplate (see Part A)

### V6 engine
Piston rings
Crankshaft and main bearings
Piston/connecting rod assemblies
Rear main oil seal/retainer
Oil pump (see Part B)
Oil pan (see Part B)
Cylinder heads (see Part B)
Camshafts and valve components (see Part B)
Timing belt and sprockets (see Part B)
Timing belt covers (see Part B)
Intake and exhaust manifolds (see Part B)
Cylinder head covers (see Part B)
Flywheel/driveplate (see Part B)

---

## 21  Piston rings – installation

Refer to illustrations 21.3, 21.4, 21.9a, 21.9b and 21.12

1  Before installing the new piston rings, the ring end gaps must be checked. It's assumed that the piston ring side clearance has been checked and verified correct (see Section 17).

2  Lay out the piston/connecting rod assemblies and the new ring sets so the ring sets will be matched with the same piston and cylinder during the end gap measurement and engine assembly.

3  Insert the top (number one) ring into the first cylinder and square it up with the cylinder walls by pushing it in with the top of the piston (see illustration). The ring should be near the bottom of the cylinder, at the lower limit of ring travel.

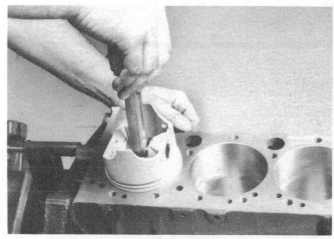

**21.3** When checking piston ring end gap, the ring must be square in the cylinder bore (this is done by pushing the ring down with the top of a piston as shown)

## 2C-24    Chapter 2 Part C   General engine overhaul procedures

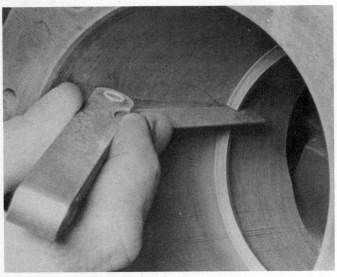

**21.4  With the ring square in the cylinder, measure the end gap with a feeler gauge**

**21.9b  DO NOT use a piston ring installation tool when installing the oil ring side rails**

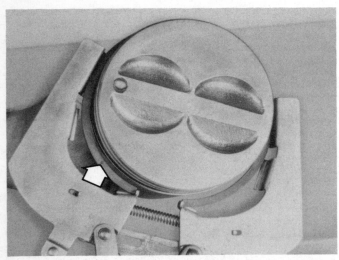

**21.12  Installing the compression rings with a ring expander – the mark (arrow) must face up**

**21.9a  Installing the spacer/expander in the oil control ring groove**

4  To measure the end gap, slip feeler gauges between the ends of the ring until a gauge equal to the gap width is found **(see illustration)**. The feeler gauge should slide between the ring ends with a slight amount of drag. Compare the measurement to this Chapter's Specifications. If the gap is larger or smaller than specified, double-check to make sure you have the correct rings before proceeding.

5  If the gap is too small, replace the rings – DO NOT file the ends to increase the clearance.

6  Excess end gap isn't critical unless it's greater than 0.040-inch. Again, double-check to make sure you have the correct rings for your engine.

7  Repeat the procedure for each ring t1hat will be installed in the first cylinder and for each ring in the remaining cylinders. Remember to keep rings, pistons and cylinders matched up.

8  Once the ring end gaps have been checked/corrected, the rings can be installed on the pistons.

9  The oil control ring (lowest one on the piston) is usually installed first. It's composed of three separate components. Slip the spacer/expander into the groove **(see illustration)**. If an anti-rotation tang is used, make sure it's inserted into the drilled hole in the ring groove. Next, install the lower side rail. Don't use a piston ring installation tool on the oil ring side rails, as they may be damaged. Instead, place one end of the side rail into the groove between the spacer/expander and the ring land, hold it firmly in place and slide a finger around the piston while pushing the rail into the groove **(see illustration)**. Next, install the upper side rail in the same manner.

10  After the three oil ring components have been installed, check to make sure that both the upper and lower side rails can be turned smoothly in the ring groove.

11  The number two (middle) ring is installed next. It's usually stamped with a mark which must face up, toward the top of the piston. **Note:** *Always follow the instructions printed on the ring package or box – different manufacturers may require different approaches.* Do not mix up the top and middle rings, as they have different cross-sections.

12  Use a piston ring installation tool and make sure the identification mark is facing the top of the piston, then slip the ring into the middle groove on the piston **(see illustration)**. Don't expand the ring any more than necessary to slide it over the piston.

13  Install the number one (top) ring in the same manner. Make sure the mark is facing up. Be careful not to confuse the number one and number two rings.

14  Repeat the procedure for the remaining pistons and rings.

## 22  Crankshaft – installation and main bearing oil clearance check

*Refer to illustrations 22.10, 22.12 and 22.14*

1  Crankshaft installation is the first major step in engine reassembly. It's

# Chapter 2 Part C  General engine overhaul procedures

22.10  Lay the Plastigage strips (arrow) on the main bearing journals, parallel to the crankshaft centerline

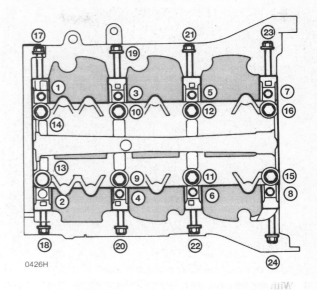

22.12  Bolt tightening sequence for the main bearing caps on Legend models

assumed at this point that the engine block and crankshaft have been cleaned, inspected and repaired or reconditioned.
2  Position the engine with the bottom facing up.
3  Remove the main bearing cap bolts and lift out the caps and bridge. Lay the caps out in the proper order to ensure correct installation.
4  If they're still in place, remove the old bearing inserts from the block and the main bearing caps. Wipe the main bearing surfaces of the block and caps with a clean, lint free cloth. They must be kept spotlessly clean!

## Main bearing oil clearance check

5  Clean the back sides of the new main bearing inserts and lay the bearing half with the oil groove in each main bearing saddle in the block. Lay the other bearing half from each bearing set in the corresponding main bearing cap. Make sure the tab on each bearing insert fits into the recess in the block or cap. Also, the oil holes in the block must line up with the oil holes in the bearing insert. **Caution:** *Do not hammer the bearings into place and don't nick or gouge the bearing faces. No lubrication should be used at this time.*
6  If you're working on a V6 engine, the thrust bearings (washers) must be installed in the number three position. On four-cylinder engines, the thrust bearings (washers) must be installed in the number four position. Oil the contact surfaces.
7  Clean the faces of the bearings in the block and the crankshaft main bearing journals with a clean, lint free cloth. Check or clean the oil holes in the crankshaft, as any dirt here can go only one way – straight through the new bearings.
8  Once you're certain the crankshaft is clean, carefully lay it in position in the main bearings.
9  Before the crankshaft can be permanently installed, the main bearing oil clearance must be checked.
10  Trim several pieces of the appropriate size Plastigage (they must be slightly shorter than the width of the main bearings) and place one piece on each crankshaft main bearing journal, parallel with the journal axis **(see illustration)**.
11  Clean the faces of the bearings in the caps and install the caps in their respective positions (don't mix them up) with the arrows pointing toward the front of the engine. Carefully lay the main bearing cap bridge in place. Don't disturb the Plastigage. Apply a light coat of oil to the bolt threads and the under sides of the bolt heads, then install them.
12  Tighten the main bearing cap bolts, in three steps, to the torque listed in this Chapter's Specifications. Don't rotate the crankshaft at any time during this operation! On Legends, follow the recommended sequence **(see illustration)**.

22.14  Compare the width of the crushed Plastigage to the scale on the envelope to determine the main bearing oil clearance (always take the measurement at the widest point of the Plastigage); be sure to use the correct scale – standard and metric ones are included

13  Remove the bolts and carefully lift off the main bearing caps. Keep them in order. Don't disturb the Plastigage or rotate the crankshaft. If any of the main bearing caps are difficult to remove, tap them gently from side-to-side with a soft-face hammer to loosen them.
14  Compare the width of the crushed Plastigage on each journal to the scale printed on the Plastigage envelope to obtain the main bearing oil clearance **(see illustration)**. Check this Chapter's Specifications to make sure it's correct.
15  If the clearance is not as specified, the bearing inserts may be the wrong size (which means different ones will be required – see Section 19). Before deciding that different inserts are needed, make sure that no dirt or oil was between the bearing inserts and the caps or block when the clearance was measured. If the Plastigage is noticeably wider at one end than the other, the journal may be tapered (see Section 18).
16  Carefully scrape all traces of the Plastigage material off the main bearing journals and/or the bearing faces. Don't nick or scratch the bearing faces.

## Chapter 2 Part C  General engine overhaul procedures

23.3 Drive the old seal out using a blunt punch and a small hammer

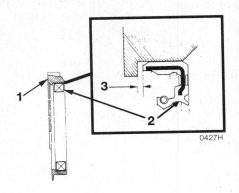

23.4 On the Integra, the depth of the oil seal in the retainer must be set as shown

1  Oil seal retainer
2  Seal
3  0.008 to 0.020 inch (0.2 to 0.5 mm)

### Final crankshaft installation

17  Carefully lift the crankshaft out of the engine. Clean the bearing faces in the block, then apply a thin, uniform layer of clean moly-base grease or engine assembly lube to each of the bearing surfaces. Coat the thrust washers as well.

18  Lubricate the crankshaft surfaces that contact the oil seals with moly-base grease, engine assembly lube or clean engine oil.

19  Make sure the journals are clean, then lay the crankshaft back in place in the block. Clean the faces of the bearings in the caps, then apply lubricant to them.

20  With the engine block positioned so the crankshaft is at the top, install the pistons and connecting rods (see Section 24).

21  Install the caps and bridge in their respective positions with the arrows pointing toward the front of the engine. **Note:** *Be sure to install the thrust washers.*

22  Apply a light coat of oil to the bolt threads and the under sides of the bolt heads, then install them. Start the bolts by hand. Tap the ends of the crankshaft forward and backward with a lead or brass hammer to line up the thrust washer and crankshaft surfaces before the bolts are tightened. Tighten all main bearing cap and bridge bolts to the torque listed in this Chapter's Specifications. On Legend models, be sure to follow the recommended sequence **(see illustration 22.12)**.

23  Rotate the crankshaft a number of times by hand to check for any obvious binding.

24  Check the crankshaft endplay with a feeler gauge or a dial indicator as described in Section 13. The endplay should be correct if the crankshaft thrust faces aren't worn or damaged and new thrust washers have been installed.

25  Install a new rear main oil seal, then bolt the retainer to the block (see Section 23).

### 23  Rear main oil seal installation

*Refer to illustrations 23.3, 23.4 and 23.5*

1  The crankshaft must be installed first and the main bearing caps and bridge bolted in place, then the new seal should be installed in the retainer and the retainer bolted to the block.

2  Check the seal contact surface on the crankshaft very carefully for scratches and nicks that could damage the new seal lip and cause oil leaks. If the crankshaft is damaged, the only alternative is a new or different crankshaft.

23.5 Drive the new seal into the retainer with a block of wood or a section of pipe, if you have one large enough – make sure that you don't cock the seal in the retainer bore

3  The old seal can be removed from the retainer by driving it out from the back side with a hammer and punch **(see illustration)**. Be sure to note how far it's recessed into the bore before removing it; the new seal will have to be recessed an equal amount. Be very careful not to scratch or otherwise damage the bore in the retainer or oil leaks could develop.

4  Make sure the retainer is clean, then apply a thin coat of engine oil to the outer edge of the new seal. The seal must be pressed squarely into the bore, so hammering it into place isn't recommended. If you don't have access to a press, sandwich the housing and seal between two smooth pieces of wood and press the seal into place with the jaws of a large vise. The pieces of wood must be thick enough to distribute the force evenly around the entire circumference of the seal. Work slowly and make sure the seal enters the bore squarely. On Integras, the seal should be installed to the depth shown **(see illustration)**.

5  If a vise in not available, the seal can be tapped into the retainer with a hammer. Use a block of wood to distribute the force evenly and make sure the seal is driven in squarely **(see illustration)**.

# Chapter 2 Part C General engine overhaul procedures

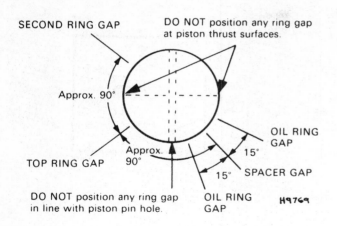

24.5 Position the piston ring gaps as shown here before installing the piston/connecting rod assemblies in the engine

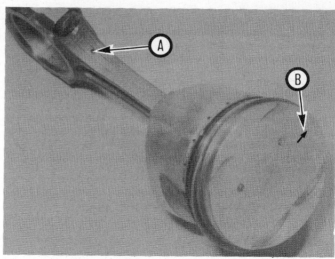

24.9 When installing pistons, the small hole in the connecting rod (A) should face the firewall side of the engine on Legends or the intake manifold side of the engine on Integras, and the arrows (B) should face the timing belt end of the engine

24.11 The piston can be driven (gently) into the cylinder bore with the end of a wooden or plastic hammer handle

6  The seal lips must be lubricated with clean engine oil or grease before the seal/retainer is slipped over the crankshaft and bolted to the block. No gasket is required. Instead, clean the surface and then apply a 2mm wide bead of sealant (Acura no. 08718-550000 or equivalent) to the retainer-to-block surface just prior to installation.

7  Tighten the bolts a little at a time to the torque listed in this Chapter's Specifications.

## 24 Pistons/connecting rods – installation and rod bearing oil clearance check

*Refer to illustrations 24.5, 24.9, 24.11, 24.13 and 24.17*

1  Before installing the piston/connecting rod assemblies, the cylinder walls must be perfectly clean, the top edge of each cylinder must be chamfered, and the crankshaft must be in place.

2  Remove the cap from the end of the number one connecting rod (refer to the marks made during removal). Remove the original bearing inserts and wipe the bearing surfaces of the connecting rod and cap with a clean, lint-free cloth. They must be kept spotlessly clean.

### Connecting rod bearing oil clearance check

3  Clean the back side of the new upper bearing insert, then lay it in place in the connecting rod. Make sure the tab on the bearing fits into the recess in the rod so the oil holes line up. Don't hammer the bearing insert into place and be very careful not to nick or gouge the bearing face. Don't lubricate the bearing at this time.

4  Clean the back side of the other bearing insert and install it in the rod cap. Again, make sure the tab on the bearing fits into the recess in the cap, and don't apply any lubricant. It's critically important that the mating surfaces of the bearing and connecting rod are perfectly clean and oil-free when they're assembled.

5  Position the piston ring gaps at staggered intervals around the piston (see illustration).

6  Slip a section of plastic or rubber hose over each connecting rod cap bolt.

7  Lubricate the piston and rings with clean engine oil and attach a piston ring compressor to the piston. Leave the skirt protruding about 1/4-inch to guide the piston into the cylinder. The rings must be compressed until they're flush with the piston.

8  Rotate the crankshaft until the number one connecting rod journal is at BDC (bottom dead center) and apply a coat of engine oil to the cylinder walls.

9  With the arrow on top of the piston (see illustration) facing the timing belt end of the engine, gently insert the piston/connecting rod assembly into the number one cylinder bore and rest the bottom edge of the ring compressor on the engine block.

10  Tap the top edge of the ring compressor to make sure it's contacting the block around its entire circumference.

11  Gently tap on the top of the piston with the end of a wooden hammer handle (see illustration) while guiding the end of the connecting rod into place on the crankshaft journal. The piston rings may try to pop out of the ring compressor just before entering the cylinder bore, so keep some downward pressure on the ring compressor. Work slowly, and if any resistance is felt as the piston enters the cylinder, stop immediately. Find out what's hanging up and fix it before proceeding. Do not, for any reason, force the piston into the cylinder – you might break a ring and/or the piston.

12  Once the piston/connecting rod assembly is installed, the connecting rod bearing oil clearance must be checked before the rod cap is permanently bolted in place.

13  Cut a piece of the appropriate size Plastigage slightly shorter than the width of the connecting rod bearing and lay it in place on the number one

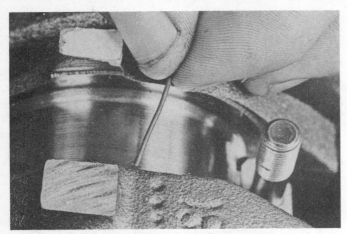

**24.13 Lay the Plastigage strips on each rod bearing journal, parallel to the crankshaft centerline**

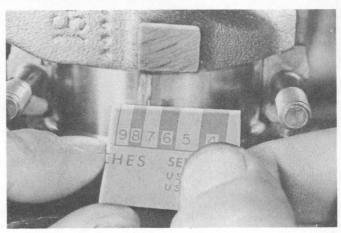

**24.17 Measuring the width of the crushed Plastigage to determine the rod bearing oil clearance (be sure to use the correct scale – standard and metric ones are included)**

connecting rod journal, parallel with the journal axis **(see illustration)**.
14   Clean the connecting rod cap bearing face, remove the protective hoses from the connecting rod bolts and install the rod cap. Make sure the mating mark on the cap is on the same side as the mark on the connecting rod. If you're working on a four cylinder engine, make sure the connecting rod oil hole is facing the intake manifold side of the engine If you're working on a V6 make sure the connecting rod oil hole is facing the firewall side of the engine.
15   Apply a light coat of oil to the under sides of the nuts, then install and tighten them to the torque listed in this Chapter's Specifications, working up to it in three steps. Use a thin-wall socket to avoid erroneous torque readings that can result if the socket is wedged between the rod cap and nut. If the socket tends to wedge itself between the nut and the cap, lift up on it slightly until it no longer contacts the cap. Do not rotate the crankshaft at any time during this operation.
16   Remove the nuts and detach the rod cap, being very careful not to disturb the Plastigage.
17   Compare the width of the crushed Plastigage to the scale printed on the Plastigage envelope to obtain the oil clearance **(see illustration)**. Compare it to the Specifications to make sure the clearance is correct.
18   If the clearance is not as specified, the bearing inserts may be the wrong size (which means different ones will be required). Before deciding that different inserts are needed, make sure that no dirt or oil was between the bearing inserts and the connecting rod or cap when the clearance was measured. Also, recheck the journal diameter. If the Plastigage was wider at one end than the other, the journal may be tapered (refer to Section 18).

### Final connecting rod installation

19   Carefully scrape all traces of the Plastigage material off the rod journal and/or bearing face. Be very careful not to scratch the bearing – use your fingernail or the edge of a credit card.
20   Make sure the bearing faces are perfectly clean, then apply a uniform layer of clean moly-base grease or engine assembly lube to both of them. You'll have to push the piston into the cylinder to expose the face of the bearing insert in the connecting rod – be sure to slip the protective hoses over the rod bolts first.
21   Slide the connecting rod back into place on the journal, remove the protective hoses from the rod cap bolts, install the rod cap and tighten the nuts to the torque listed in this Chapter's Specifications. Again, work up to the torque in three steps.
22   Repeat the entire procedure for the remaining pistons/connecting rods.
23   The important points to remember are . . .
   a) Keep the back sides of the bearing inserts and the insides of the connecting rods and caps perfectly clean when assembling them.
   b) Make sure you have the correct piston/rod assembly for each cylinder.
   c) The arrow on the piston must face the timing belt end of the engine.
   d) Be sure the stamped marks on the rod and rod cap are on the same side.
   e) Lubricate the cylinder walls with clean oil.
   f) Lubricate the bearing faces when installing the rod caps after the oil clearance has been checked.
24   After all the piston/connecting rod assemblies have been properly installed, rotate the crankshaft a number of times by hand to check for any obvious binding.
25   As a final step, the connecting rod endplay must be checked. Refer to Section 12 for this procedure.
26   Compare the measured endplay to the Specifications to make sure it's correct. If it was correct before disassembly and the original crankshaft and rods were reinstalled, it should still be right. If new rods or a new crankshaft were installed, the endplay may be inadequate. If so, the rods will have to be removed and taken to an automotive machine shop for resizing.

**25   Initial start-up and break-in after overhaul**

**Warning:** *Have a fire extinguisher handy when starting the engine for the first time.*

1   Once the engine has been installed in the vehicle, double-check the engine oil and coolant levels.
2   With the spark plugs out of the engine and the ignition system disabled (see Section 3), crank the engine until oil pressure registers on the gauge or the light goes out.
3   Install the spark plugs, connect the plug wires and restore the ignition system functions (see Section 3).
4   Start the engine. It may take a few moments for the fuel system to build up pressure, but the engine should start without a great deal of effort.
5   After the engine starts, allow it to warm up to normal operating temperature. While the engine is warming up, make a thorough check for fuel, oil and coolant leaks.
6   Shut the engine off and recheck the engine oil and coolant levels.
7   Drive the vehicle to an area with minimum traffic, accelerate at full throttle from 30 to 50 mph, then allow the vehicle to slow to 30 mph with the throttle closed. Repeat the procedure 10 or 12 times. This will load the piston rings and cause them to seat properly against the cylinder walls. Check again for oil and coolant leaks.
8   Drive the vehicle gently for the first 500 miles (no sustained high speeds) and keep a constant check on the oil level. It is not unusual for an engine to use oil during the break-in period.
9   At approximately 500 to 600 miles, change the oil and filter.
10   For the next few hundred miles, drive the vehicle normally. Do not pamper it or abuse it.
11   After 2000 miles, change the oil and filter again and consider the engine broken in.

# Chapter 3 Cooling, heating and air conditioning systems

## Contents

| | |
|---|---|
| Air conditioning and heater control assembly – removal and installation | 13 |
| Air conditioning and heating system – check and maintenance | 14 |
| Air conditioning compressor – removal and installation | 16 |
| Air conditioning condenser – removal and installation | 17 |
| Air conditioning receiver/drier – removal and installation | 15 |
| Antifreeze – general information | 2 |
| Blower unit – removal and installation | 11 |
| Coolant level check | See Chapter 1 |
| Coolant reservoir – removal and installation | 6 |
| Coolant temperature sending unit – check and replacement | 10 |
| Cooling system check | See Chapter 1 |
| Cooling system servicing (draining, flushing and refilling) | See Chapter 1 |
| Drivebelt check, adjustment and replacement | See Chapter 1 |
| Engine cooling fan – check and replacement | 4 |
| General information | 1 |
| Heater core – removal and installation | 12 |
| Oil cooler – removal and installation | 7 |
| Radiator – removal and installation | 5 |
| Thermostat – check and replacement | 3 |
| Underhood hose check and replacement | See Chapter 1 |
| Water pump – check | 8 |
| Water pump – replacement | 9 |

## Specifications

### General

| | |
|---|---|
| Coolant capacity | See Chapter 1 |
| Drivebelt tension | See Chapter 1 |
| Radiator pressure cap rating | |
|   Legend | 14 to 17 psi |
|   Integra | 11 to 15 psi |
| Thermostat rating (fully open) | 196-degrees F |

### Torque specifications

**Ft-lbs** (unless otherwise indicated)

| | |
|---|---|
| Oil cooler center bolt (Legend) | 36 |
| Oil cooler mounting bolts (Legend) | 16 |
| Oil cooler center bolt (Integra) | 54 |
| Thermostat housing bolts | |
|   Legend | |
|     M6 | 108 in-lbs |
|     M8 | 16 |
|   Integra | 108 in-lbs |
| Water pump attaching bolts | |
|   Legend | |
|     M6 | 108 in-lbs |
|     M8 | 16 |
|   Integra | 108 in-lbs |

# 3–2  Chapter 3  Cooling, heating and air conditioning systems

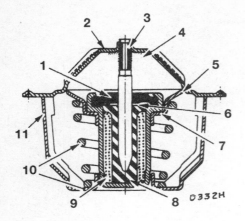

**1.2  Pellet type thermostat**

| | | | |
|---|---|---|---|
| 1 Flange seal | 5 Valve seat | 9 Wax pellet |
| 2 Flange | 6 Teflon seal | 10 Coil spring |
| 3 Piston | 7 Valve | 11 Frame |
| 4 Nut | 8 Rubber diaphragm | |

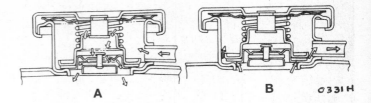

**1.3  Pressure-type radiator cap**

## 1  General information

*Refer to illustrations 1.2 and 1.3*

### Engine cooling system

All vehicles covered by this manual employ a pressurized engine cooling system with thermostatically controlled coolant circulation. An impeller-type water pump mounted on the engine block pumps coolant through the engine. The coolant flows around each cylinder and toward the rear of the engine. Cast-in coolant passages direct coolant around the intake and exhaust ports, near the spark plug areas and in close proximity to the exhaust valve guides.

A wax pellet type thermostat controls engine coolant temperature. During warm up, the closed thermostat prevents coolant from circulating through the radiator. As the engine nears normal operating temperature, the thermostat opens and allows hot coolant to travel through the radiator, where it's cooled before returning to the engine **(see illustration)**.

The cooling system is sealed by a pressure type radiator cap, which raises the boiling point of the coolant and increases the cooling efficiency of the radiator. If the system pressure exceeds the cap pressure relief value, the excess pressure in the system forces the spring-loaded valve inside the cap off its seat and allows the coolant to escape through the overflow tube into a coolant reservoir. When the system cools the excess coolant is automatically drawn from the reservoir back into the radiator **(see illustration)**.

The coolant reservoir does double duty as both the point at which fresh coolant is added to the cooling system to maintain the proper fluid level and as a holding tank for overheated coolant.

This type of cooling system is known as a closed design because coolant that escapes past the pressure cap is saved and reused.

### Heating system

The heating system consists of a blower fan and heater core located in the heater box, the hoses connecting the heater core to the engine cooling system and the heater/air conditioning control head on the dashboard. Hot engine coolant is circulated through the heater core. When the heater mode is activated, a flap door opens to expose the heater box to the passenger compartment. A fan switch on the control head activates the blower motor, which forces air through the core, heating the air.

### Air conditioning system

The air conditioning system consists of a condenser mounted in front of the radiator, an evaporator mounted adjacent to the heater core, a compressor mounted on the engine, a receiver-drier which contains a high pressure relief valve and the plumbing connecting all of the above components.

A blower fan forces the warmer air of the passenger compartment through the evaporator core (sort of a radiator-in-reverse), transferring the heat from the air to the refrigerant. The liquid refrigerant boils off into low pressure vapor, taking the heat with it when it leaves the evaporator.

## 2  Antifreeze – general information

**Warning:** *Do not allow antifreeze to come in contact with your skin or painted surfaces of the vehicle. Rinse off spills immediately with plenty of water. Antifreeze is highly toxic if ingested. Never leave antifreeze lying around in an open container or in puddles on the floor; children and pets are attracted by it's sweet smell and may drink it. Check with local authorities about disposing of used antifreeze. Many communities have collection centers which will see that antifreeze is disposed of safely.*

The cooling system should be filled with a water/ethylene glycol based antifreeze solution, which will prevent freezing down to at least -20-degrees F, or lower if local climate requires it. It also provides protection against corrosion and increases the coolant boiling point.

The cooling system should be drained, flushed and refilled at the specified intervals (see Chapter 1). Old or contaminated antifreeze solutions are likely to cause damage and encourage the formation of rust and scale in the system. Use distilled water with the antifreeze.

Before adding antifreeze, check all hose connections, because antifreeze tends to leak through very minute openings. Engines don't normally consume coolant, so if the level goes down, find the cause and correct it.

The exact mixture of antifreeze-to-water which you should use depends on the relative weather conditions. The mixture should contain at least 50-percent antifreeze, but should never contain more than 70-percent antifreeze. Consult the mixture ratio chart on the antifreeze container before adding coolant. Hydrometers are available at most auto parts stores to test the coolant. Use antifreeze which meets the vehicle manufacturer's specifications.

## 3  Thermostat – check and replacement

**Warning:** *Do not remove the radiator cap, drain the coolant or replace the thermostat until the engine has cooled completely. Read the warning at the beginning of Section 2.*

### Check

1  Before assuming the thermostat is to blame for a cooling system problem, check the coolant level, drivebelt tension (see Chapter 1) and temperature gauge operation.

2  If the engine seems to be taking a long time to warm up (based on heater output or temperature gauge operation), the thermostat is probably stuck open. Replace the thermostat with a new one.

3  If the engine runs hot, use your hand to check the temperature of the upper radiator hose. If the hose isn't hot, but the engine is, the thermostat is probably stuck closed, preventing the coolant inside the engine from escaping to the radiator. Replace the thermostat. **Caution:** *Don't drive the vehicle without a thermostat. The computer may stay in open loop and emissions and fuel economy will suffer.*

# Chapter 3  Cooling, heating and air conditioning systems

3.7  Follow the lower radiator hose to locate the thermostat – on the Legend (shown) it's located below the throttle body (intake air duct removed)

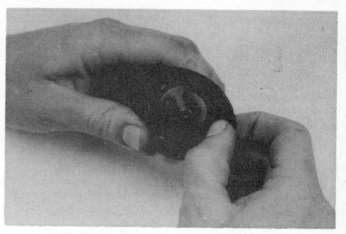

3.13  Fit a new rubber seal over the thermostat

4   If the upper radiator hose is hot, it means that the coolant is flowing and the thermostat is open. Consult the Troubleshooting section at the front of this manual for cooling system diagnosis.

### Replacement

*Refer to illustrations 3.7, 3.13 and 3.14*

5   Disconnect the negative battery cable from the battery.
6   Drain the cooling system (see Chapter 1). If the coolant is relatively new or in good condition (see Chapter 1), save it and reuse it.
7   Follow the lower radiator hose to the engine to locate the thermostat housing **(see illustration)**. On the Integra, it's located below and adjacent to the distributor. On Legends, it's located below the throttle body; it may be necessary to remove the intake air duct for access.
8   Loosen the hose clamp, then detach the hose from the fitting. If it's stuck, grasp it near the end with a pair of adjustable pliers and twist it to break the seal, then pull it off. If the hose is old or deteriorated, cut it off and install a new one.
9   If the outer surface of the large fitting that mates with the hose is deteriorated (corroded, pitted, etc.) it may be damaged further by hose removal. If it is, the thermostat housing cover will have to be replaced.
10  Remove the bolts and detach the housing outlet. Note the position of any brackets for correct reinstallation. If the cover is stuck, tap it with a soft-face hammer to jar it loose. Be prepared for some coolant to spill as the gasket seal is broken.
11  Note how it's installed (with the jiggle pin up), then remove the thermostat.

12  Remove all traces of old gasket material and/or sealant from the housing and outlet.
13  Install a new rubber seal over the thermostat **(see illustration)**.
14  Install the new thermostat in the housing without using sealant. Make sure the jiggle pin is at the top and the correct end faces out – the spring end is directed into the engine **(see illustration)**.
15  Install the outlet cover and bolts. Tighten the bolts to the torque listed in this Chapter's Specifications.
16  Reattach the hose and tighten the hose clamp securely. Install all components that were removed for access.
17  Refill the cooling system (see Chapter 1).
18  Start the engine and allow it to reach normal operating temperature, then check for leaks and proper thermostat operation (as described in Steps 2 through 4).

## 4   Engine cooling fan – check and replacement

**Warning:** *To avoid possible injury or damage, DO NOT operate the engine with a damaged fan. Do not attempt to repair fan blades – replace a damaged fan with a new one.*

### Check

*Refer to illustrations 4.1 and 4.2*

1   To test the fan motor, unplug the electrical connector at the motor **(see illustration)** and use jumper wires to connect the fan directly to the bat-

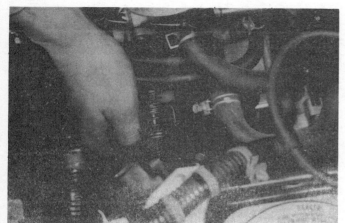

3.14  Install the thermostat with the spring facing in and the jiggle pin at the top (Legend shown, Integra similar)

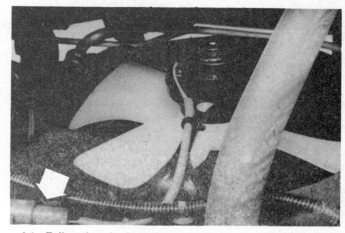

4.1  Follow the wire from the motor to the connector (arrow), unplug the connector and power the motor directly from the battery with jumper wires (viewed from below)

4.2 The fan switch (thermosensor) is located in the bottom of the radiator (arrow) – unplug and bridge the terminals in the connector

4.5 Unplug the fan wiring connector(s) (the model shown has two fans)

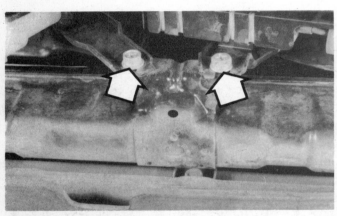

4.6 Remove the fan lower mounting bolts (arrows) (Legend shown, Integra similar)

4.7 Remove the upper fan mounting bolts, the inner brackets (arrows) are shown here

4.9a Pull the radiator toward the front of the vehicle, then tilt the fan back until it clears the filler neck and lift it out

4.9b The fan on the passenger's side provides additional cooling for the air conditioning condenser

tery. If the fan still doesn't work, replace the motor.

2 If the motor tested OK, the fault lies in the fuses, coolant temperature (thermosensor) switch, resistor, relay or the wiring which connects the components. To test the switch, remove the connector plug (see illustra-tion) and bridge the terminals in the plug together. With the ignition on (but engine off for safety) the fan should come on. Carefully check all fuses, wiring and connections. If no obvious problems are found, further diagno-sis should be done by a dealer service department or other repair shop.

## Chapter 3 Cooling, heating and air conditioning systems

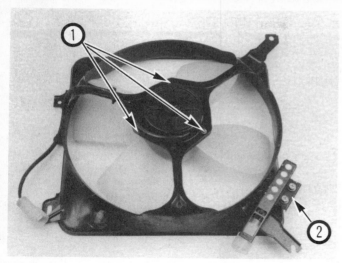

**4.11a Engine cooling fan**
1 Motor mounting screws  2 Resistor

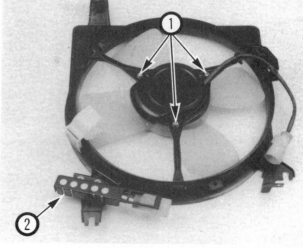

**4.11b Condenser fan**
1 Motor mounting screws  2 Resistor

### Replacement

#### Legend

*Refer to illustrations 4.5, 4.6, 4.7, 4.9a, 4.9b, 4.11a and 4.11b*

**Note:** *The Legend has two fans; both are removed in the same manner.*

3   Disconnect the negative battery cable from the battery.
4   Set the parking brake and block the rear wheels to prevent the vehicle from rolling. Raise the front of the vehicle and support it securely with jackstands. Remove the lower splash pan under the radiator.
5   Insert a small screwdriver into the connector to lift the lock tab and unplug the fan wiring **(see illustration)**.
6   Remove the fan lower mounting bolts **(see illustration)**.
7   Unbolt the fan upper brackets **(see illustration)** and the condenser top mountings.
8   Remove the upper radiator support bar (see Section 5).
9   Tilt the radiator forward slightly, then carefully lift the fan out of the engine compartment **(see illustrations)**.
10  To detach the fan from the motor, remove the motor shaft nut or screws.
11  To remove the bracket from the fan motor, remove the mounting screws **(see illustrations)**.
12  Installation is the reverse of removal.

#### Integra

*Refer to illustrations 4.14*

**Note:** *Some Integra are equipped with two fans; both are removed in the same manner.*

13  Disconnect the negative battery cable from the battery.
14  Remove the intake air duct **(see illustration)**.
15  Set the parking brake and block the rear wheels to prevent the vehicle from rolling. Raise the front of the vehicle and support it securely with jackstands. Remove the lower splash pan under the radiator.
16  Insert a small screwdriver into the connector to lift the lock tab and unplug the fan wiring.
17  Remove the fan mounting bolts.
18  Carefully lift the fan out of the engine compartment.
19  To detach the fan from the motor, remove the motor shaft nut.
20  To remove the bracket from the fan motor, remove the mounting screws.
21  Installation is the reverse of removal.

**4.14 Unbolt (arrows) the intake air duct**

### 5  Radiator – removal and installation

*Refer to illustrations 5.4, 5.6, 5.8, 5.9 and 5.13*

**Warning:** *Wait until the engine is completely cool before beginning this procedure. Read the warning at the beginning of Section 2.*

#### Removal

1   Disconnect the negative battery cable from the battery.
2   Set the parking brake and block the rear wheels. Raise the front of the vehicle and support it securely on jackstands. Remove the splash pan beneath the radiator.
3   Drain the cooling system (see Chapter 1). If the coolant is relatively new or in good condition, save it and reuse it.
4   If the vehicle is equipped with an automatic transaxle, disconnect the

## Chapter 3 Cooling, heating and air conditioning systems

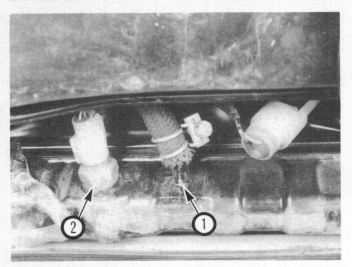

**5.4 Disconnect the cooling lines and wiring connectors (Integra shown, Legend similar)**

1. Transaxle cooling line fitting
2. Cooling fan switch and connector

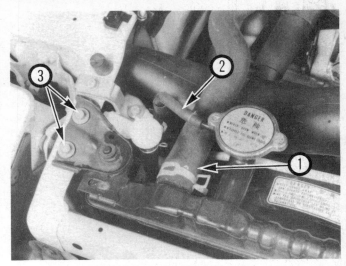

**5.6 Radiator mounting details**

1. Upper radiator hose
2. Overflow hose
3. Mounting bolts (Integra only)

cooler lines from the radiator **(see illustration)**. Use a drip pan to catch spilled fluid and plug the lines and fittings.
5   Unplug wiring connector for the cooling fan switch.
6   Loosen the hose clamps, then detach the radiator hoses from the fittings **(see illustration)**. If they're stuck, grasp each hose near the end with a pair of slip joint pliers and twist it to break the seal, then pull it off – be careful not to damage the radiator fittings! If the hoses are old or deteriorated, cut them off and install new ones.
7   Remove the engine cooling fan(s) (see Section 4).
8   On Legend models, unbolt the front bulkhead crossmember. On Integra models, remove the air intake duct and mounting bolts **(see illustration)**.
9   Carefully lift out the radiator **(see illustration)**. Don't spill coolant on the vehicle or scratch the paint.
10  With the radiator removed, it can be inspected for leaks and damage. If it needs repair, have a radiator shop or dealer service department perform the work as special techniques are required.
11  Bugs and dirt can be removed from the radiator by spraying with a garden hose from the back side.
12  Check the radiator mounts for deterioration and replace if necessary.

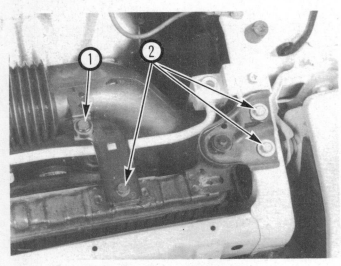

**5.8 Integra radiator details**

1. Intake air duct bolt
2. Mounting bolts

### Installation

13  Installation is the reverse of the removal procedure. Guide the radiator into the mounts until they seat properly **(see illustration)**.
14  After installation, fill the cooling system with the proper mixture of antifreeze and water. Refer to Chapter 1 if necessary.
15  Start the engine and check for leaks. Allow the engine to reach normal operating temperature, indicated by the upper radiator hose becoming hot. Recheck the coolant level and add more if required.
16  If you're working on an automatic transaxle equipped vehicle, check and add fluid as needed.

### 6   Coolant reservoir – removal and installation

*Refer to illustration 6.2, 6.3 and 6.4*
**Warning:** *The engine must be completely cool before removing the reservoir. Read the warning at the beginning of Section 2.*

1   The coolant reservoir is mounted adjacent to the radiator in the corner of the engine compartment.
2   On Legend models, follow the overflow hose from the radiator neck to the top of the coolant reservoir. Remove the cap with the hose still attached. Lift the reservoir straight up out of the bracket **(see illustration)**.

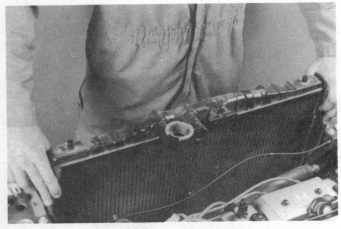

**5.9 Lift the radiator straight up (Legend shown)**

# Chapter 3  Cooling, heating and air conditioning systems

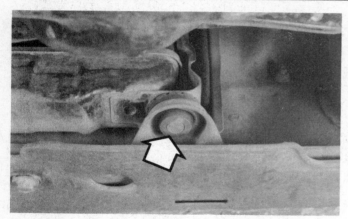

5.13  Be sure the posts on the bottom of the radiator fit into the mounts (arrow)

6.2  On Legend models, remove the cap, then lift the reservoir straight out of the bracket (arrow)

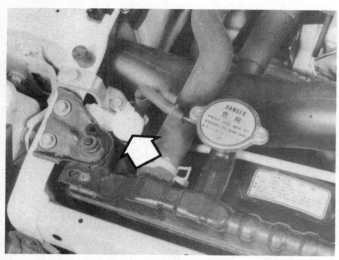

6.3  On Integra models, the filler neck (arrow) is near the radiator cap – a vertical hose connects to the reservoir

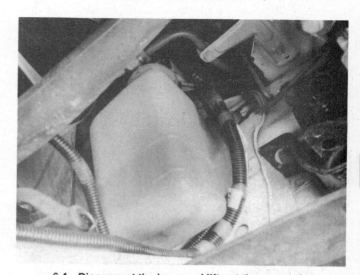

6.4  Disconnect the hose and lift out the reservoir

3   On Integra models, the filler neck is adjacent to the radiator cap **(see illustration)**, but the coolant reservoir is accessed from under the vehicle. Set the parking brake and block the rear wheels. Raise the front of the vehicle and support it securely on jackstands.
4   Remove the lower splash shield for access. Disconnect the hose from the top of the reservoir, then lower the reservoir out **(see illustration)**.
5   On all models temporarily pour the coolant into a container.
6   Wash out and inspect the reservoir for cracks and chafing. Examine the reservoir closely. If it's damaged, replace it.
7   Installation is the reverse of removal.

## 7   Oil cooler – removal and installation

*Refer to illustrations 7.4 and 7.6*

### Removal

1   Allow the engine to cool completely.
2   The oil cooler is mounted between the oil filter and engine block. Remove the oil filter and drain the coolant (see Chapter 1).
3   Detach the two coolant lines from the oil cooler. Be prepared for coolant to escape from the open fittings. Cap or plug the open fittings.
4   On Legend models only, disconnect the wire from the oil pressure sending unit **(see illustration)**.
5   Unbolt the oil cooler and separate it from the engine.

7.4  Disconnect the wire from the oil pressure sending unit (arrow)

# Chapter 3 Cooling, heating and air conditioning systems

7.6 On the Legend (shown) the seal (arrow) fits around both tubes – on the Integra a round seal is used

8.4 The weep hole (arrow) is located on the underside of the pump (Legend pump shown, Integra pump similar) – pump removed for clarity

## Installation

6  Installation is the reverse of removal. Be sure to use a new seal **(see illustration)** between the block and oil cooler and tighten the coolant hoses securely at the fittings. On Legends, apply sealant to the threads of the mounting bolts.
7  Install a new oil filter and change the engine oil (see Chapter 1).
8  Add coolant and oil as needed.
9  Start the engine and check for oil and coolant leaks.
10  Recheck the coolant and oil levels.

## 8  Water pump – check

*Refer to illustration 8.4*
1  A failure in the water pump can cause serious engine damage due to overheating.
2  There are three ways to check the operation of the water pump while it's installed on the engine. If the pump is defective, it should be replaced with a new or rebuilt unit.
3  With the engine running at normal operating temperature, squeeze the upper radiator hose. If the water pump is working properly, a pressure surge should be felt as the hose is released. **Warning:** *Keep your hands away from the drivebelts and fan blades!*
4  Water pumps are equipped with weep (or vent) holes. If a failure oc-

curs in the pump seal, coolant will leak from the hole. With the timing belt cover removed, you'll need a flashlight and small mirror to find the hole on the water pump from underneath to check for leaks **(see illustration)**.
5  If the water pump shaft bearings fail there may be a howling sound at the pump while it's running. Shaft wear can be felt with the timing belt removed if the water pump pulley is rocked up and down (with the engine off). Don't mistake drivebelt slippage, which causes a squealing sound, for water pump bearing failure.

## 9  Water pump – replacement

*Refer to illustrations 9.6 and 9.11*
**Warning:** *Wait until the engine is completely cool before beginning this procedure. Read the warning at the beginning of Section 2.*
1  Disconnect the negative battery cable from the battery.
2  Drain the cooling system (see Chapter 1). If the coolant is relatively new or in good condition, save it and reuse it.
3  Remove the drivebelts (see Chapter 1).
4  Remove the timing belt (see Chapter 2, Part A or B).
5  Remove any accessory brackets from the water pump.
6  Remove the bolts and detach the water pump from the engine **(see illustration)**. Note the locations of the different diameters of bolts as they're removed.

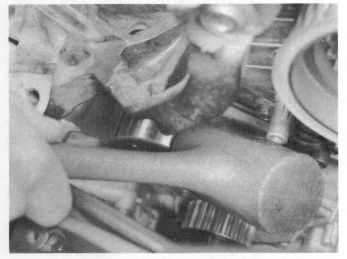

9.6 Use a soft-face hammer to break the pump loose

9.11 Press the new O-ring seal into the groove in the pump (arrow)

# Chapter 3  Cooling, heating and air conditioning systems

**10.1  On the Legend, the coolant temperature sending unit (arrow) is located near the bleeder screw adjacent to the upper radiator hose**

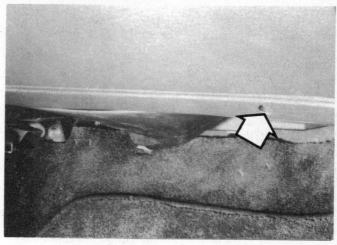

**11.22  Remove the lower dash cover screws (arrow)**

7  Clean the bolt threads and the threaded holes in the engine to remove corrosion and sealant.
8  Compare the new pump to the old one to make sure they're identical.
9  Remove all traces of old gasket sealer and O-ring from the engine.
10  Clean the engine and new water pump mating surfaces with lacquer thinner or acetone.
11  Apply a thin layer of RTV sealant to the O-ring groove of the new pump, then carefully set a new O-ring in the groove **(see illustration)**.
12  Carefully attach the pump to the engine and thread the bolts into the holes finger tight.
13  Install the remaining bolts (if they also hold an accessory bracket in place, be sure to reposition the bracket at this time). Tighten them to the torque listed in this Chapter's Specifications in 1/4-turn increments. Don't overtighten them or the pump may be distorted.
14  Reinstall all parts removed for access to the pump.
15  Refill and bleed the cooling system and check the drivebelt tension (see Chapter 1). Run the engine and check for leaks.

## 10  Coolant temperature sending unit – check and replacement

*Refer to illustration 10.1*

**Warning:** *Wait until the engine is completely cool before beginning this procedure.*

1  The coolant temperature indicator system is composed of a temperature gauge mounted in the instrument panel and a coolant temperature sending unit mounted on the engine **(see illustration)**. Some vehicles have more than one sending unit, but only one is used for the indicator system. **Warning:** *This vehicle is equipped with an electric cooling fan. Stay clear of the fan blades, which can come on even when the engine is not running.*
2  If an overheating indication occurs even when the engine is cold, check the wiring between the dash and the sending unit for a short circuit to ground.
3  If the gauge is inoperative, test the circuit by briefly grounding the wire to the sending unit while the ignition is on (engine not running for safety). If the gauge deflects full scale, replace the sending unit.
4  If the gauge doesn't respond in the test outlined in Step 3, check for an open circuit in the gauge wiring.
5  If the sending unit must be replaced, simply unscrew it from the engine and quickly install the replacement. Use sealant on the threads. Make sure the engine is cool before removing the defective sending unit. There will be some coolant loss as the unit is removed, so be prepared to catch it. Check the level after the replacement part has been installed.

## 11  Blower unit – removal and installation

### Integra models without air conditioning

1  Disconnect the negative cable from the battery.
2  The blower unit is located in the passenger compartment above the right front (passenger's) footwell.
3  Remove the glove compartment and glove compartment frame.
4  Detach the blower duct.
5  Label and disconnect the electrical connector from the blower unit.
6  Remove the blower unit retaining screws and lower the unit from the housing.
7  If the motor is being replaced, transfer the fan to the new motor prior to installation.
8  Installation is the reverse of removal. Check for proper operation.

### Integra models with air conditioning

9  Disconnect the negative cable from the battery.
10  The blower unit is located in the passenger compartment above the right front footwell.
11  Remove the glove compartment, glove compartment frame and side frame.
12  Unbolt the retractor control unit.
13  Remove the center console (see Chapter 11).
14  Remove the center console bracket.
15  Unbolt the lower dashboard bracket and insert a screwdriver. Loosen the sealing strip toward the right (passenger's) side.
16  Disconnect the electrical connector from the blower unit.
17  Remove the blower unit retaining screws and lower the unit from the vehicle.
18  Unbolt the fan and motor from the housing. If the motor is being replaced, transfer the fan to the new motor prior to installation.
19  Installation is the reverse of removal. Check for proper operation.

### Legend models

*Refer to illustrations 11.22, 11.23 and 11.25*

20  Disconnect the negative cable from the battery.
21  The blower unit is located in the passenger compartment above the right front footwell.
22  Remove the lower dash cover **(see illustration)**.

# Chapter 3  Cooling, heating and air conditioning systems

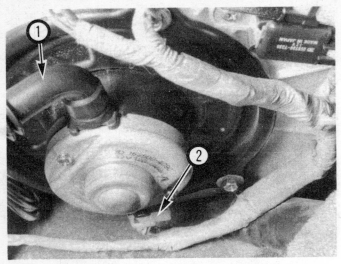

**11.23  Disconnect the hose and wiring**
1  Hose    2  Wiring

**11.25  Remove the mounting nut (arrow) to detach the fan from the motor shaft**

23  Detach the hose and wiring from the motor **(see illustration)**.
24  Remove the three mounting screws and lower the blower from the dash.
25  Remove the fan from the blower motor by removing the nut **(see illustration)** and lifting the fan off the shaft.
26  Installation is the reverse of removal. Check for proper operation.

## 12  Heater core – removal and installation

### Removal

1  Disconnect the negative cable from the battery.
2  Drain the cooling system (see Chapter 1).
3  Working in the engine compartment, disconnect the heater hoses and valve where they enter the firewall.
4  Remove the instrument panel and the center console (see Chapter 11).
5  Remove the heater controls (see Section 13).
6  Label and detach the air ducts, wiring and controls still attached to the heater assembly.
7  Unbolt the heating unit and lift it from the vehicle.
8  Remove the screws and retainers, take out the old heater core and install the new unit.

### Installation

9  Reassemble the heater unit and check the operation of the air control flaps. If any parts bind, correct the problem before installation.
10  Reinstall the remaining parts in the reverse order of removal.
11  Refill and bleed the cooling system (see Chapter 1), reconnect the battery and run the engine. Check for leaks and proper system operation.

## 13  Air conditioning and heater control assembly – removal and installation

### Legend models
Refer to illustration 13.4

1  Disconnect the negative cable from the battery.
2  Remove the front console and the radio (see Chapters 11 and 12).
3  Remove the four mounting screws.

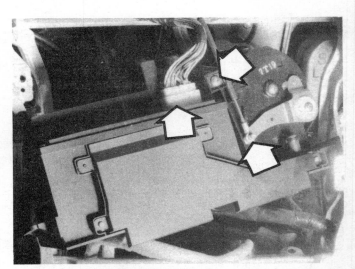

**13.4  Disconnect the heater control cable and unplug the electrical connector (arrows)**

4  Detach the heater control cable **(see illustration)**.
5  Pull the control assembly out of the dash far enough to unplug the electrical connector, then pull it the rest of the way out of the dash.
6  Installation is the reverse of removal.
7  Run the engine and check for proper functioning of the heater and air conditioning.

### Integra models

8  Disconnect the negative cable from the battery.
9  Remove the glove compartment (see Chapter 11).
10  Remove the hole covers and the mounting screws.
11  Pull the control out slightly. Mark the air mix cable at the clamp to ensure installation in the same position, then disconnect it.
12  Detach the wiring from the control assembly and lift the assembly from the dash.
13  To install the unit, reverse the above procedure.
14  To adjust the cable, align the mark you made in Step 11. Fasten the clip and check for stiffness or binding through the full range of operation.
15  Run the engine and check for proper functioning of the heater (and air conditioning, if equipped).

# Chapter 3  Cooling, heating and air conditioning systems

14.5a  Check the sight glass (arrow) for bubbles (Legend shown)

14.5b  On Integra models, the sight glass (arrow) is located near the headlight on the driver's side

## 14  Air conditioning and heating system – check and maintenance

### Air conditioning system

Refer to illustrations 14.5a, 14.5b, 14.6 and 14.9

**Warning:** *The air conditioning system is under high pressure. Do not loosen any hose fittings or remove any components until after the system has been discharged by a dealer service department or an automotive air conditioning shop. Always wear eye protection when adding refrigerant or disconnecting air conditioning system fittings.*

1  The following maintenance checks should be performed on a regular basis to ensure that the air conditioner continues to operate at peak efficiency.
   a) Inspect the condition of the compressor drivebelt. If it is worn or deteriorated, replace it (see Chapter 1).
   b) Check the drivebelt tension and, if necessary, adjust it (see Chapter 1).
   c) Inspect the system hoses. Look for cracks, bubbles, hardening and deterioration. Inspect the hoses and all fittings for oil bubbles or seepage. If there is any evidence of wear, damage or leakage, replace the hose(s).
   d) Inspect the condenser fins for leaves, bugs and any other foreign material that may have embedded itself in the fins. Use a "fin comb" or compressed air to remove debris from the condenser.
   e) Make sure the system has the correct refrigerant charge.
   f) If you hear water sloshing around in the dash area or have water dripping on the carpet, slip the evaporative housing condensation drain tube off and insert a piece of wire into both openings to check for blockage.

2  It's a good idea to operate the system for about ten minutes at least once a month. This is particularly important during the winter months because long term non-use can cause hardening, and subsequent failure, of the seals. Note that using the Defrost function operates the compressor.

3  Because of the complexity of the air conditioning system and the special equipment necessary to service it, in-depth troubleshooting and repairs are beyond the scope of this manual. However, simple component replacement procedures are provided in this Chapter.

4  The most common cause of poor cooling is simply a low system refrigerant charge. If a noticeable drop in system cooling ability occurs, one of the following quick checks will help you determine whether the refrigerant level is low.

5  With the air conditioning operating inspect the sight glass **(see illustrations)**. On Legends, the sight glass is located on the top of the receiver-drier near the front corner of the engine compartment on the passenger's side. On Integras, the sight glass is located in the refrigerant line near the front corner of the engine compartment on the driver's side. If the refrigerant looks foamy, it's low. Charge the system (see below).

### Adding refrigerant

6  Buy an automotive "charging kit" at an automotive parts store. A charging kit includes a 14-ounce can of refrigerant, a can tap valve and a short section of hose which can be attached between the tap valve and the system low side service valve. **Warning:** *Do not connect to the "high side" of the system!* **(see illustration)**. Because one can of refrigerant may not be sufficient to bring the system charge up to its proper level, it's a good idea to buy a couple additional cans. Make sure that the first can contains red refrigerant dye. If the system is leaking, the red dye will leak out with the refrigerant and help you pinpoint the location of the leak. **Warning:** *Wear eye protection while performing this Step.*

7  Connect the charging kit in accordance with the manufacturer's instructions.

8  Warm up the engine and operate the system.

14.6  Connect the charging kit to the low-pressure side of the air conditioning system – it has larger diameter hoses than the high-pressure side

# 3–12  Chapter 3  Cooling, heating and air conditioning systems

**14.9  Place a thermometer in the center dash vent to monitor the temperature of the air entering the passenger compartment**

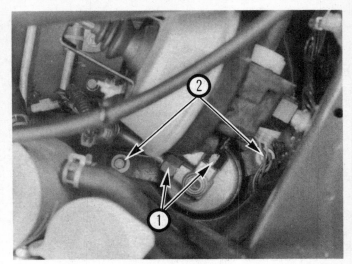

**15.3a  Disconnect the refrigerant lines and remove the mounting bolts (Legend shown)**

1  Refrigerant line fittings     2  Mounting bolts

**15.3b  On the Integra, the receiver-drier is located below the headlight on the driver's side (viewed from below)**

9  Place a thermometer in the center dashboard vent **(see illustration)** and add refrigerant until the indicated temperature is around 40 to 45-degrees F. When adding refrigerant, never exceed the system's capacity.

## Heating systems

10  If the air coming out of the heater vents isn't hot, the problem could stem from any of the following causes:
   a) The thermostat is stuck open, preventing the engine coolant from warming up enough to carry heat to the heater core. Replace the thermostat (see Section 3).
   b) A heater hose is blocked, preventing the flow of coolant through the heater core. Feel both heater hoses at the firewall. They should be hot. If one of them is cold, there is an obstruction in one of the hoses or in the heater core, or the heater control valve is shut. Detach the hoses and back flush the heater core with a water hose. If the heater core is clear but circulation is impeded, remove the two hoses and flush them out with a water hose.
   c) If flushing fails to remove the blockage from the heater core, the core must be replaced.

11  If the blower motor speed does not correspond to the setting selected on the blower switch, the problem could be a bad fuse, circuit, switch, blower motor resistor or motor.
   a) Before checking an inoperative blower motor or circuit, always check the fuse first.
   b) Using a test light or voltmeter, check the voltage at the motor.
   c) Pull the heating/air conditioning control assembly (see Section 13) far enough from the dash to verify – with a test light or voltmeter – that current is reaching the blower switch on the control assembly. If the switch is not getting current, troubleshoot the circuit between the battery and the switch (see wiring diagrams at the end of this manual).
   d) Using a test light or voltmeter, verify that the blower motor is getting current. If the blower motor is not getting current, check the circuit between the switch and motor, including the power transistor and blower relays. These are located on the blower housing under the dash.

12  If there isn't any air coming out of the vents:
   a) Turn the ignition ON and activate the fan control. Place your ear at the heating/air conditioning register (vent) and listen. Most motors are audible. Can you hear the motor running?
   b) If you can't (and have already verified that the blower switch and the blower motor circuit are good), the blower motor itself is probably bad (see Section 11). You can verify the motor's condition by hooking up a fused jumper wire directly between battery positive terminal and the blower motor.

13  If the carpet under the heater core is damp, or if antifreeze vapor or steam is coming through the vents, the heater core is leaking. Remove it (see Section 12) and install a new unit (most radiator shops will not repair a leaking heater core).

## 15  Air conditioning receiver/drier – removal and installation

*Refer to illustrations 15.3a and 15.3b*

**Warning:** *For this operation, the system must be discharged by an air conditioning technician. Do not attempt to do this by yourself. The refrigerant is under high pressure and can cause serious injury.*

1  Have the refrigerant discharged and recycled by an air conditioning technician.
2  Disconnect the ground cable from the negative terminal of the battery.
3  Disconnect the refrigerant lines **(see illustrations)** from the receiver/drier and cap the open fittings to prevent dirt and moisture entry. On Inte-

# Chapter 3  Cooling, heating and air conditioning systems

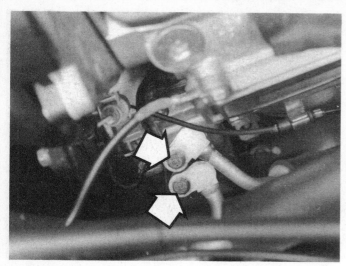

**16.6  Remove the bolts (arrows) and detach the refrigerant lines (Legend shown, Integra similar)**

**16.8a  On the Legend, the lower mounting bolts (arrows) are visible here**

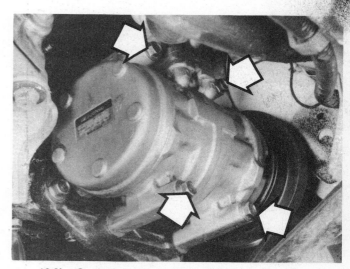

**16.8b  On the Integra, the mounting bolts (arrows) are readily visible**

gras, it will be necessary to remove the lower splash pan on the driver's side for access.
4  Remove the mounting bolts and lift the receiver/drier out of the vehicle.
5  Installation is the reverse of removal.
6  Have the system evacuated, charged and leak tested by the shop that discharged it. If the receiver was replaced, have them add 1/3 ounce of refrigeration oil.

## 16  Air conditioning compressor – removal and installation

Refer to illustration 16.6, 16.8a and 16.8b
**Warning:** *The air conditioning system is under high pressure. DO NOT disassemble any part of the system (hoses, compressor, line fittings, etc.) until after the system has been discharged by a dealer service department or service station.*
**Note:** *The receiver-drier should be replaced whenever the compressor is replaced.*

### Removal
1  Have the air conditioning system refrigerant discharged and recycled (see Warning above).
2  Disconnect the negative battery cable from the battery.
3  Set the parking brake, block the rear wheels and raise the front of the vehicle, supporting it securely on jackstands. Remove the lower splash guard.
4  Remove the drivebelt (see Chapter 1). On Integra models, remove the intake air duct.
5  Disconnect the compressor clutch wiring harness.
6  Disconnect the refrigerant lines from the compressor **(see illustration)**. Plug the open fittings to prevent entry of dirt and moisture.
7  On Legend models, remove the front bumper, grille, spoiler and right radius rod (see Chapters 10 and 11).
8  Unbolt the compressor from the mounting brackets **(see illustrations)** and remove it from the vehicle.
9  The clutch may have to be transferred from the original to the new compressor.

### Installation
10  Installation is the reverse of removal. Replace all O-rings with new ones specifically made for air conditioning system use and lubricate them with refrigerant oil.

11  Have the system evacuated, recharged and leak tested by the shop that discharged it. If the compressor is being replaced, pour out the oil from the old compressor into a graduated container and add that amount of new refrigerant oil to the new compressor. Also, add one ounce of refrigerant oil when the system is recharged with refrigerant.

## 17  Air conditioning condenser – removal and installation

**Warning:** *For this operation, the system must be discharged by an air conditioning technician. Do not attempt to do this by yourself. The refrigerant is under high pressure and can cause serious injury.*

1  Have the refrigerant discharged and recycled by an air conditioning technician.
2  Disconnect the negative cable from the battery.
3  On Legend models, remove the grille and front bulkhead crossmember for access (see Chapter 11 and Section 5). On Integra models, remove the front bumper (see Chapter 11).

4  Remove the center vertical brace and latch assembly, if necessary.
5  Disconnect the inlet and outlet fittings. On the Legend, the fittings are both on the passenger's side. On Integras, there is one fitting on each side. Cap the open fittings immediately to keep moisture and dirt out of the system.
6  Remove the condenser mounting bolts/nuts and lift the condenser out.
7  Install the condenser, brackets and bolts, making sure the rubber cushions fit on the mounting points properly.
8  Reconnect the refrigerant lines, using new O-rings where needed.
9  Reinstall the remaining parts in the reverse order of removal.
10  Have the system evacuated, charged and leak tested by the shop that discharged it. If the condenser is being replaced, add 1/3 ounce of refrigeration oil.

# Chapter 4  Fuel and exhaust systems

## Contents

| | | | |
|---|---|---|---|
| Accelerator cable – replacement | 8 | Fuel pressure relief procedure | 2 |
| Air cleaner housing – removal and installation | 7 | Fuel pump/fuel pressure – check | 3 |
| Air filter replacement | See Chapter 1 | Fuel pump – removal and installation | 4 |
| Bypass Control System (Legend models only) – general information and check | 15 | Fuel system check | See Chapter 1 |
| | | Fuel tank cleaning and repair – general information | 6 |
| Exhaust system check | See Chapter 1 | Fuel tank – removal and installation | 5 |
| Exhaust system servicing – general information | 16 | General information | 1 |
| Fuel filter replacement | See Chapter 1 | Idle speed check and adjustment | See Chapter 1 |
| Fuel injection system – check | 10 | Injector resistor – check and replacement | 14 |
| Fuel injection system – general information | 9 | Throttle body – component check, removal and installation | 11 |
| Fuel injectors – check, removal and installation | 13 | Throttle linkage check | See Chapter 1 |
| Fuel pressure regulator – check and replacement | 12 | | |

## Specifications

### General

| | |
|---|---|
| Fuel pressure | 35 to 41 psi (with vacuum hose disconnected from pressure regulator) |
| Fuel injector resistance | 1.5 to 2.5 ohms |
| Injector resistor resistance | 5.0 to 7.0 ohms |

### Torque specifications                Ft-lbs (unless otherwise indicated)

| | |
|---|---|
| Fuel injection service bolt | 108 in-lbs |
| Throttle body mounting nuts | 16 |
| Fuel rail mounting nuts | 108 in-lbs |

## 1  General information

These vehicles are equipped with the Programmed Fuel Injection (PGM-FI) system. This system uses timed impulses to sequentially inject the fuel directly into the intake ports of each cylinder. The injectors are controlled by the Electronic Control Unit (ECU). The ECU monitors various engine parameters and delivers the exact amount fuel, in the correct sequence, into the intake ports.

All Integra models are equipped with an electric fuel pump, mounted near the fuel tank behind the left rear wheel. The electric fuel pump on Legend models is mounted in the fuel tank.

The exhaust system consists of a header pipe, a catalytic converter, an exhaust pipe and a muffler. Each of these components is replaceable. For further information regarding the catalytic converter, refer to Chapter 6.

## 2  Fuel pressure relief procedure

*Refer to illustration 2.2*

**Warning:** *Gasoline is extremely flammable, so take extra precautions when you work on any part of the fuel system. Don't smoke or allow open flames or bare light bulbs near the work area, and don't work in a garage where a natural gas-type appliance (such as a water heater or clothes dryer) with a pilot light is present. If you spill any fuel on your skin, rinse it off immediately with soap and water. When you perform any kind of work on the fuel system, wear safety glasses and have a Class B type fire extinguisher on hand.*

1   Detach the cable from the negative battery terminal. Unscrew the fuel filler cap to relieve pressure built up in the fuel tank.

# Chapter 4  Fuel and exhaust systems

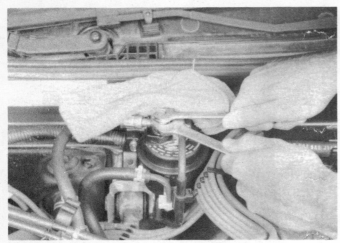

**2.2  To relieve the fuel pressure, you'll need one wrench to loosen the service bolt on top of the fuel filter and another wrench to hold the special banjo bolt into which the service bolt is installed. Once the wrenches are in place, place the rag over the fittings before loosening**

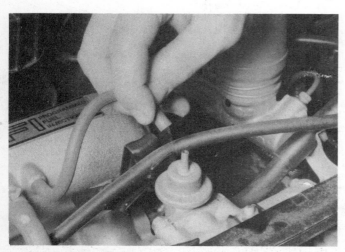

**3.5a  Location of the pressure regulator on Integra models**

**3.5b  Location of the fuel presssure regulator on Legend models**

**3.4  Remove the service bolt from the top of the fuel filter and attach a fuel pressure gauge**

2  You'll need two wrenches for this procedure – one to loosen the service bolt at the top of the fuel filter and another to hold the special banjo bolt into which the service bolt is installed **(see illustration)**.
3  Place a shop rag over the service bolt.
4  While holding the special banjo bolt, slowly loosen the service bolt one complete turn.
5  Always replace the washer between the service bolt and the special banjo bolt whenever the service bolt is loosened to relieve fuel pressure. Tighten the service bolt to the torque listed in this Chapter's Specifications.

### 3  Fuel pump/fuel pressure – check

**Warning:** *Gasoline is extremely flammable, so take extra precautions when you work on any part of the fuel system. Don't smoke or allow open flames or bare light bulbs near the work area, and don't work in a garage where a natural gas-type appliance (such as a water heater or clothes dryer) with a pilot light is present. If you spill any fuel on your skin, rinse it off immediately with soap and water. When you perform any kind of work on the fuel system, wear safety glasses and have a Class B type fire extinguisher on hand.*

#### Preliminary check

1  If you suspect insufficient fuel delivery, first inspect all fuel lines to ensure that the problem is not simply a leak in a line.

#### Fuel pump operational check

**Note:** *On Integra models, the fuel pump is located underneath the vehicle, immediately ahead of the left rear wheel. On Legend models, it is located inside the fuel tank (see Section 4).*

2  Set the parking brake and have an assistant turn the ignition switch to the On position while you listen to the fuel pump. You should hear a whirring sound, lasting for a couple of seconds. Start the engine. The whirring sound should now be continuous (although harder to hear with the engine running). If there is no whirring sound, either the fuel pump or the fuel main relay circuit is defective.

#### Pressure check

*Refer to illustrations 3.4, 3.5a and 3.5b*

3  Relieve the fuel pressure (see Section 2).
4  Remove the service bolt from the top of the fuel filter and attach a fuel pressure gauge **(see illustration)**, using a special adapter which may be purchased at a tool store or dealer service department.
5  Start the engine. Detach the vacuum hose from the pressure regulator **(see illustrations)**. With the engine idling, measure the fuel pressure. It should be as listed in this Chapter's Specifications.

# Chapter 4  Fuel and exhaust systems

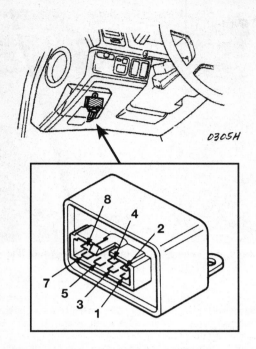

3.16  The main relay is located under the dash near the fuse box – refer to the terminal numbers when testing the relay

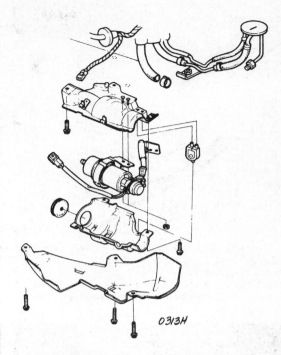

4.5  Details of the Integra electric fuel pump

6   If the fuel pressure is not within specification, check the following:
   a) If the pressure is higher than specified, check for a faulty regulator (see Section 12) or a pinched or clogged fuel return hose or pipe.
   b) If the pressure is lower than specified:
      1) Inspect the fuel filter – make sure it's not clogged.
      2) Look for a pinched or clogged fuel hose between the fuel tank and the fuel pump.
      3) Check the pressure regulator for a malfunction (see Section 12).
      4) Look for leaks in the fuel line.
      5) Look for a pinched, broken or disconnected regulator vacuum hose.
7   If there are no problems with any of the above-listed components, check the fuel pump (see below).

## Fuel pump check

8   If you suspect a problem with the fuel pump, verify the pump actually runs. Have an assistant turn the ignition switch to On – you should hear a brief whirring noise as the pump comes on and pressurizes the system. Have the assistant start the engine. This time you should hear a constant whirring sound from the pump (but it's more difficult to hear with the engine running).
9   If the pump does not come on (makes no sound), proceed to the next step.
10  Raise the rear of the vehicle and place it securely on jackstands (Integra models only).
11  Remove the left rear wheel (Integra models only).
12  On Integra models, remove the fuel pump cover (see Section 4) and detach the black and yellow wires (make sure the ignition switch is turned off before disconnecting the wires) from the fuel pump. On Legend models, the fuel pump connector is directly above the fuel tank, and access to it can be gained from the luggage area.
13  Touch the positive probe of a voltmeter to the black/yellow (Legend) or the yellow (Integra) wire and the negative probe to the black wire, then turn on the ignition switch and verify there is voltage available.
14  If voltage is available, replace the fuel pump (see Section 4).
15  If no voltage is available, check the main relay (see below).

## Main relay check

*Refer to illustration 3.16*

16  To test the main relay, first remove it from its location next to the under-dash fuse panel **(see illustration)**.
17  Using a pair of jumper wires, connect battery voltage to the no. 4 relay terminal, ground the no. 8 terminal, then check for continuity between the no. 5 and no. 7 terminals. If there's no continuity, replace the relay.
18  Connect battery voltage to the no. 5 relay terminal, ground the no. 2 terminal and verify there's continuity between the no. 1 and no. 3 terminals. If there isn't, replace the relay.
19  Connect battery voltage to the no. 3 relay terminal, ground the no. 8 terminal. Verify there's continuity between the no. 5 and no. 7 terminals. If there is no continuity, replace the relay.

---

4   Fuel pump – removal and installation

**Warning:** *Gasoline is extremely flammable, so take extra precautions when you work on any part of the fuel system. Don't smoke or allow open flames or bare light bulbs near the work area, and don't work in a garage where a natural gas-type appliance (such as a water heater or clothes dryer) with a pilot light is present. If you spill any fuel on your skin, rinse it off immediately with soap and water. When you perform any kind of work on the fuel system, wear safety glasses and have a Class B type fire extinguisher on hand.*

### Integra models

*Refer to illustrations 4.5 and 4.7*

1   Detach the cable from the negative battery terminal.
2   Relieve the fuel system pressure (see Section 2).
3   Loosen the lug nuts of the left rear wheel. Raise the rear of the vehicle and place it securely on jackstands.
4   Remove the left rear wheel.
5   Remove the fuel pump shield retaining bolts and detach the shield **(see illustration)**.

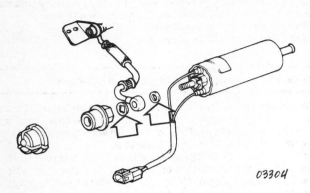

**4.7 Fuel hose-to-pump connection details (Integra models) – be sure to replace the sealing washers when installing the fuel pump (arrows)**

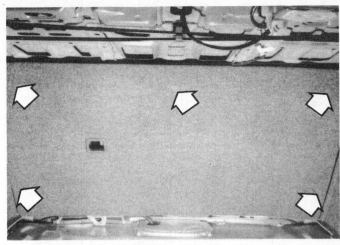

**4.12 Remove the clips (arrows) from the maintenance cover**

6  Detach the fuel lines and electrical connectors from the fuel pump.
7  Unscrew the pulsation damper from the end of the pump **(see illustration)**. Be prepared for fuel spillage. **Warning:** *Do not disassemble the pump.*
8  Unbolt the pump from its mounting bracket and remove it.
9  Installation is the reverse of removal. Be sure to replace the sealing washers **(see illustration 4.7)**.
10  After you have installed the new pump, have an assistant turn the ignition switch to the On position two or three times while you watch for any leaks where the fuel lines are attached to the pump.

### Legend models

*Refer to illustrations 4.12, 4.13 and 4.14*

11  Detach the cable from the negative battery terminal. Relieve the fuel pressure from the system (see Section 2).
12  Remove the maintenance access cover in the luggage compartment **(see illustration)**.
13  Remove the four bolts that retain the fuel pump cover plate **(see illustration)**.
14  Unplug the electrical connector from the fuel pump and sending unit **(see illustration)**.
15  Remove the mounting nuts from the fuel pump.
16  Remove the fuel pump from the tank.
17  Remove the sock filter from the end of the pump.
18  Detach the pump from its bracket.
19  Installation is the reverse of removal.

### 5  Fuel tank – removal and installation

*Refer to illustrations 5.9*

**Warning:** *Gasoline is extremely flammable, so take extra precautions when you work on any part of the fuel system. Don't smoke or allow open flames or bare light bulbs near the work area, and don't work in a garage where a natural gas-type appliance (such as a water heater or clothes dryer) with a pilot light is present. If you spill any fuel on your skin, rinse it off immediately with soap and water. When you perform any kind of work on the fuel system, wear safety glasses and have a Class B type fire extinguisher on hand.*

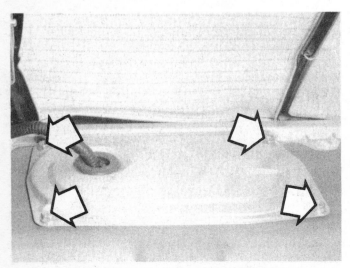

**4.13 Remove the four bolts (arrows) from the fuel pump cover plate**

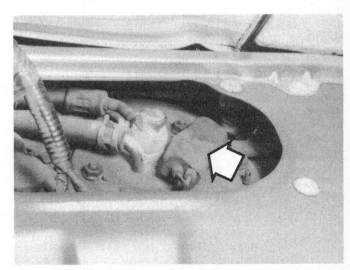

**4.14 Peel back the rubber boot (arrow) and unplug the electrical connector**

# Chapter 4  Fuel and exhaust systems

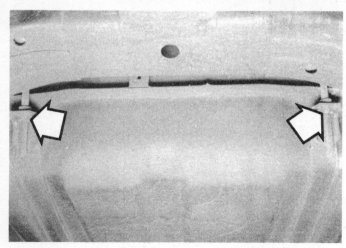

5.9  Remove the nuts (arrows) and drop the fuel tank retaining straps down

9   Disconnect both fuel tank retaining straps and pivot them down until they are hanging out of the way (see illustration).
10  Lower the tank enough to disconnect the electrical connectors and ground strap from the fuel pump and fuel gauge sending unit.
11  Remove the tank from the vehicle.
12  Installation is the reverse of removal.

## 6  Fuel tank cleaning and repair – general information

1   All repairs to the fuel tank or filler neck should be carried out by a professional who has experience in this critical and potentially dangerous work. Even after cleaning and flushing of the fuel system, explosive fumes can remain and ignite during repair of the tank.
2   If the fuel tank is removed from the vehicle, it should not be placed in an area where sparks or open flames could ignite the fumes coming out of the tank. Be especially careful inside garages where a natural gas-type appliance is located, because the pilot light could cause an explosion.

## 7  Air cleaner housing – removal and installation

*Refer to illustrations 7.4, 7.6 and 7.7*

1   Detach the cable from the negative battery terminal.
2   Remove the air cleaner cover and filter element (see Chapter 1).
3   Remove the battery (see Chapter 5).
4   Loosen the clamps that retain the air intake hoses to the air cleaner housing (see illustration).
5   Remove the clamps that hold the wire harness to the air cleaner housing.
6   Remove the bolts that hold the air cleaner housing to the battery tray (see illustration).
7   Label and detach all hoses from the air cleaner housing (see illustration). Lift the housing from the engine compartment.
8   Installation is the reverse of removal.

## 8  Accelerator cable – replacement

*Refer to illustrations 8.3, 8.4, 8.5 and 8.8*

1   Detach the cable from the negative battery terminal.
2   On Legend models, remove the control box from the firewall (directly above accelerator cable).

**Note:** *The following procedure is much easier to perform if the fuel tank is empty. Some tanks have a drain plug for this purpose. If the tank does not have a drain plug, the fuel can be siphoned from the tank using a siphoning kit, available at most auto parts stores. NEVER start the siphoning action with your mouth!*

1   Remove the fuel tank filler cap to relieve fuel tank pressure.
2   Relieve the fuel system pressure (see Section 2).
3   Detach the cable from the negative terminal of the battery.
4   If the tank has a drain plug, remove it and drain the fuel into an approved gasoline container. If it doesn't have a drain plug, siphon the fuel into an approved gasoline container, using a siphoning kit (available at most auto parts stores).
5   On Legend models, remove the cover in the luggage compartment and disconnect the fuel gauge and fuel pump connector (see Section 4). On Integra models, disconnect the fuel gauge connector from under the vehicle (see Section 4).
6   Raise the rear of the vehicle and place it securely on jackstands.
7   Label, then disconnect the fuel lines and any hoses or brackets.
8   Support the fuel tank with a floor jack. Position a piece of wood between the jack head and the fuel tank to protect the tank.

7.4  Lift up on the bail to release the clamp (Legend shown)

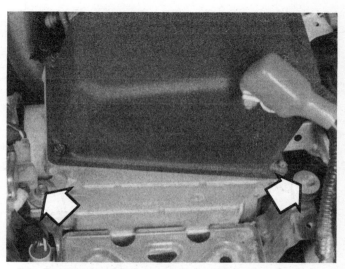

7.6  Remove the bolts (arrows) that retain the housing to the battery tray (Legend shown)

## Chapter 4 Fuel and exhaust systems

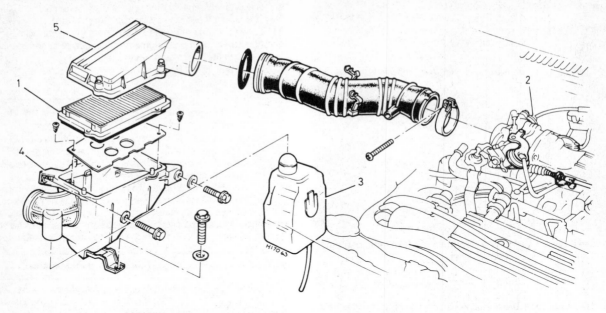

**7.7 Exploded view of the air cleaner housing and related components (Integra models)**

1. Air filter element
2. Throttle body
3. Resonator
4. Air cleaner housing
5. Air cleaner cover

3 Loosen the locknut and remove the accelerator cable from its bracket **(see illustration)**.
4 Rotate the throttle shaft bellcrank until the cable is out of its guide groove in the bellcrank **(see illustration)** and detach the cable from the bellcrank.
5 Working underneath the dash, detach the cable from the accelerator pedal **(see illustration)**.
6 Pull the grommet from the firewall and pull the cable through the firewall from the engine compartment side.
7 Installation is the reverse of removal.
8 To adjust the cable **(see illustration)**:
 a) Lift up on the cable to remove any slack.
 b) Turn the adjusting nut until it is 1/8-inch (3 mm) away from the cable bracket.
 c) Tighten the locknut and check cable deflection at the throttle linkage. Deflection should be 3/8 to 1/2-inch. If deflection is not within specifications, loosen the locknut and turn the adjusting nut until the deflection is as specified.
 d) After you have adjusted the throttle cable, have an assistant help you verify that the throttle valve opens all the way when you depress the accelerator pedal to the floor and that it returns to the idle position when you release the accelerator. Verify the cable operates smoothly. It must not bind or stick.
 e) If the vehicle is equipped with an automatic transaxle, adjust the transaxle throttle valve cable (see Chapter 7B).
 f) If the vehicle is equipped with cruise control, have the cruise control cable adjusted by a dealer service department or other repair shop.

**8.3 Loosen the locknut with an open end wrench and remove the cable from the bracket (Legend shown)**

**8.4 Rotate the throttle shaft until the cable is out of the guide groove in the bellcrank, then detach the cable end**

# Chapter 4 Fuel and exhaust systems

8.5 Detach the cable end (arrow) from the accelerator pedal linkage

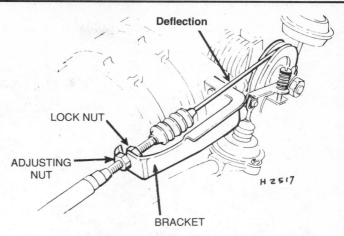

8.8 Lift up the cable to remove the slack, then turn the adjusting nut until it is 3 mm (1/8-inch) from the cable bracket, then tighten the locknut

## 9 Fuel injection system – general information

*Refer to illustrations 9.1a and 9.1b*

The Programmed Fuel Injection (PGM-FI) system **(see illustrations)** consists of three sub-systems: air intake, electronic control and fuel delivery. The intake manifold absolute pressure and engine speed are used to determine the correct air/fuel ratio that is to be injected into the combustion chambers. This is known as the speed-density method of fuel control. The system uses an Electronic Control Unit (ECU) along with the sensors (coolant temperature sensor, throttle position sensor, manifold absolute pressure (MAP) sensor etc.) to determine the proper air/fuel ratio under all operating conditions.

The fuel injection system and the emission control system are closely linked in function and design. For additional information, refer to Chapter 6.

### Air intake system

The air intake system consists of the air cleaner, the air intake pipe, the throttle body, the idle control system and the intake manifold. A resonator in the air intake tube provides silencing as air is drawn into the system.

The throttle body is either a single or dual barrel, side-draft design with the primary air horn at the top. The lower portion of the throttle body is heated by engine coolant to prevent icing in cold weather. The throttle body on the Legend Sedan (1988 through 1990), Legend Coupe and Integra is a single barrel, side draft design. The idle adjusting screw is located on top of the throttle body. The throttle body on the 1986 and 1987 Legend Sedan is a dual barrel, side draft design with the primary air horn at the top. The idle adjusting screw is located at the top of the throttle body also. On both systems, a throttle angle sensor is attached to the throttle shaft to monitor changes in the throttle opening. To slow movement of the throttle valve as it closes, a dashpot is added to vehicles equipped with a manual transaxle.

When the engine is idling, the air/fuel ratio is controlled by the idle control system, which consists of the Electronic Control Unit (ECU), the fast idle valve and the Electronic Air Control Valve (EACV) or the Electronic Idle Control Valve (EICV). All 1988 through 1990 models are equipped with the EACV, while all 1986 and 1987 models are equipped with the EICV.

This valve changes the amount of air into the intake manifold as controlled by the ECU. The ECU receives information from the sensors (vehicle speed, coolant temperature, air conditioning, power steering mode etc.) and adjusts the idle according to the demands of the engine and driver. Finally, to prevent rough running after the engine starts, the valve is opened during cranking and immediately after starting to provide additional air into the intake manifold. Both the EACV and the EICV have similiar functions and differ only in the design and current load rating.

### Electronic control system

The Electronic Control System is explained in detail in Chapter 6 Emission Control Systems.

### Fuel delivery system

The fuel delivery system consists of these components: The fuel pump, the pressure regulator, the fuel injectors, the injector resistor and the main relay.

The fuel pump is an in-line, direct drive type. Fuel is drawn through a filter into the pump, flows past the armature through the one-way valve, passes through another filter and is delivered to the injectors. A relief valve prevents excessive pressure build-up by opening in the event of a blockage in the discharge side and allowing fuel to flow from the high to the low pressure side.

The pressure regulator maintains a constant fuel pressure to the injectors. The spring chamber of the pressure regulator is connected to the intake manifold to constantly maintain the fuel pressure at 36 psi higher than the pressure in the manifold. When the difference between the fuel pressure and manifold pressure exceeds this figure, the diaphragm is pushed up and excess fuel is fed back to the fuel tank through the return line.

The injectors are solenoid-actuated, constant stroke, pintle types consisting of a solenoid, plunger, needle valve and housing. When current is applied to the solenoid coil, the needle valve raises and pressurized fuel fills the injector housing and squirts out the nozzle. The needle valve lift and the fuel pressure are constant, so the injection quantity is determined by the length of time the valve is open, i.e. the length of time during which current is supplied to the solenoid coils.

Because it determines opening and closing intervals – which in turn determine the air-fuel mixture ratio – injector timing must be quite accurate. To attain the best possible injector response, the current rise time, when voltage is being applied to each injector coil, must be as short as possible. The number of windings in the coil has therefore been reduced to lower the inductance in the coil. However, this creates low coil resistance, which could compromise the durability of the coil. The flow of current in the coil is therefore restricted by a resistor installed in the injector wire harness.

The main relay, which is installed adjacent to the fuse box, is a direct coupler type which contains the relays for the electronic control unit power supply and the fuel pump power supply.

9.1a Fuel injection components on the Integra (1987 shown, others similar)

1 Air cleaner housing
2 Air intake duct
3 Fuel filter
4 Throttle body unit
5 Dashpot diaphragm
6 Fuel rail
7 Fuel pressure regulator
8 Fuel injector
9 Injector resistor

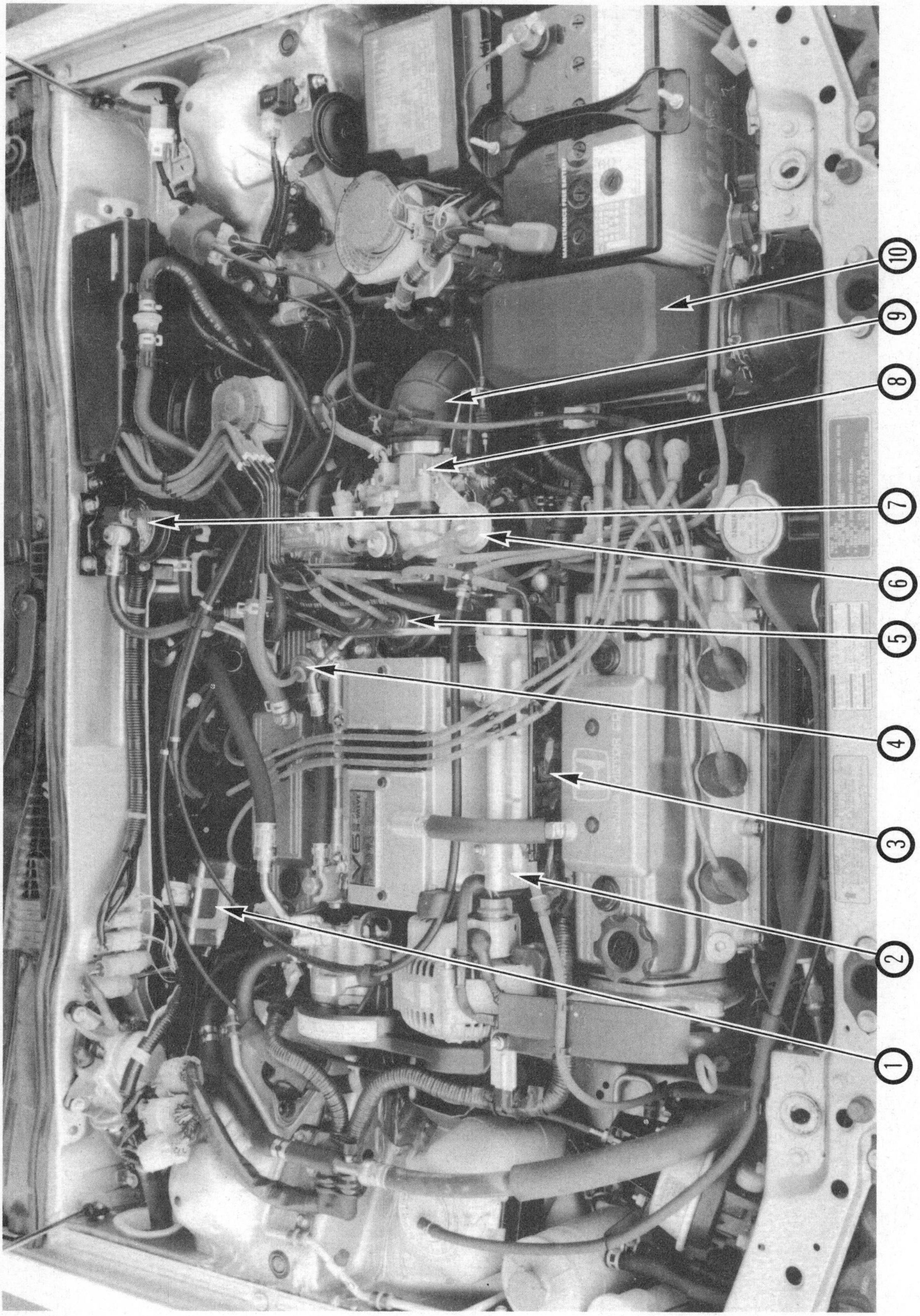

Chapter 4 Fuel and exhaust systems    4–9

9.1b Fuel injection components on the Legend (1988 shown, others similar)

1  Injector resistor
2  Fuel rail
3  Fuel injector
4  Fuel pressure regulator
5  Air bypass valve
6  Dashpot diaphragm
7  Fuel filter
8  Throttle body unit
9  Air intake duct
10 Air cleaner housing

11.1 Detach the hose that connects the throttle body with the charcoal canister and connect a vacuum gauge

11.18 Apply vacuum to the number 10 hose (1988 Legend Sedan shown) and watch for a slow but steady release of vacuum

## 10  Fuel injection system – check

**Note:** *the following procedure is based on the assumption that the fuel pressure is adequate (see Section 3).*

1  Check the ground wire connections on the intake manifold for tightness. Check all wiring harness connectors that are related to the system. Loose connectors and poor grounds can cause many problems that resemble more serious malfunctions.
2  Check to see that the battery is fully charged, as the control unit and sensors depend on an accurate supply voltage in order to properly meter the fuel.
3  Check the air filter element – a dirty or partially blocked filter will severly impede performance and economy (see Chapter 1).
4  If a blown fuse is found, replace it and see if it blows again. If it does, search for a grounded wire in the harness to the fuel pump.
5  Check the air intake duct to the intake manifold for leaks, which will result in an excessively lean mixture. Also check the condition of all vacuum hoses connected to the intake manifold.
6  Remove the air intake duct from the throttle body and check for dirt, carbon or other residue build-up. If it's dirty, clean it with carburetor cleaner and a toothbrush.
7  With the engine running, place a screwdriver against each injector, one at a time, and listen through the handle for a clicking sound, indicating operation.
8  The remainder of the system checks can be found in the following Sections.

## 11  Throttle body – component check, removal and installation

### Throttle body

#### Check

*Refer to illustration 11.1*

1  On top of the throttle body, locate the vacuum hose that goes to the canister. Detach it from the throttle body and attach a vacuum gauge in its place **(see illustration)**.
2  Start the engine and warm it to its normal operating temperature (wait until the cooling fan comes on twice). Verify the gauge indicates no vacuum.
3  Open the throttle slightly from idle and verify that the gauge indicates vacuum. If the gauge indicates no vacuum, check the port to make sure it is not clogged. Clean it with carburetor cleaner if necessary.

4  Stop the engine and verify the throttle cable and valve operate smoothly without binding or sticking.
5  If the throttle cable or valve binds or sticks, check for a build-up of sludge on the cable or throttle shaft.
6  If a build-up of sludge is evident, try removing it with carburetor cleaner or a similar solvent.
7  If cleaning fails to remedy the problem, replace the throttle body.

#### Removal and installation

8  Detach the cable from the negative battery terminal.
9  Remove the air duct that connects the air cleaner assembly to the throttle body.
10  Label the throttle angle sensor connector and disconnect it from the throttle body unit. Also label and detach all vacuum hoses from the throttle body.
11  Detach the accelerator cable (see Section 8) and, if equipped, the transaxle throttle valve cable (see Chapter 7B).
12  Detach the coolant hoses from the throttle body.
13  Remove the four mounting nuts and detach the throttle body and gasket.
14  Installation is the reverse of removal. Be sure to adjust the accelerator cable (see Section 8) and, if equipped, the throttle valve cable (see Chapter 7B).

### Throttle control (dashpot) system

#### All Legend Sedans, 1986 and 1987 Legend Coupes and 1986 and 1987 Integras

*Refer to illustrations 11.18, 11.19 and 11.20*

15  The dashpot slows the closing of the throttle valve during gear shifting or deceleration.
16  Check all the vacuum lines for leaks, damage or for a disconnected hose.
17  Disconnect the number 8 hose (1986 and 1987 Legend sedans only), or the number 10 hose (all other Legend models) or the number 6 hose (1986 and 1987 Integra models) from the dashpot diaphragm.
18  Apply vacuum **(see illustration)** and watch as the vacuum releases (the throttle should close slowly).
19  If the vacuum holds or releases very quickly, replace the dashpot check valve **(see illustration)** and retest the system.
20  Connect a vacuum pump to the dashpot diaphragm **(see illustration)** and apply vacuum. The rod should move and hold vacuum (remain stationary).
21  If the vacuum does not hold, replace the dashpot diaphragm with a new unit.

# Chapter 4  Fuel and exhaust systems

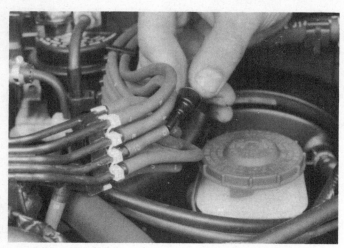

11.19  Follow the number 10 hose back toward the control box to find the check valve

11.20  Apply vacuum directly to the dashpot diaphragm

### 1988 and 1989 Integras
*Refer to illustration 11.25*

22  Start the engine and allow it to warm up.
23  Connect a tachometer in accordance with the manufacturer's instructions.
24  Disconnect the number 23 vacuum hose from the dashpot and check the engine speed. It should be approximately 2500 ± 500 rpm.
25  If the engine speed is too high, adjust it by bending the tab on the throttle arm **(see illustration)**.
26  If the engine speed didn't change when the hose was disconnected, connect a vacuum gauge to the number 23 hose and check for vacuum (the gauge should register vacuum). If the gauge doesn't indicate vacuum, check the hose for a blockage or cracks and repair it as necessary.
27  If the gauge does register vacuum, but the engine speed didn't change when the hose was disconnected in Step 24, replace the dashpot.
28  Check and, if necessary, adjust the idle speed (see Chapter 1).

## 12  Fuel pressure regulator – check and replacement

**Warning:** *Gasoline is extremely flammable, so take extra precautions when you work on any part of the fuel system. Don't smoke or allow open flames or bare light bulbs near the work area, and don't work in a garage where a natural gas-type appliance (such as a water heater or clothes dryer) with a pilot light is present. If you spill any fuel on your skin, rinse it off immediately with soap and water. When you perform any kind of work on the fuel system, wear safety glasses and have a Class B type fire extinguisher on hand.*

The fuel pressure regulator maintains constant fuel pressure to the injection system. When the difference between the fuel pressure and the manifold pressure exceeds 36 psi, the diaphragm in the fuel pump is pushed up and the excess fuel returns to the fuel tank through the fuel return line. On Legend models only, a pressure regulator cut-off valve controls the amount of vacuum to the pressure regulator.

### Check

#### Vacuum check (Legend models only)
*Refer to illustration 12.2*

1  Inspect the fuel system for pinched or broken vacuum hoses and fuel lines.
2  Disconnect the vacuum line from the pressure regulator and connect a vacuum gauge to the hose **(see illustration)**.
3  The coolant temperature must be below 221-degrees F (preferably not started for a few hours). Start the engine and allow it to idle. Check for vacuum. There should be vacuum.

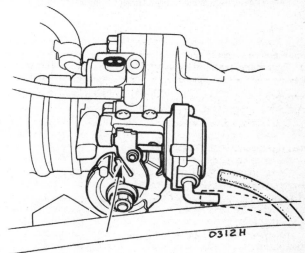

11.25  Bend this tab (arrow) to adjust the engine rpm when the dashpot hose is disconnected (1988 and 1989 Integra models only)

12.2  Detach the fuel pressure regulator hose from the metal vacuum line and attach a vacuum gauge

# Chapter 4  Fuel and exhaust systems

**12.10  Cover the line with a rag to prevent damage while pinching with pliers**

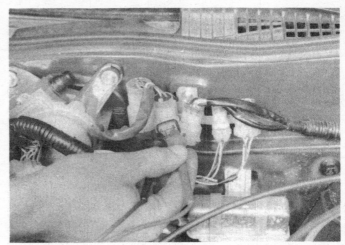

**12.12  Attach the positive end of the voltmeter to the BLK/YEL terminal and the negative end of the probe to the LT GRN terminal of the 6P connector**

4   If there is no vacuum, check the pressure regulator solenoid valve (see Steps 11 through 19)
5   Start the engine and allow it to warm up until the coolant is above 221-degrees F. Check for vacuum. There should be no vacuum.
6   If there is vacuum, check the pressure regulator solenoid valve (see Steps 11 through 19)
7   If all the pressure checks are correct, replace the fuel pressure regulator with a new unit.

### Pressure check
*Refer to illustration 12.10*

8   Inspect the fuel system for pinched or broken vacuum hoses and fuel lines.
9   Connect a fuel pressure gauge following the procedure outlined in Section 3.
10  Check the fuel pressure. Detach the vacuum hose from the regulator (see Section 3) and verify the fuel pressure rises. If the fuel pressure doesn't rise, carefully pinch the fuel return line with a pair of pliers (place a rag over the hose first, so as not to damage it) **(see illustration)**. If the pressure now rises, replace the fuel pressure regulator.

### Solenoid valve check (Legend models only)
11  On Legend Sedan and Coupe models, when the coolant temperature exceeds 221-degrees F or the temperature of the intake air exceeds 176-degrees F, the Pressure Regulator Cut-off Solenoid Valve (PRCSV) closes off vacuum to the pressure regulator, thereby allowing high fuel pressure to the injection system and preventing vapor lock.

### Test Number 1
*Refer to illustration 12.12*

12  Make sure the coolant temperature exceeds 221-degrees F (engine thoroughly warmed up). Disconnect the 6P connector **(see illustration)** and attach the positive end of the voltmeter to the BLK/YEL terminal and the negative end to the LT GRN terminal. Measure the voltage within the first 60 seconds of running time.
13  If there is no voltage, repair the open circuit in the BLK/YEL wire between the solenoid valve and the Number 11 fuse (Legend Sedan) or the Number 9 fuse (Legend Coupe).
14  If there are no obvious open circuits in the wire harness, take the vehicle to a dealership service department for diagnosis.

### Test Number 2
15  Start the engine and allow it to idle.
16  Disconnect the number 3 vacuum hose on Legend Coupe and 1988 through 1990 Legend Sedan or the number 2 vacuum hose on 1986 and 1987 Legend Sedan from the intake manifold. Use a vacuum gauge and check for vacuum.
17  If there is no vacuum, inspect the vacuum port for blockage.

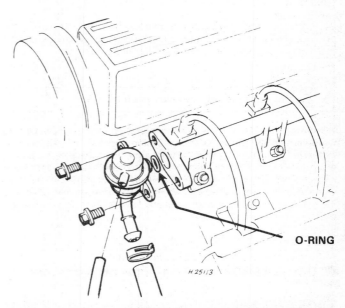

**12.22  To replace the fuel pressure regulator, detach the vacuum hose and fuel return hose, then remove the two retaining bolts – be sure to replace the O-ring**

18  Disconnect the 6P connector **(see illustration 12.12)** and attach the positive end of the voltmeter to the BLK/YEL terminal and the negative end to the LT GRN terminal.
19  If there is voltage, inspect for a short in the light green wire between the solenoid valve and the ECU. If there is no obvious sign of damage, have the vehicle tested at a dealer service department or other repair shop.

### Replacement
*Refer to illustration 12.22*

20  Detach the cable from the negative battery terminal.
21  Relieve the system fuel pressure (see Section 2).
22  Detach the vacuum hose and fuel return hose from the regulator **(see illustration)**.
23  Remove the two bolts and detach the regulator.

# Chapter 4  Fuel and exhaust systems

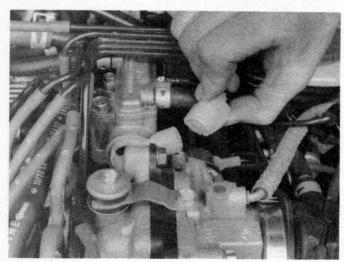

13.2 Disconnect the EACV on Legend models before testing the injectors

13.3a Before unplugging an injector connector, use a scribe (shown) or a small screwdriver to pry the spring clip loose

24  Installation is the reverse of removal. Be sure to use a new O-ring. Apply clean engine oil to the O-ring and install it in its proper position. Be sure you don't damage the O-ring when you install the regulator.

## 13  Fuel injectors – check, removal and installation

**Warning:** *Gasoline is extremely flammable, so take extra precautions when you work on any part of the fuel system. Don't smoke or allow open flames or bare light bulbs near the work area, and don't work in a garage where a natural gas-type appliance (such as a water heater or clothes dryer) with a pilot light is present. If you spill any fuel on your skin, rinse it off immediately with soap and water. When you perform any kind of work on the fuel system, wear safety glasses and have a Class B type fire extinguisher on hand.*

### Check

*Refer to illustrations 13.2, 13.3a, 13.3b and 13.4*

1  Start the engine and warm it to its normal operating temperature.

2  On Legend models, disconnect the EACV 2P connector **(see illustration)**.
3  With the engine idling, unplug each injector one-at-a-time **(see illustrations)**, note the change in idle speed then reconnect the injector. If the idle speed drop is almost the same for each cylinder, the injectors are operating correctly. If unplugging a particular injector fails to change the idle speed, proceed to the next step.
4  Turn the engine off. Remove the connector from the injector, and measure the resistance between the two terminals of the injector **(see illustration)**.
5  The resistance should be 1.5 to 2.5 ohms. If not, replace it with a new one.
6  If the resistance is as specified, connect a high-impedance voltmeter or a special injector harness test light, (available at some auto parts stores) to the electrical connector.
   a) If the voltage fluctuates between zero and two volts, the injector is receiving proper voltage.
   b) If there is no voltage, check the injector resistor (see Section 14).
   c) If the injector resistor is operating normally, check the wiring between the resistor and the injector(s) and between the injector(s) and the ECU for a short circuit, break in the wire or bad connection.

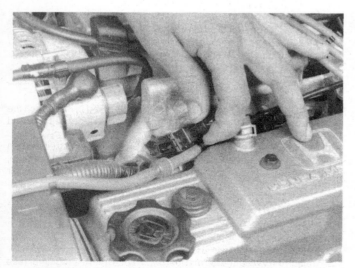

13.3b Detach each injector connector, one at a time, while carefully listening for any change in engine performance

13.4 Check the resistance value of each injector

# Chapter 4  Fuel and exhaust systems

13.13  Remove the fuel lines from the fuel rail (Legend shown)

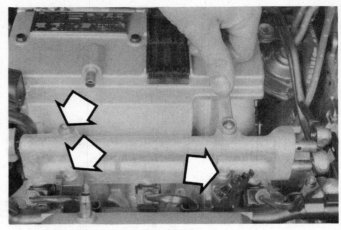

13.14a  Remove the fuel rail mounting nuts (arrows) . . .

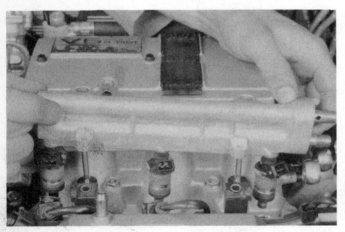

13.14b  . . . then lift the fuel rail up

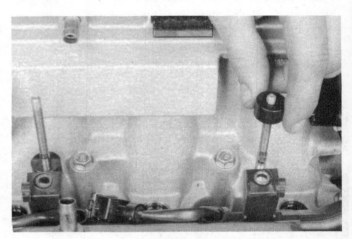

13.14c  If necessary, remove the phenolic spacers

## Removal

*Refer to illustrations 13.13, 13.14a, 13.14b, 13.14c and 13.15*

7  Detach the cable from the negative battery terminal.
8  Relieve the fuel pressure (see Section 2).
9  Remove the air intake duct, if necessary.
10  Unplug the injector connectors **(see illustration 13.2)**.
11  Detach the vacuum hose and fuel return hose from the fuel pressure regulator (see Section 12).
12  Detach any ground cables from the fuel rail.
13  Detach the fuel lines from the fuel rail **(see illustration)**.
14  Remove the mounting nuts **(see illustration)** and detach the fuel rail and injectors **(see illustration)**. On Legend models remove the phenolic spacers, if necessary **(see illustration)**.
15  Remove the injector(s) from the fuel rail and remove and discard the seal ring(s) **(see illustration)**. Note the location of the injector O-ring and cushion ring, then remove and discard them. **Note:** *Whether you're replacing an injector or a leaking O-ring, it's a good idea to remove all the injectors from the fuel rail and replace all the O-rings, seal rings and cushion rings.*

## Installation

*Refer to illustration 13.19*

16  Coat the new cushion rings with clean engine oil and slide them onto the injectors.
17  Coat the new O-rings with clean engine oil and install them on the injector(s), then insert each injector into its corresponding bore in the fuel rail.

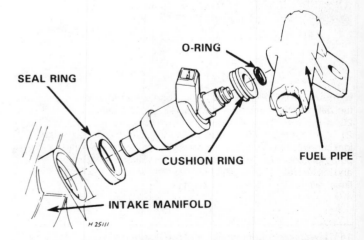

13.15  Injector installation details (Integra shown, Legend similar)

18  Coat the new seal rings with clean engine oil and press them into the injector bore(s) in the intake manifold.
19  Install the injector and fuel rail assembly on the intake manifold. Make sure the centerline of the electrical connector on each injector is aligned

# Chapter 4  Fuel and exhaust systems

4 – 15

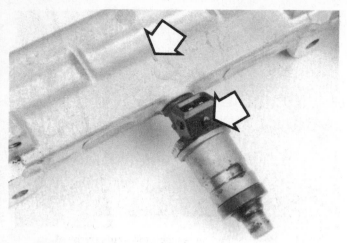

**13.19  Be sure the mark on each injector connector is aligned with the mark on the fuel rail (Legend shown, Integra similar)**

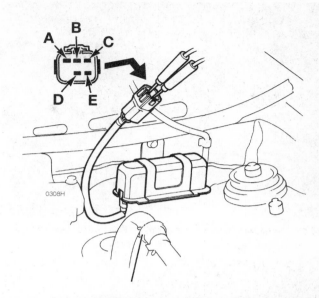

**14.4a  To check the injector resistor on the Integra, measure the resistance between the power supply terminal (A) and the other 4 terminals in the connector (arrows) – resistance should be about 5 to 7 ohms**

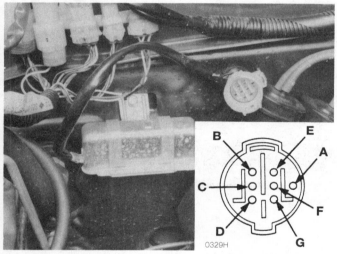

**14.4b  Repeat the same test for the Legend but measure all 6 terminals**

with its corresponding mark on the fuel rail **(see illustration)**. Tighten the fuel rail mounting nuts to the torque listed in this Chapter's Specifications.
20  The remainder of installation is the reverse of removal.
21  After the injector/fuel rail assembly installation is complete, turn the ignition switch to On, but don't operate the starter (this activates the fuel pump for about two seconds, which builds up fuel pressure in the fuel lines and the fuel rail). Repeat this about two or three times, then check the fuel lines, rail and injectors for fuel leakage.

## 14  Injector resistor – check and replacement

*Refer to illustrations 14.4a and 14.4b*

### Check

1  Detach the cable from the negative battery terminal.
2  Locate the injector resistor **(see illustrations 9.1a and 9.1b)**. It's on the firewall in the engine compartment.
3  Trace the wire harness from the resistor back to its connector and unplug it.
4  Check the resistance between the power supply terminal (A) and each of the other six (Legend) or four (Integra) terminals in the connector **(see illustrations)**. Resistance for each of the checks should be about 5 to 7 ohms.
5  If the indicated resistance isn't within specification, replace the resistor.

### Replacement

6  Detach the cable from the negative battery terminal.
7  Disconnect the electrical connector from the injector resistor.
8  Remove the bolt that attaches the resistor to the body and remove the unit.
9  Installation is the reverse of removal.

## 15  Bypass Control System (Legend models only) – general information and check

### General information

1  Legend models are equipped with a Bypass Control System (see Chapter 2, Part B) to divert the path of intake air into the combustion chamber. Two air intake paths are provided in the intake manifold to allow the option of the intake path length most favorable for the particular engine speed. Optimum performance is achieved by switching the paths from either the long intake path (for high torque at low RPM) or the short intake path (for maximum horsepower at high RPM).
2  The bypass control solenoids (switching mechanisms) are controlled by the sensors and the ECU. Any failure with the solenoids can lead to problems with the system, resulting in poor driveability. Follow the simple checks to help diagnose any system defects. **Note:** *The Bypass Control System does not have any self diagnostic codes directly relating to this system.*

15.3 Connect a vacuum gauge to the number 2 hose and check for vacuum

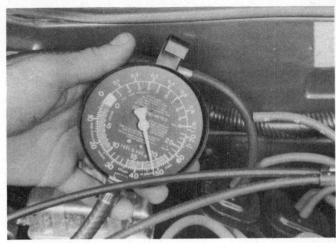

15.4 Connect a vacuum gauge to the vacuum tank and check for vacuum – the tank(s) are located on the firewall.

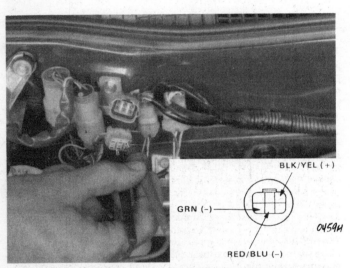

15.5 Check for voltage at the BLK/YEL (+) terminal and the RED/BLU (-) terminal

15.8 Connect a vacuum gauge to the number 8 hose and check for vacuum

## Check

*Refer to illustrations 15.3, 15.4, 15.5 and 15.8*

3   Start the engine and allow it to idle. Remove the number 2 hose from the bypass control diaphragm and connect a vacuum gauge **(see illustration)**.

4   If there is vacuum, continue with Steps 8 through 11. If there is no vacuum, remove the number 12 hose from the vacuum tank A **(see illustration)** and check for vacuum at the tank.

5   If there is no vacuum, check the line between the intake manifold and vacuum tank A for any damage or obstructions. If there is vacuum, disconnect the 6P connector at the firewall **(see illustration)**. Measure voltage between the BLK/YEL (+) terminal and the RED/BLU (-) terminal. There should NOT be battery voltage. If there is, replace the bypass control solenoid A with a new unit and retest.

6   If there is no battery voltage, measure the voltage between the BLK/YEL (+) and ground. There should be voltage.
   a) If there is no voltage, check for an open circuit condition in the BLK/YEL (+) wire between the 6P connector and the number 9 (fuel pump) fuse.

   b) If there is voltage, have the wire harness and ECU checked at a dealer service department or other repair shop.

7   If there was vacuum initially (see Step 2), raise the engine rpm to 4000 and check for vacuum at the number 2 hose **(see illustration 15.3)**. There should be no vacuum.

8   Reconnect the number 2 hose and check for vacuum at the number 8 hose **(see illustration)** at idle. There should be vacuum.

9   Increase the engine speed to to 3500 rpm and check for vacuum at the number 8 hose. There should be no vacuum. If all the tests are correct, the Bypass Control System is OK.

## Bypass valve test

*Refer to illustrations 15.11 and 15.12*

10   Check the bypass valve shaft and linkage for binding or any obvious damage.

11   Check that A **(see illustration)** of the bypass valve is in contact with the stopper when the number 2 and the number 8 hoses are disconnected from the diaphragm.

12   Check that B **(see illustration)** of the bypass valve is in contact with

# Chapter 4  Fuel and exhaust systems

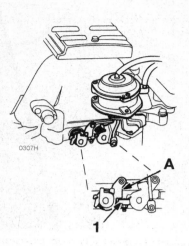

**15.11  Observe the location of the stopper (1) and the cam (A) when the hoses are disconnected – they should be contacting each other**

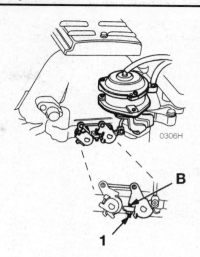

**15.12  Observe the location of the full close screw – when vacuum is applied to both diaphragms the cam should contact the screw**

1   Full close screw               B   Cam

**16.1  Check for any broken or missing rubber hangers**

the full close screw when manifold vacuum is applied to both diaphragms.
13   If there are any problems, remove the intake manifold and clean the bypass valves and shafts with carburetor cleaner. If the shafts still bind, replace the assembly with a new one (see Chapter 2, Part B).
14   If either diaphragm doesn't hold vacuum, replace the diaphragm assembly. It's held on with two screws.

## 16  Exhaust system servicing – general information

*Refer to illustration 16.1*

**Warning:** *Inspection and repair of exhaust system components should be done only after enough time has elapsed after driving the vehicle to allow the system components to cool completely. Also, when working under the vehicle, make sure it is securely supported on jackstands.*

1   The exhaust system consists of the exhaust manifold(s), the catalytic converter, the muffler, the tailpipe and all connecting pipes, brackets, hangers and clamps. The exhaust system is attached to the body with mounting brackets and rubber hangers **(see illustration)**. If any of the parts are improperly installed, excessive noise and vibration will be transmitted to the body.

### Muffler and pipes

2   Conduct regular inspections of the exhaust system to keep it safe and quiet. Look for any damaged or bent parts, open seams, holes, loose connections, excessive corrosion or other defects which could allow exhaust fumes to enter the vehicle. Also check the catalytic converter when you inspect the exhaust system (see below). Deteriorated exhaust system components should not be repaired; they should be replaced with new parts.
3   If the exhaust system components are extremely corroded or rusted together, welding equipment will probably be required to remove them. The convenient way to accomplish this is to have a muffler repair shop remove the corroded sections with a cutting torch. If, however, you want to save money by doing it yourself (and you don't have a welding outfit with a cutting torch), simply cut off the old components with a hacksaw. If you have compressed air, special pneumatic cutting chisels can also be used. If you do decide to tackle the job at home, be sure to wear safety goggles to protect your eyes from metal chips and work gloves to protect your hands.
4   Here are some simple guidelines to follow when repairing the exhaust system:

   a) Work from the back to the front when removing exhaust system components.
   b) Apply penetrating oil to the exhaust system component fasteners to make them easier to remove.
   c) Use new gaskets, hangers and clamps when installing exhaust systems components.
   d) Apply anti-seize compound to the threads of all exhaust system fasteners during reassembly.
   e) Be sure to allow sufficient clearance between newly installed parts and all points on the underbody to avoid overheating the floor pan and possibly damaging the interior carpet and insulation. Pay particularly close attention to the catalytic converter and heat shield.

### Catalytic converter

**Warning:** *The converter gets very hot during operation. Make sure it's cooled down before you touch it.*
**Note:** *See Chapter 6 for more information on the catalytic converter.*

5   Periodically, inspect the heat shield for cracks, dents and loose or

missing fasteners.

6 Remove the heat shield and inspect the converter for cracks or other damage.

7 If the converter must be replaced, remove the mounting nuts from the flanges at each end, detach the rubber mounts and separate the converter from the exhaust system (you should be able to push the exhaust pipes at each end out of the way to clear the converter studs.

8 Installation is the reverse of removal. Be sure to use new gaskets.

# Chapter 5  Engine electrical systems

## Contents

| | |
|---|---|
| Alternator – removal and installation | 17 |
| Battery cables – check and replacement | 3 |
| Battery check, maintenance and charging | See Chapter 1 |
| Battery – emergency jump starting | 2 |
| Battery – removal and installation | 4 |
| Centrifugal advance mechanism (1986 and 1987 models only) – check | 14 |
| Charging system – check | 16 |
| Charging system – general information and precautions | 15 |
| Distributor – removal and installation | 10 |
| Drivebelt check, adjustment and replacement | See Chapter 1 |
| General information | 1 |
| Igniter – check (1986 and 1987 models only) | 8 |
| Ignition coil – check and replacement | 7 |
| Ignition system – check | 6 |
| Ignition system – general information | 5 |
| Ignition timing check and adjustment | See Chapter 1 |
| Radio condenser – check | 9 |
| Reluctor air gap (1986 and 1987 models only) – check and adjustment | 11 |
| Reluctor (1986 and 1987 models only) – replacement | 12 |
| Spark plug replacement | See Chapter 1 |
| Spark plug wire, distributor cap and rotor check and replacement | See Chapter 1 |
| Starter motor – in-vehicle check | 20 |
| Starter motor – removal and installation | 21 |
| Starter solenoid – removal and installation | 22 |
| Starting system – general information and precautions | 19 |
| Vacuum advance mechanism – check and replacement | 13 |
| Voltage regulator and alternator brushes – replacement | 18 |

## Specifications

### Ignition system
Coil
  Primary resistance (between terminals A and D)
    Legend ............................................. 0.3 to 0.4 ohms
    Integra (1986 and 1987) ............................ 1.2 to 1.5 ohms
    Integra (1988 and 1989) ............................ 0.3 to 0.5 ohms
  Secondary resistance
    Legend (except 1989 and 1990 A/T) .................. 9,040 to 13,560 ohms
    Legend A/T (1989 and 1990 only) .................... 14,400 to 21,600 ohms
    Integra (1986 and 1987) ............................ 9,040 to 13,560 ohms
    Integra (1988 and 1989) ............................ 9,440 to 14,160 ohms
  Resistance between terminals B and D
    Legend ............................................. 2,090 to 2,310 ohms
    Integra (1986 and 1987) ............................ Approximately 2,200 ohms
Igniter resistance
  1986 ................................................. 50,000 ohms or more at 70-degrees F
  1987 ................................................. 500 ± 50 ohms or more at 70-degrees F
Pick-up coil resistance (1986 and 1987 Integra and Legend) ... 750 ± 100 ohms
Radio condenser capacitance .............................. 0.47 ± 0.09 microfarads
Alternator brush length (minimum) ........................ 1/4-inch

## 1  General information

The engine electrical systems include all ignition, charging and starting components. Because of their engine-related functions, these components are discussed separately from chassis electrical devices such as the lights, the instruments, etc. (which are included in Chapter 12).

Always observe the following precautions when working on the electrical systems:

a) Be extremely careful when servicing engine electrical compo-

nents. They are easily damaged if checked, connected or handled improperly.
b) Never leave the ignition switch on for long periods of time with the engine off.
c) Don't disconnect the battery cables while the engine is running.
d) Maintain correct polarity when connecting a battery cable from another vehicle during jump starting.
e) Always disconnect the negative cable first and hook it up last or the battery may be shorted by the tool being used to loosen the cable clamps.

It's also a good idea to review the safety-related information regarding the engine electrical systems located in the Safety First section near the front of this manual before beginning any operation included in this Chapter.

## 2  Battery – emergency jump starting

Refer to the Booster battery (jump) starting procedure at the front of this manual.

## 3  Battery cables – check and replacement

1  Periodically inspect the entire length of each battery cable for damage, cracked or burned insulation and corrosion. Poor battery cable connections can cause starting problems and decreased engine performance.
2  Check the cable-to-terminal connections at the ends of the cables for cracks, loose wire strands and corrosion. The presence of white, fluffy deposits under the insulation at the cable terminal connection is a sign that the cable is corroded and should be replaced. Check the terminals for distortion, missing mounting bolts and corrosion.
3  When removing the cables, always disconnect the negative cable first and hook it up last or the battery may be shorted by the tool used to loosen the cable clamps. Even if only the positive cable is being replaced, be sure to disconnect the negative cable from the battery first (see Chapter 1 for further information regarding battery cable removal).
4  Disconnect the old cables from the battery, then trace each of them to their opposite ends and detach them from the starter solenoid and ground terminals. Note the routing of each cable to ensure correct installation.
5  If you are replacing either or both of the old cables, take them with you when buying new cables. It is vitally important that you replace the cables with identical parts. Cables have characteristics that make them easy to identify: positive cables are usually red and larger in cross-section; ground cables are usually black and smaller in cross-section.
6  Clean the threads of the solenoid or ground connection with a wire brush to remove rust and corrosion. Apply a light coat of battery terminal corrosion inhibitor, or petroleum jelly, to the threads to prevent future corrosion.
7  Attach the cable to the solenoid or ground connection and tighten the mounting nut/bolt securely.
8  Before connecting a new cable to the battery, make sure that it reaches the battery post without having to be stretched.
9  Connect the positive cable first, followed by the negative cable.

## 4  Battery – removal and installation

Refer to illustrations 4.1, 4.2 and 4.3

1  **Caution:** *Always disconnect the negative cable first and hook it up last or the battery may be shorted by the tool being used to loosen the cable clamps.* Disconnect both cables from the battery terminals **(see illustration)**.
2  Remove the battery hold-down clamp **(see illustration)**.
3  Lift out the battery. Be careful – it's heavy. **Note:** *Battery straps and*

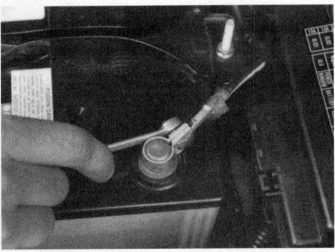

4.1  To remove the battery, detach the negative, then the positive cable clamps from their respective terminals

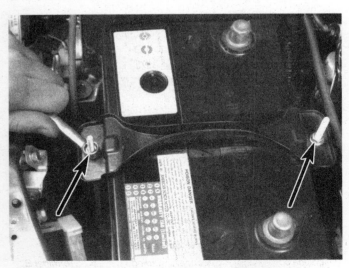

4.2  Remove the two nuts (arrows) and detach the hold-down clamp

4.3  If available, attach a battery strap and lift the battery straight up

# Chapter 5   Engine electrical systems

handlers (see illustration) *are available at most auto parts stores for a reasonable price. They make it easier to remove and carry the battery.*
4   While the battery is out, remove and inspect the carrier (tray) for corrosion.
5   If corrosion has leaked down to the battery support, remove the bolts and lift the support out. Clean the deposits from the metal to prevent the support from further oxidation.
6   If you are replacing the battery, make sure you get one that's identical, with the same dimensions, amperage rating, cold cranking rating, etc.
7   Installation is the reverse of removal.

## 5   Ignition system – general information

**Warning:** *The transistorized electronic ignition systems used on these models generate considerably higher voltage than conventional systems. Be extra careful when servicing these ignition systems.*

The Programmed Ignition (PGM-IG) system provides complete control of the ignition timing by determining the optimum timing using a microcomputer in response to engine speed and vacuum pressure in the intake manifold. These parameters are relayed to the ECU by the CRANK/CYL Sensor, TDC Sensor, Throttle Angle Sensor, Coolant Temperature Sensor and MAP Sensor. Ignition timing is altered during warm-up, idling and warm running conditions by the PGM-IG system. This electronic ignition system also consists of the ignition switch, battery, coil, distributor, spark plug wires and spark plugs.

All distributors are driven by the camshaft. Distributors on early models employ centrifugal and vacuum advance systems; later model distributors are advanced and retarded by the Electronic Control Unit (ECU). Several distributors are used on the vehicles covered by this manual: On 1986 and 1987 Integras, the distributor is separated from the crank angle sensor, which is attached to the opposite camshaft. On 1988 and 1989 Integras, the distributor and the crank angle sensor are in the same housing, but this unit does not disassemble completely. On 1986 and 1987 Legends, the pick-up coil and igniter are replaceable items. 1988 through 1990 Legend models employ a crank angle sensor which is located inside the distributor; these distributors must be replaced as a single unit if they become defective. Refer to the local dealership parts department for any questions concerning the availability of the distributor parts and assemblies.

## 6   Ignition system – check

Refer to illustration 6.3
**Warning:** *Because of the very high voltage generated by the ignition system, extreme care should be taken whenever an operation is performed involving ignition components. This not only includes the coil, igniter and spark plug wires, but related items connected to the system as well, such as the plug connections, tachometer and any test equipment.*
1   With the ignition switch turned to the "on" position, a "Battery" light or an "Oil Pressure" light is a basic check for ignition and battery supply to the ECU.
2   Check all ignition wiring connections for tightness, cuts, corrosion or any other signs of a bad connection.
3   Use a calibrated ignition tester to verify adequate secondary voltage (25,000 volts) at each spark plug (see illustration). A faulty or poor connection at that plug could also result in a misfire. Also check for carbon deposits inside the spark plug boot.
4   Check for carbon tracking on the coil. If carbon tracking is evident, replace the coil and be sure the secondary wires related to that coil are clean and tight. Excessive wire resistance or faulty connections could cause damage to the coil.
5   Using an ohmmeter, check the resistance between the coil terminals. If an open is found (verified by an infinite reading), replace the coil.
6   Using an ohmmeter, check the resistance of the spark plug wires. Each wire should measure less than 25,000 ohms.
7   Additional checks should be performed by a dealer service department or an automotive repair shop.

**6.3   To use a calibrated ignition tester, simply disconnect a spark plug wire, connect it to the tester, clip the tester to a convenient ground and operate the starter with the ignition on – if there's enough power to fire the plug, sparks will be visible between the electrode tip and the tester body**

## 7   Ignition coil – check and replacement

### Check
1   Make sure the ignition switch is turned Off for the following checks.
2   Remove the rubber boot (cover) from the coil (if equipped) and, on Legend and 1986 and 1987 Integra models, detach the high tension lead from the secondary terminal (coil tower). On 1988 and 1989 Integra models, remove the distributor cap (see Chapter 1). Mark and disconnect the wires from the primary terminals.

### 1988 and 1989 Integra
Refer to illustration 7.3
3   Using an ohmmeter, touch the probes to the primary terminals (A and B) of the coil, measure the resistance and compare your reading to the value listed in this Chapter's Specifications (see illustration).

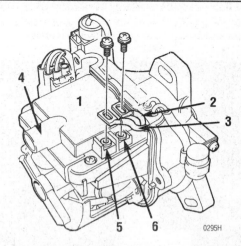

**7.3   On 1988 and 1989 Integra models, after the screws are removed and wires disconnected, measure the resistance between terminals A and B of the coil, then between the secondary winding terminal and terminal A**

| 1   Ignition coil | 4   Secondary winding | 5   Terminal B |
| 2   Blk/yel wire  |     terminal          | 6   Terminal A |
| 3   Wht/blu wire  |                       |                |

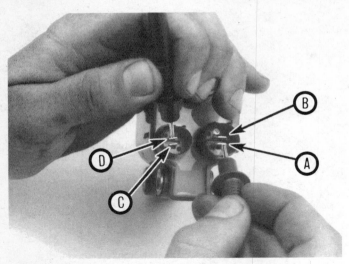

**7.8a Checking the resistance between the coil primary terminals (Legend model shown, 1986 and 1987 Integra similar)**

4  Touch the probes to the secondary winding terminal and the positive primary terminal (A) **(see illustration 7.3)**, measure the resistance and compare your reading to the secondary resistance value listed in this Chapter's Specifications.
5  The figures in the specifications will vary somewhat with the temperature of the coil. The specified resistance values are for a coil temperature of about 70-degrees F.
6  If the coil fails either check, replace it.

### 1986 and 1987 Integra and all Legends
*Refer to illustrations 7.8a, 7.8b, 7.9 and 7.10*
**Note:** *The coil tests for the 1986 and 1987 Integra and all Legend models are similar, but the specifications differ. Also, on 1989 and 1990 Legends with automatic transmission, the coil primary pin designations are different* **(see illustration 7.8b)**, *but the tests are the same.*
7  Unplug the primary and secondary connectors and the coil high tension lead.
8  Using an ohmmeter, touch the probes to primary terminals A and D **(see illustrations)**, measure the resistance and compare your reading to the values listed in this Chapter's Specifications.

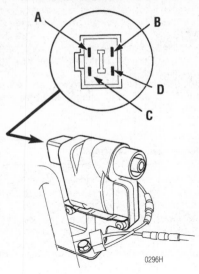

**7.8b On 1989 and 1990 Legend automatic transmission models, the connector and terminal locations are different, but the tests are the same**

9  Touch the probes to terminal A and the secondary terminal (coil tower) **(see illustration)**, measure the resistance and compare your reading to the secondary resistance values listed in this Chapter's Specifications.
10  Touch the probes to terminals B and D **(see illustration)**, measure the resistance and compare your reading to the value listed in this Chapter's Specifications.
11  The above figures will vary somewhat with coil temperature. The specified resistance values are for a coil temperature of about 70-degrees.
12  If the coil passes all three checks, it's okay. Plug in the connectors. If it fails any of the above checks, replace it.

### *Replacement*
#### 1988 and 1989 Integra
13  Detach the cable from the negative terminal of the battery.

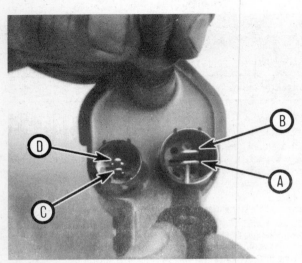

**7.9 Checking the resistance between the coil positive terminal and the high-tension terminal (Legends and 1986 and 1987 Integra)**

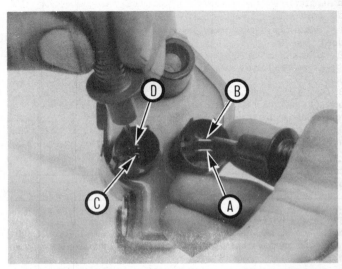

**7.10 Checking the resistance between terminals B and D (Legends and 1986 and 1987 Integra)**

# Chapter 5  Engine electrical systems

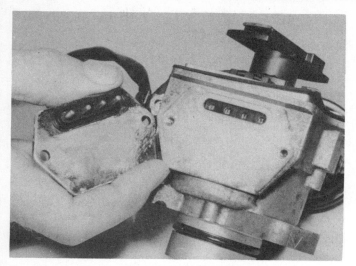

8.1 Unplug the igniter from the distributor case

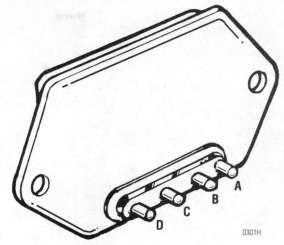

8.2 Igniter terminal identification

14  Remove the distributor cap (see Chapter 1) and leak cover (if equipped). Remove the screws and detach the wires from the primary terminals.
15  Remove the four screws and slide the coil out.
16  Installation is the reverse of removal.

### 1986 and 1987 Integra and all Legends
17  Detach the cable from the negative terminal of the battery.
18  Using pieces of numbered tape, mark the positions of the primary wires. Unplug the primary connectors and detach the high-tension lead from the coil.
19  Remove the two mounting bolts and detach the coil from its mounting bracket.
20  Installation is the reverse of removal.

## 8  Igniter – check (1986 and 1987 models only)

*Refer to illustrations 8.1 and 8.2*
**Note:** *Checking the igniter on 1988 and later models is beyond the scope of the home mechanic. Take the vehicle to a dealer service department for diagnosis.*
1  Remove the igniter cover and pull out the igniter unit **(see illustration)**.
2  Check for continuity in both directions between A and B terminals on the igniter **(see illustration)**. There should be continuity in only one direction.
3  Connect an ohmmeter positive probe to terminal D and the negative probe to the igniter unit (ground). The resistance should be within the range listed in this Chapter's Specifications.
4  If the igniter fails either of the above checks, replace it. **Note:** *When installing the igniter, pack silicone grease in the connector housing.*

## 9  Radio condenser – check

*Refer to illustration 9.1*
**Note:** *The radio condenser is a device that reduces ignition noise in the radio. It is included in this Chapter because it can prevent the engine from running if it fails.*
1  If you own or have access to a condenser tester, check the capacitance of the condenser **(see illustration)** and compare your reading with the value listed in this Chapter's Specifications. If you don't have a condenser tester or access to one, take the condenser to a television repair

9.1  The radio condenser (arrow) is located under the coil on Legend models

shop and have it tested.
2  If the indicated capacitance isn't within specification, replace the condenser.

## 10  Distributor – removal and installation

### Removal
*Refer to illustrations 10.6a and 10.6b*
1  Detach the cable from the negative battery terminal.
2  Detach the vacuum hose(s) from the vacuum advance diaphragm (if equipped), any clamps and wire harness connectors on the distributor. Mark the wires and hoses so they can be returned to their original locations.
3  Detach the primary wires from the coil (except 1988 and 1989 Integra).
4  Look for a raised number or letter on the distributor cap. This marks the location for the number one cylinder spark plug wire terminal. If the cap does not have a mark for the number one terminal, locate the number one spark plug and trace the wire back to the terminal on the cap.
5  Remove the distributor cap (see Chapter 1) and turn the engine over until the rotor is pointing toward the number one spark plug terminal (see the locating TDC procedure in Chapter 2).

# Chapter 5 Engine electrical systems

**10.6a** Make one mark directly underneath the rotor tip . . .

**10.6b** . . . and another between the distributor base and the cylinder head (arrow)

6  Make a mark on the edge of the distributor base directly below the rotor tip and in line with it (if the rotor on your engine has more than one tip, use the center one for reference). Also, mark the distributor base and the cylinder head to ensure the distributor is installed correctly **(see illustrations)**.
7  If not already done, unplug the igniter wires.
8  Remove the distributor hold-down bolt(s) and pull out the distributor.
**Caution:** *DO NOT turn the crankshaft while the distributor is out of the engine, or the alignment marks will be useless.*

## Installation
*Refer to illustrations 10.9 and 10.10*

**Note:** *If the crankshaft has been moved while the distributor is out, the number one piston must be repositioned at TDC. This can be done by feeling for compression pressure at the number one plug hole as the crankshaft is turned. Once compression is felt, align the ignition timing zero mark with the pointer.*

9  Install a new O-ring on the distributor housing **(see illustration)**.
10  Insert the distributor into the cylinder head in exactly the same relationship to the head that it was when removed. **Note:** *The lugs on the end of the distributor and the corresponding grooves in the camshaft end are offset to eliminate the possibility of installing the distributor 180-degrees out of phase* **(see illustration)**.
11  Recheck the alignment marks between the distributor base and the cylinder head to verify the distributor is in the same position it was in before removal. Also check the rotor to see if it's aligned with the mark you made on the distributor.
12  Loosely install the hold-down bolt(s).
13  The remainder of installation is the reverse of removal. Check the ignition timing (see Chapter 1) and tighten the distributor hold-down bolt(s) securely.

## 11  Reluctor air gap (1986 and 1987 models only) – check and adjustment

*Refer to illustrations 11.3a and 11.3b*

1  Detach the cable from the negative battery terminal.
2  Remove the distributor cap and rotor (see Chapter 1).

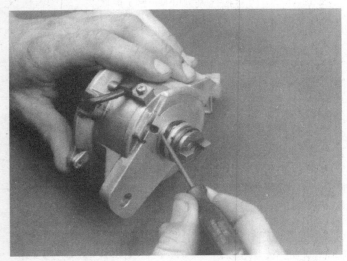

**10.9** Removing the O-ring – put a new O-ring on the bottom of the distributor housing before installing the distributor

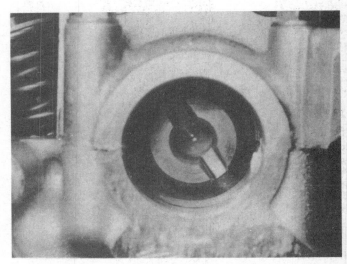

**10.10** The end of the camshaft has an offset groove which matches the lugs on the distributor shaft – this ensures you won't install the distributor 180-degrees out of phase

# Chapter 5 Engine electrical systems

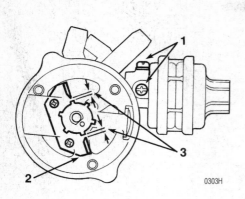

11.3a  Reluctor air gap adjustment on 1986 and 1987 Legends

1  Diaphragm screws   2  Stator   3  Air gaps

11.3b  Reluctor air gap adjustment on 1986 and 1987 Integras

1  Air gaps   2  Stator   3  Reluctor   4  Screws

3  Using a non-magnetic feeler gauge, verify the air gaps between the stator and reluctor are equal **(see illustrations)**.
4  If the gaps aren't equal, loosen the screws **(see illustration 11.3a and 11.3b)** and move the stator until the air gaps are equal. Tighten the screws, then recheck the gaps to make sure they are still equal. **Note:** *On 1986 and 1987 Legend distributors, loosen the diaphragm screws before loosening the stator screws.*

## 12  Reluctor (1986 and 1987 models only) – replacement

*Refer to illustration 12.3*
1  Detach the cable from the negative battery terminal.
2  Remove the distributor cap (see Chapter 1).
3  Remove the reluctor by prying it off with a pair of small screwdrivers **(see illustration)**. Be careful – using excessive force to pry off the reluctor may result in a damaged stator.
4  Installation is the reverse of removal. Make sure you install the new reluctor with the number or letter manufacturing code facing up and the gap in the pin facing away from the shaft.

## 13  Vacuum advance mechanism – check and replacement

**Note:** *Some models have electronically controlled advance and no vacuum advance mechanism. If your distributor does not have a vacuum advance diaphragm on the side, there is no vacuum advance mechanism.*

### Check
1  Detach the cable from the negative battery terminal.
2  Remove the distributor cap.
3  Detach the vacuum hose(s) from the vacuum advance diaphragm on the distributor and attach a vacuum pump to the fitting. If there are two hoses, attach the pump where the outer hose connects.
4  Turn the breaker plate right and left to check for freedom of movement.
5  Apply a gradual vacuum while watching the breaker plate. Verify the breaker plate operates smoothly – there should be no binding.
  a) If there's binding, find the source of the binding and free the breaker plate.

### Replacement
6  Remove the E-clip from the diaphragm arm.
7  Remove the diaphragm mounting screws.
  b) If the vacuum pump gauge indicates a loss of vacuum, the diaphragm is defective and must be replaced.

8  Detach the diaphragm arm, then pull the diaphragm out of the distributor.
9  Installation is the reverse of removal.

12.3  To remove the reluctor from the distributor shaft, pry it off with a pair of screwdrivers – be sure to use a couple of rags under the screwdrivers to protect the stator assembly from damage (Legend shown, Integra similar)

## 14  Centrifugal advance mechanism (1986 and 1987 models only) – check

**Note:** *This procedure applies to 1986 and 1987 models only. 1988 and later models have electronically controlled advance and no centrifugal advance.*

1  Detach the vacuum hose(s) from the vacuum advance diaphragm and plug it (them).
2  Connect a timing light in accordance with the manufacturer's instructions.
3  Start the engine and increase the engine speed from idle to about 2500 RPM. Observe the timing marks as described in the ignition timing procedure in Chapter 1. As the engine accelerates, the timing should gradually increase farther and farther Before Top Dead Center (BTDC). If it doesn't, check the centrifugal advance mechanism for sticking or binding.

## 15 Charging system – general information and precautions

The charging system includes the alternator, an internal voltage regulator, a charge indicator, the battery, a fusible link and the wiring between all the components. The charging system supplies electrical power for the ignition system, the lights, the radio, etc. The alternator is driven by a drivebelt at the one end of the engine.

The purpose of the voltage regulator is to limit the alternator's voltage to a preset value. This prevents power surges, circuit overloads, etc., during peak voltage output.

The fusible link is a short length of wire integral with the engine compartment wiring harness. The link is four wire gauges smaller in diameter than the circuit it protects. See Chapter 12 for additional information regarding fusible links.

The charging system doesn't ordinarily require periodic maintenance. However, the drivebelt, battery and wires and connections should be inspected at the intervals outlined in Chapter 1.

The dashboard warning light should come on when the ignition key is turned to On, but it should go off immediately after the engine is started. If it remains on, there is a malfunction in the charging system (see Section 16). Some vehicles are also equipped with a voltmeter. If the voltmeter indicates abnormally high or low voltage, check the charging system (see Section 16).

Be very careful when making electrical circuit connections to a vehicle equipped with an alternator and note the following:
  a) When reconnecting wires to the alternator from the battery, be sure to note the polarity.
  b) Before using arc welding equipment to repair any part of the vehicle, disconnect the wires from the alternator and the battery terminals.
  c) Never start the engine with a battery charger connected.
  d) Always disconnect both battery leads before using a battery charger.
  e) The alternator is turned by an engine drivebelt which could cause serious injury if your hands, hair or clothes become entangled in it with the engine running.
  f) Because the alternator is connected directly to the battery, it could arc or cause a fire if overloaded or shorted out.
  g) Wrap a plastic bag over the alternator and secure it with rubberbands before steam cleaning the engine.

## 16 Charging system – check

1   If a malfunction occurs in the charging circuit, don't automatically assume that the alternator is causing the problem. First check the following items:
  a) Check the drivebelt tension and condition (Chapter 1). Replace it if it's worn or deteriorated.
  b) Make sure the alternator mounting and adjustment bolts are tight.
  c) Inspect the alternator wiring harness and the connectors at the alternator and voltage regulator. They must be in good condition and tight.
  d) Check the fusible link (if equipped) located between the starter solenoid and the alternator. If it's burned, determine the cause, repair the circuit and replace the link (the vehicle won't start and/or the accessories won't work if the fusible link blows). Sometimes a fusible link may look good, but still be bad. If in doubt, remove it and check for continuity.
  e) Start the engine and check the alternator for abnormal noises (a shrieking or squealing sound indicates a bad bearing).
  f) Check the specific gravity of the battery electrolyte. If it's low, charge the battery (doesn't apply to maintenance-free batteries).
  g) Make sure the battery is fully charged (one bad cell in a battery can cause overcharging by the alternator).
  h) Disconnect the battery cables (negative first, then positive). Inspect the battery posts and the cable clamps for corrosion. Clean them thoroughly if necessary (see Chapter 1). Reconnect the cable to the negative terminal.
  i) With the key off, connect a test light between the negative battery post and the disconnected negative cable clamp.
     1) If the test light does not come on, reattach the clamp and proceed to the next step.
     2) If the test light comes on, there is a short (drain) in the electrical system of the vehicle. The short must be repaired before the charging system can be checked.
     3) Disconnect the alternator wiring harness.
        (a) If the light goes out, the alternator is bad.
        (b) If the light stays on, pull each fuse until the light goes out (this will tell you which component is shorted).

2   Using a voltmeter, check the battery voltage with the engine off. It should be approximately 12-volts.

3   Start the engine and check the battery voltage again. It should now be approximately 14-to-15 volts.

4   Turn on the headlights. The voltage should drop, and then come back up, if the charging system is working properly.

5   If the voltage reading is more than the specified charging voltage, replace the voltage regulator (refer to Section 18). If the voltage is less, the alternator diode(s), stator or rectifier may be bad or the voltage regulator may be malfunctioning.

## 17 Alternator – removal and installation

*Refer to illustration 17.3*

1   Detach the cable from the negative terminal of the battery.
2   Mark and detach the electrical connectors from the alternator.
3   Loosen the alternator adjusting bolt and pivot bolt, then detach the drivebelt **(see illustration)**.
4   On Integra models equipped with air conditioning, remove the left driveaxle (see chapter 8). Remove the adjusting and pivot bolts and separate the alternator from the engine.
5   If you are replacing the alternator, take the old one with you when purchasing a replacement unit. Make sure the new/rebuilt unit looks identical to the old alternator. Look at the terminals - they should be the same in number, size and location as the terminals on the old alternator. Finally, look at the identification numbers - they will be stamped into the housing or printed on a tag attached to the housing. Make sure the numbers are the same on both alternators.
6   Many new/rebuilt alternators DO NOT have a pulley installed, so you may have to switch the pulley from the old unit to the new/rebuilt one. When buying an alternator, find out the shop's policy regarding pulleys some shops will perform this service free of charge.

**17.3  Here's a typical alternator – to remove it, loosen the pivot bolt (arrow) and the adjusting bolt at the rear of the alternator (not shown), then remove the drivebelt, remove the bolts and lift out the alternator**

# Chapter 5 Engine electrical systems

18.2a Remove the three nuts from the rear cover

18.2b Take the nut, washer and terminal insulator off terminal B and remove the alternator rear cover

18.3 Once the rear cover is removed, remove the five screws (arrows) that retain the voltage regulator and the brush holder

18.4a Remove the brush holder, . . .

18.4b . . . then remove the regulator

7 Installation is the reverse of removal.
8 After the alternator is installed, adjust the drivebelt tension (see Chapter 1).
9 Check the charging voltage to verify proper operation of the alternator (see Section 16).

## 18 Voltage regulator and alternator brushes – replacement

Refer to illustrations 18.2a, 18.2b, 18.3, 18.4a, 18.4b, 18.5 and 18.7
**Note:** *Don't attempt to overhaul the alternator. If replacing the brushes and regulator does not solve the alternator problem, take the alternator to a dealer and have it rebuilt or exchange it as a core for a rebuilt unit.*

1 Remove the alternator (see Section 17) and place it on a clean workbench.
2 Remove the three rear cover nuts, the nut and terminal insulator and the rear cover **(see illustrations)**.
3 Remove the five voltage regulator and brush holder retaining screws **(see illustration)**.
4 Remove the brush holder and the regulator from the rear end frame **(see illustrations)**. If you are only replacing the regulator, proceed to Step 6, install the new unit, reassemble the alternator and install it on the engine (see Section 17). If you are going to replace the brushes, proceed with the next Step.

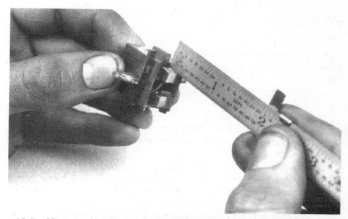

18.5 Measure the exposed length of the brushes and compare your measurements to the specified minimum length to determine whether they should be replaced

5 Measure the exposed length of each brush **(see illustration)** and compare it to the specified minimum length. If the length of either brush is less than the minimum listed in this Chapter's Specifications, replace the brushes.
6 Make sure that each brush moves smoothly in the brush holder.

# Chapter 5  Engine electrical systems

**18.7  To install the brush holder, depress each brush with a small screwdriver to clear the shaft**

7  Install the brush holder by depressing each brush with a small screwdriver to clear the shaft **(see illustration)**.
8  Install the voltage regulator and brush holder screws into the rear frame.
9  Install the rear cover and tighten the three nuts securely.
10  Install the terminal insulator and tighten it with the nut.
11  Install the alternator (see Section 17).

## 19  Starting system – general information and precautions

The sole function of the starting system is to turn over the engine quickly enough to allow it to start.

The starting system consists of the battery, the starter motor, the starter solenoid and the wires connecting them. The solenoid is mounted directly on the starter motor.

The solenoid/starter motor assembly is installed on the upper part of the engine, next to the transmission bellhousing.

When the ignition key is turned to the Start position, the starter solenoid is actuated through the starter control circuit. The starter solenoid then connects the battery to the starter. The battery supplies the electrical energy to the starter motor, which does the actual work of cranking the engine.

The starter motor on some vehicles equipped with manual transaxles can only be operated when the clutch pedal is depressed; the starter on all vehicles equipped with an automatic transaxles can only be operated when the selector lever is in Park or Neutral.

Always observe the following precautions when working on the starting system:
a) Excessive cranking of the starter motor can overheat it and cause serious damage. Never operate the starter motor for more than 15 seconds at a time without pausing to allow it to cool for at least two minutes.
b) The starter is connected directly to the battery and could arc or cause a fire if mishandled, overloaded or shorted out.
c) Always detach the cable from the negative terminal of the battery before working on the starting system.

## 20  Starter motor – in-vehicle check

**Note:** *Before diagnosing starter problems, make sure the battery is fully charged.*

1  If the starter motor does not turn at all when the switch is operated, make sure the shift lever is in Neutral or Park (automatic transmission) or the clutch pedal is depressed (manual transmission).
2  Make sure the battery is charged and all cables, both at the battery and starter solenoid terminals, are clean and secure.
3  If the starter motor spins but the engine is not cranking, the overrunning clutch in the starter motor is slipping and the starter motor must be replaced. Also, the ring gear on the flywheel or driveplate may be worn.
4  If, when the switch is actuated, the starter motor does not operate at all but the solenoid clicks, the problem lies with either the battery, the main solenoid contacts or the starter motor itself (or the engine is seized).
5  If the solenoid plunger cannot be heard when the switch is actuated, the battery is bad, the fusible link is burned (the circuit is open) or the solenoid itself is defective.
6  To check the solenoid, connect a jumper lead between the battery (+) and the ignition switch wire terminal (the small terminal) on the solenoid. If the starter motor now operates, the solenoid is OK and the problem is in the ignition switch, neutral start switch or the wiring.
7  If the starter motor still does not operate, remove the starter/solenoid assembly for disassembly, testing and repair.
8  If the starter motor cranks the engine at an abnormally slow speed, first make sure that the battery is charged and that all terminal connections are tight. If the engine is partially seized, or has the wrong viscosity oil in it, it will crank slowly.
9  Run the engine until normal operating temperature is reached, then disconnect the coil wire from the distributor cap and ground it on the engine.
10  Connect a voltmeter positive lead to the positive battery post and connect the negative lead to the negative post.
11  Crank the engine and take the voltmeter readings as soon as a steady figure is indicated. Do not allow the starter motor to turn for more than 15 seconds at a time. A reading of nine volts or more, with the starter motor turning at normal cranking speed, is normal. If the reading is nine volts or more but the cranking speed is slow, the motor, solenoid contacts or circuit connections are faulty. If the reading is less than nine volts and the cranking speed is slow, the starter motor is probably bad.

## 21  Starter motor – removal and installation

1  Detach the cable from the negative terminal of the battery.
2  Clearly label, then disconnect the wires from the terminals on the starter motor solenoid. Disconnect any clips securing the wiring to the starter.
3  Remove the mounting bolts and detach the starter.
4  Installation is the reverse of removal.

## 22  Starter solenoid – removal and installation

1  Disconnect the cable from the negative terminal of the battery.
2  Remove the starter motor (see Section 21).
3  Disconnect the large wire from the solenoid to the starter motor terminal.
4  Remove the screws which secure the solenoid to the starter motor gear housing and detach the solenoid from the gear housing.
5  While the solenoid is removed, check the overrunning clutch by sliding it along its shaft. If it doesn't move freely, or if the clutch slips when you rotate the armature while holding the drive gear, replace the clutch assembly. If the gear is worn or damaged, replace the complete overrunning clutch assembly (the gear isn't available separately). If the starter gear teeth are damaged, you should also inspect the flywheel or driveplate ring gear for damage.
6  Installation is the reverse of removal.

# Chapter 6  Emissions control systems

## Contents

| | |
|---|---|
| Air injection system (Legend models only) ................. 8 | General information ............................................... 1 |
| Catalytic converter ....................................... 9 | Information sensors ............................................ 4 |
| Electronic Air/Idle Control Valve (EACV/EICV) – check | Programmed Fuel Injection (PGM-FI) system – |
| and replacement ....................................... 5 | general information ......................................... 2 |
| Exhaust Gas Recirculation (EGR) system | Positive Crankcase Ventilation (PCV) system .............. 6 |
| (Legend models only) .................................. 10 | Self diagnosis system – description and code access ......... 3 |
| Fuel evaporative control system ........................ 7 | |

## Specifications

### General

Oxygen sensor resistance
   Closed throttle deceleration ............................ Less than 0.4 volts
   Cruising (4,000 rpm) ..................................... More than 0.6 volts
CRANK circuit resistance
   Legend ..................................................... 500 to 1,000 ohms
   Integra (1988 and 1989) ................................. 350 to 550 ohms
CYL circuit resistance
   Legend ..................................................... 500 to 1,000 ohms
   Integra (1986 and 1987) ................................. 650 to 850 ohms
Electronic Air Control Valve (EACV) resistance ............... 8 to 15 ohms
Electronic Idle Control Valve (EICV) resistance .............. 6 to 20 ohms
Intake air temperature sensor resistance ...................... 1 to 4 k-ohms at room temperature

## 1  General information

*Refer to illustrations 1.6a and 1.6b*

To prevent pollution of the atmosphere from incompletely burned and evaporating gases, and to maintain good driveability and fuel economy, a number of emission control systems are incorporated. They include the:

*Self diagnosis system*
*Electronic engine controls*
*Exhaust Gas Recirculation (EGR) system*
*Air injection system*
*Fuel evaporative control system*
*Positive Crankcase Ventilation (PCV) system*
*Catalytic converter*

The Sections in this Chapter include general descriptions, checking procedures within the scope of the home mechanic and component replacement procedures (when possible) for each of the systems listed above.

Before assuming that an emissions control system is malfunctioning, check the fuel and ignition systems carefully. The diagnosis of some emission control devices requires specialized tools, equipment and training. If checking and servicing become too difficult or if a procedure is beyond your ability, consult a dealer service department or other repair shop. Remember, the most frequent cause of emissions problems is simply a loose or broken vacuum hose or wire, so always check the hose and wiring connections first.

This doesn't mean, however, that emissions control systems are particularly difficult to maintain and repair. You can quickly and easily perform many checks and do most of the regular maintenance at home with com-

## Chapter 6  Emissions control systems

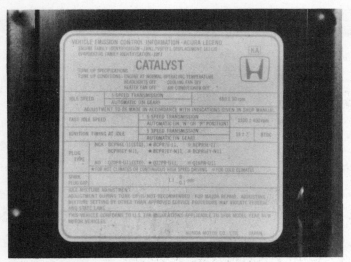

1.6a  The Vehicle Emission Control Information (VECI) label provides essential tune-up specifications like idle speed and fast idle, spark plug types, etc.

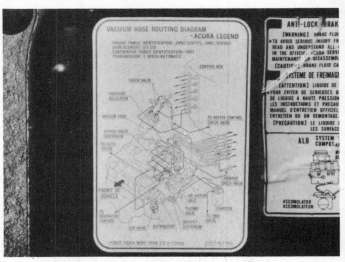

1.6b  The Vacuum Hose Routing Diagram tells you what emission control devices the vehicle is equipped with, gives you their approximate locations and provides a vacuum hose routing schematic, which is helpful when you're looking for leaks and disconnected or misrouted hoses

mon tune-up and hand tools. **Note:** *Because of a Federally mandated extended warranty which covers the emissions control system components, check with your dealer about warranty coverage before working on any emissions-related systems. Once the warranty has expired, you may wish to perform some of the component checks and/or replacement procedures in this Chapter to save money.*

Pay close attention to any special precautions outlined in this Chapter. It should be noted that the illustrations of the various systems may not exactly match the system installed on your vehicle because of changes made by the manufacturer during production or from year-to-year.

A Vehicle Emissions Control Information (VECI) label is attached to the underside of the hood **(see illustration)**. This label contains important emissions specifications and adjustment information. A second label, the Vacuum Hose Routing Diagram, **(see illustration)** provides a vacuum hose schematic with emissions components identified. When servicing the engine or emissions systems, the VECI label and the vacuum hose routing diagram in your particular vehicle should always be checked for up-to-date information.

## 2  Programmed Fuel Injection (PGM-FI) system – general information

*Refer to illustrations 2.1a and 2.1b*

The Programmed Fuel Injection (PGM-FI) system **(see illustrations)** consists of three sub-systems: air intake, electronic control and fuel delivery. The intake manifold absolute pressure and the engine speed are used to determine the correct air/fuel ratio that is to be injected into the combustion chambers. This is known as the speed density method of fuel control. The PGM-FI system uses an Electronic Control Unit (ECU) along with the sensors (coolant temperature sensor, throttle position sensor, manifold absolute pressure (MAP) sensor etc.) to determine the proper fuel/air ratio under all operating conditions.

The fuel injection system and the emission control system are closely linked in function and design. For additional information, refer to Chapter 4.

### Electronic control system

The electronic control system consists of an eight-bit microprocessor (computer), output actuators and various information sensors:

The crank angle sensor, which is an integral part of the distributor assembly, consists of two rotors (TDC and CYL) and a pickup for each rotor. Depending on the year and model, some Acuras are equipped with the crank angle sensor separated from the distributor while others are equipped with a single combined unit (refer to Chapter 5). The distributor is driven off the end of the camshaft, and the rotors are coupled to the distributor shaft, so they turn together as a unit as the cam rotates. The CYL pickup detects the position of the no. 1 cylinder as the base for sequential injection; the TDC pickup determines the injection timing for each cylinder. The TDC pickup also monitors engine speed to help determine the basic discharge duration for different operating conditions.

The Manifold Absolute Pressure (MAP) sensor converts manifold pressure readings into electrical voltage signals and sends them to the ECU. This data, along with the data from the TDC and CYL sensors, enables the ECU to determine the duration during which fuel is injected.

The atmospheric pressure (PA) sensor converts atmospheric pressures into voltage signals and sends them to the ECU. These signals enable the ECU to modify the basic fuel discharge duration to compensate for changes in the atmospheric pressure.

The coolant temperature (TW) sensor uses a temperature dependent diode (thermistor) to measure differences in the coolant temperature. The resistance of the thermistor decreases with a rise in coolant temperature. The ECU uses this input to increase or decrease the fuel discharge duration.

The intake air temperature (TA) sensor, which is located in the intake manifold, is also a thermistor. In operation, it's similar to the TW sensor but has a lower thermal capacity for quicker response time.

The throttle angle sensor is a variable resistor. The sensor is mounted on the end of the throttle valve shaft. As the throttle valve is rotated, the resistance varies, altering the output voltage to the control unit, which in turn alters the fuel discharge duration.

The oxygen sensor monitors the oxygen content in the exhaust gas and sends a variable voltage signal to the ECU, which alters the fuel discharge duration.

When the ignition key is turned to Start, the starter switch sends a signal to the ECU, which increases the amount of fuel injected, in accordance with the engine temperature. The amount of fuel injected is gradually reduced once the engine is started.

Refer to Chapter 4 for additional information on the fuel injection system and diagnosing the components.

2.1a Emission control components on Integra models (typical)

1 EACV (not visible)
2 Intake air temperature (TA) sensor (not visible)
3 Oxygen sensor
4 Fuel injector
5 Manifold Absolute Pressure (MAP) sensor
6 Throttle angle sensor (not visible)
7 CRANK/CYL sensor
8 Distributor (TDC sensor)

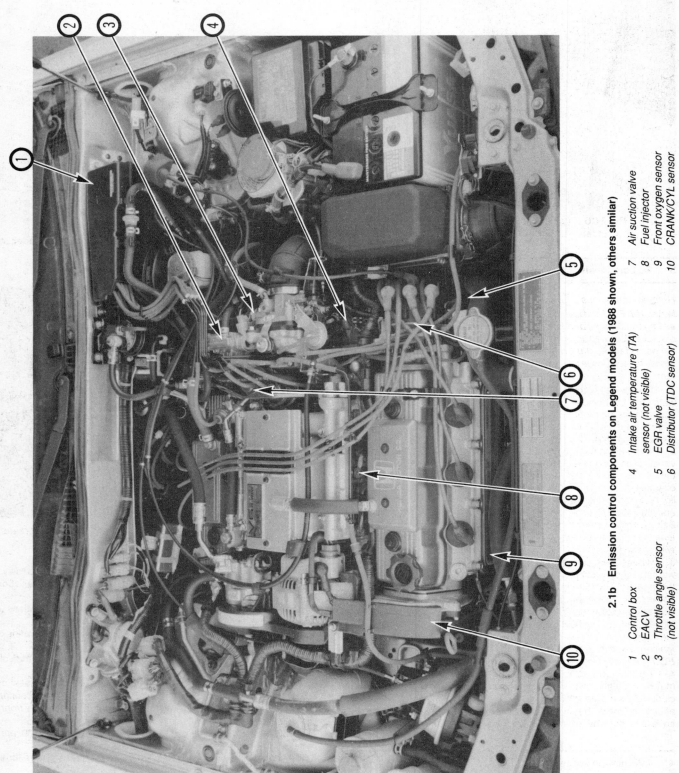

2.1b Emission control components on Legend models (1988 shown, others similar)

1  Control box
2  EACV
3  Throttle angle sensor (not visible)
4  Intake air temperature (TA) sensor (not visible)
5  EGR valve
6  Distributor (TDC sensor)
7  Air suction valve
8  Fuel injector
9  Front oxygen sensor
10 CRANK/CYL sensor

# Chapter 6  Emissions control systems

**3.1a  The ECU on all Legend Sedans and Integra models is located under the passenger front seat**

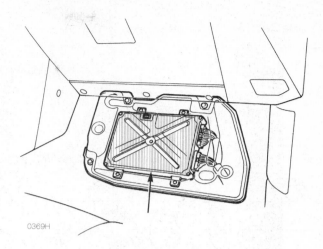

**3.1b  The ECU (arrow) on Legend Coupes is located under the carpet at the front of the passenger footwell**

## 3  Self diagnosis system – description and code access

*Refer to illustrations 3.1a and 3.b*

**Note:** *The ECU and LED display are located under the passenger seat on all Legend Sedans and Integras. On Legend Coupes, the ECU and LED display are located under the the dashboard, behind the carpet on the passenger side (see illustration). If there are two LEDs, the yellow one is strictly for setting the idle speed. The red LED is for the self-diagnosis system.*

1  To view self-diagnosis information from the ECU memory, you must watch the LED display on the ECU **(see illustrations)**. The codes are stored in the memory of the ECU and when accessed, they blink a sequence on the LED to relay a number or code that represents a system component failure.

2  With the ignition ON, the ECU will display one LED that flashes in a variety of combinations. On 1986 through 1989 Legend models and all Integras, the LED will blink the sum total representing the code number (for example, 14 short blinks for the code 14). On 1990 Legend models, the light will hold a longer blink to represent the first digit of a two digit number and then will blink short for the second digit (for example, 1 long blink then 8 short blinks for the code 18). **Note:** *If the system has more than one problem, the codes will be displayed in sequence then a pause and the codes will repeat.*

3  When the ECU sets a trouble code, the Check Engine light or the S warning light (if equipped) will come on and a trouble code will be stored in the memory. The trouble code will stay in the ECU memory until the voltage to the ECU is interrupted. To clear the memory, remove the HAZARD fuse at the battery positive terminal for 10 seconds on all Integras. Remove the ALTERNATOR SENSE fuse in the under-hood relay box to clear the memory on all Legends. **Caution:** *To prevent damage to the ECU, the ignition switch must be off when disconnecting or connecting power to the ECU (this includes disconnecting and connecting the battery).*

4  The table on the following page is a list of the typical trouble codes which may be encountered while diagnosing the computerized system. Also included are simplified troubleshooting procedures. If the problem persists after these checks have been made, more detailed service procedures will have to be done by a dealer service department or other repair shop.

## 4  Information sensors

### Oxygen sensor

#### Check

*Refer to illustration 4.8*

1  Turn the ignition switch to OFF.

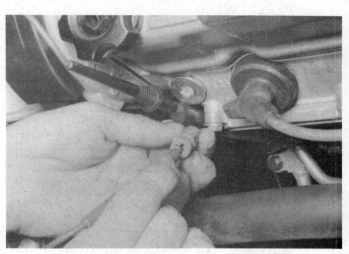

**4.8  When checking the oxygen sensor, use a digital voltmeter to receive accurate low voltage readings**

2  Remove the ALTERNATOR SENSE fuse in the under hood relay box for 10 seconds on all Legends or the HAZARD fuse in the main fuse box on Integra (see Chapter 12) for ten seconds to reset the ECU.

3  Check the fuel pressure to make sure it's within specifications (see Chapter 4).

4  If the fuel pressure is okay, warm the engine to normal operating temperature.

5  Block the rear wheels and set the parking brake. Raise the front of the vehicle and support it securely on jackstands.

6  Locate the oxygen sensor, which is screwed into the exhaust manifold. Follow the sensor wire back and unplug the electrical connector. **Note:** *Legend models are equipped with an oxygen sensor in the front and the rear exhaust manifolds.*

7  Start the engine and allow it to run for fifteen minutes at 1700 rpm. Do not close the throttle completely at this time.

8  Disconnect the wire harness from the oxygen sensor **(see illustration)**.

9  Raise the engine speed to 4000 rpm for approximately 10 seconds and allow the throttle to snap shut.

10  Using a digital voltmeter, promptly measure the voltage between the oxygen sensor side of the electrical connector and ground **(see illustration 4.8)**. It should be below 0.4-volts during closed throttle deceleration.

## Chapter 6 Emissions control systems

| Trouble codes | Circuit or system | Corrective action |
|---|---|---|
| Code 0 | Faulty ECU | Inspect the number 5 fuse (1988 through 1990 Legend), the number 13 fuse (1986 and 1987 Legend) or the number 3 fuse (Integra). Replace if necessary. Also check for open circuit in the YEL wire between the #5 (1988 through 1990 Legend), #13 (1986 and 1987 Legend) or the #3 (Integra) fuse and the combination meter. If no open circuit is found, have the wiring harness checked at a dealer service department or other qualified repair shop. |
| Code 1 | Oxygen content (front manifold on Legend) | Refer to Section 4 |
| Code 2 | Oxygen content (rear manifold, Legend only) | Refer to Section 4 |
| Code 3 and 5 | Manifold Absolute Pressure | Refer to Section 4 |
| Code 4 | Crank angle sensor | Refer to Section 4 |
| Code 6 | Coolant temperature | Have the vehicle checked at a dealer service department or other repair shop |
| Code 7 | Throttle angle | Refer to Section 4 |
| Code 8 | TDC Position (crank angle) | Refer to Section 4 |
| Code 9 | No. 1 cylinder position (crank angle) | Refer to Section 4 |
| Code 10 | Intake air temperature | Refer to Section 4 |
| Code 11 | IMA Sensor (1986 and 1987 Legends only) | Have the vehicle checked at a dealer service department or other repair shop |
| Code 12 | Exhaust Gas Recirculation System (Legend only) | Refer to Section 10 |
| Code 13 | Atmospheric Pressure | Have the vehicle checked at a dealer service department or other repair shop |
| Code 14 | Electronic Air/Idle Control (EACV or EICV) | Refer to Section 5 |
| Code 15 | Ignition output signal | Possible faulty igniter – see Chapter 5 |
| Code 16 | Fuel injector (Integra only) | See Chapter 4 |
| Code 17 | Vehicle speed sensor | Have the vehicle checked at a dealer service department or other repair shop |
| Code 18 | Ignition timing adjustment | See Chapter 5 |
| Code 19 | Lock-up control solenoid valve (automatic trans-axle vehicles) (Integra only) | Have the vehicle checked at a dealer service department or other repair shop |
| Code 20 | Electric load (Integra only) | Have the vehicle checked at a dealer service department or other repair shop |
| Code 30 | A/T FI signal A (automatic transaxle vehicles) (Legend only) | Have the vehicle checked at a dealer service department or other repair shop |
| Code 31 | A/T FI signal B (automatic transaxle vehicles) (Legend only) | Have the vehicle checked at a dealer service department or other repair shop |

# Chapter 6  Emissions control systems

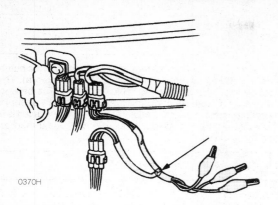

4.23a  Detach the electrical connector (located on the right side of the engine firewall on all Legends) from the MAP sensor and install the test harness (arrow)

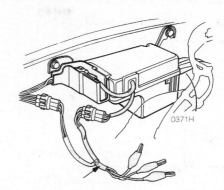

4.23b  On 1986 and 1987 Integra models, the connector is located just outside of the control box – unplug the connector and install the test harness

11  Raise the engine speed to 4000 rpm and hold it for 10 seconds. Using a digital voltmeter, promptly measure the voltage between the oxygen sensor side of the electrical connector and ground. It should be above 0.6-volts during cruising rpm.
12  If the test results are not correct, replace the oxygen sensor with a new unit.
13  If the test results are correct, have the wire harness checked at a dealer service department.

### Replacement
**Warning:** *The electric cooling fan can activate at any time, even when the ignition is in the OFF position. Disconnect the negative battery cable when working in the vicinity of the fan.*

14  Unplug the oxygen sensor electrical connector.
15  Unscrew the oxygen sensor.
16  Apply a small amount of anti-seize compound to the threads of the new oxygen sensor and install it into the exhaust manifold.
17  Plug in the electrical connector.

## Manifold Absolute Pressure (MAP) sensor
### Check
18  Check the vacuum hose from the throttle body to the MAP sensor for cracking and general deterioration. Replace it if necessary.
19  Check the electrical connector at the sensor for a snug fit. Check the terminals in the connector and the wires leading to it for looseness and breaks. Repair as required.

### All Legend models and 1986 and 1987 Integra models
*Refer to illustrations 4.23a and 4.23b*

20  Turn the ignition switch to OFF.
21  Remove the ALTERNATOR SENSE fuse in the under hood relay box for 10 seconds on all Legends or the HAZARD fuse in the main fuse box on Integra (see Chapter 12) for ten seconds to reset the ECU.
22  Warm the engine to operating temperature and check to make sure the CHECK ENGINE warning light continues to flash.
23  Turn the ignition Off and disconnect the electrical connector from the MAP sensor **(see illustrations)**. Insert the test harness into the circuit.
**Note:** *The test harness can be purchased at a dealer parts department or you can fabricate your own.*
24  With the ignition switch ON, measure the voltage between the RED (+) terminal and the GRN (-) terminal. There should be about five volts.
25  Measure the voltage between the WHITE (+) terminal and the GREEN (-) terminal. There should be about three volts.
26  If the voltage readings check out okay, replace the MAP sensor.
27  If these tests don't pinpoint the problem, have the MAP sensor circuit diagnosed by a dealer service department or other repair shop, as special tools are required from this point.

### 1988 and 1989 Integra models
*Refer to illustration 4.31*

28  Turn the ignition switch to OFF.
29  Remove the HAZARD fuse in the main fuse box (see Chapter 12) for ten seconds to reset the ECU.
30  Warm the engine to operating temperature and check to make sure the CHECK ENGINE warning light continues to flash.
31  Turn the ignition Off and disconnect the 3P connector from the MAP sensor **(see illustration)**. Insert the test harness into the circuit.
32  With the ignition switch ON, measure the voltage between the RED (+) terminal and the GRN (-) terminal. There should be about five volts.
33  Measure the voltage between the WHITE (+) terminal and the GREEN (-) terminal. There should be about three volts.
34  Measure the voltage between the WHT (+) and BRN/WHT (-) terminals. There should be about five volts.
35  If the voltage readings check out okay, replace the MAP sensor.
36  If these tests don't pinpoint the problem, have the MAP sensor circuit diagnosed by a dealer service department or other repair shop, as special tools are required from this point.

### Replacement
37  Disconnect the electrical connector and the vacuum hose from the MAP sensor.
38  Remove the bolts that retain the sensor to the firewall and remove the MAP sensor.
39  Installation is the reverse of removal.

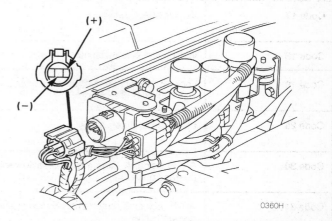

4.31  On 1988 and 1989 Integra models, the connector is located in the middle of the firewall

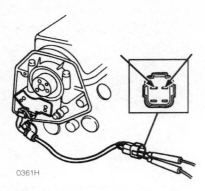

4.46 Measure the resistance between these two terminals in the connector for the CRANK circuit (Legend models)

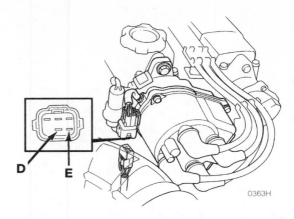

4.70 Pin designations in the connector for the CRANK circuit on 1988 and 1989 Integra models

## Crank angle sensor

40  On these models, the crank angle sensor is made up of two individual sensors – a CRANK/CYL sensor and a TDC sensor. The CRANK/CYL sensor detects the position of the number one piston as the base for the sequential fuel injection system. The TDC sensor determines the injector timing for each cylinder, as well as sending an engine speed signal to the ECU.
41  On Legend models, the CRANK/CYL sensor is located on the right end (passenger side) of the front cylinder head, just below the end of the camshaft. The TDC sensor is located inside the distributor. If the code indicates the TDC sensor is defective, have the distributor tested by a dealer service department or other repair shop.
42  On early Integra models (1986 and 1987), the CRANK/CYL sensor is a separate unit that is attached to the end of the cylinder head opposite the distributor while the TDC sensor is located inside the distributor. If the code indicates the TDC sensor is defective, have the distributor tested by a dealer service department or other repair shop. On late model Integras (1988 and 1989), both the TDC and CYL sensor are located inside the distributor. Refer to Chapter 5 for additional information on the ignition system.

### Legend models
#### CRANK circuit check
*Refer to illustration 4.46*

43  Turn the ignition switch OFF.
44  Remove the ALTERNATOR SENSE fuse in the under hood relay box for 10 seconds (see Chapter 12) to reset the ECU.
45  Start the engine and make sure the CHECK ENGINE light continues to flash.
46  Stop the engine and disconnect the electrical connector in the CRANK/CYL sensor harness **(see illustration)**.
47  Measure the resistance between terminals ORANGE/BLUE and WHITE/BLUE on 1986 and 1987 Legend or BLU/YEL and BLU/GRN on 1988 through 1990 Legend. The resistance should be between 500 and 1,000 ohms.
48  Check for continuity to ground from each of the previous designated terminals. Continuity should not exist.
49  If any of the tests are incorrect, replace the CRANK/CYL sensor with a new unit.
50  If the test results are correct and the problem still exists, have the system checked at a dealership service department.

#### CYL circuit check
51  Turn the ignition switch OFF.
52  Remove the ALTERNATOR SENSE fuse in the under hood relay box for 10 seconds (see Chapter 12) to reset the ECU.
53  Start the engine and make sure the CHECK ENGINE light continues to flash.
54  Stop the engine and disconnect the electrical connector in the CRANK/CYL sensor harness **(see illustration 4.46)**.
55  Measure the resistance between terminals WHT and ORN on all Legends. The resistance should be between 500 and 1,000 ohms.
56  Check for continuity to ground from each of the previously designated terminals. Continuity should not exist.
57  If any of the tests are incorrect, replace the CRANK/CYL sensor with a new unit.
58  If the test results are correct and the problem still exists, have the system checked at a dealership service department.

#### Replacement
59  Remove the cruise control actuator.
60  Remove the front upper timing belt cover (see Chapter 2B).
61  Remove the timing belt (see Chapter 2B).
62  Remove the three bolts and detach the pulley.
63  Remove the four bolts and remove the front upper cover back plate.
64  Unbolt the sensor from the head and install the new one.
65  Align the cam pulley pin with the camshaft hole and install the cam pulley.
66  The remainder of installation is the reverse of removal.

### 1988 and 1989 Integra models
#### Check and replacement
*Refer to illustration 4.70*

67  Turn the ignition switch OFF.
68  Remove the HAZARD fuse in the main fuse box (see Chapter 12) for ten seconds to reset the ECU.
69  Start the engine and make sure the CHECK ENGINE light continues to flash.
70  Stop the engine and disconnect the electrical connector in the CRANK/CYL sensor harness **(see illustration)**.
71  Measure the resistance between terminals D and E **(see illustration 4.70)**. The resistance should be between 350 and 550 ohms.
72  Check for continuity to ground from each of the previous designated terminals. Continuity should not exist.
73  If any of the tests are incorrect, replace the distributor with a new unit.
74  If the test results are correct and the problem still exists, have the system checked at a dealer service department or other repair shop.

### 1986 and 1987 Integra models
#### Check
*Refer to illustration 4.78*

75  Turn the ignition switch OFF.

# Chapter 6  Emissions control systems

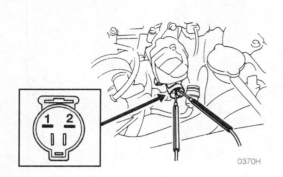

4.78  Pin designations in the connector for the crank angle sensor circuit on 1986 and 1987 Integra models

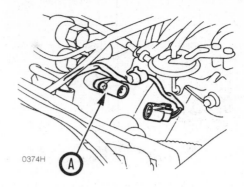

4.108  The TA sensor (A) is located in the intake manifold on all models

76  Remove the ALTERNATOR SENSE fuse in the under hood relay box for 10 seconds on all Legends or the HAZARD fuse in the main fuse box on Integra (see Chapter 12) for ten seconds to reset the ECU.
77  Start the engine and make sure the CHECK ENGINE light continues to flash.
78  Stop the engine and disconnect the electrical connector from the crank angle sensor (see illustration).
79  Measure the resistance between the RED and WHITE terminals (see illustration 4.78). The resistance should be between 650 and 850 ohms.
80  If the test is incorrect, replace the crank angle sensor with a new unit.
81  If the test results are correct and the problem still exists, have the system checked at a dealer service department or other repair shop.

### Replacement
82  Remove the bolts and pull the crank angle sensor from the cylinder head.
83  Install a new O-ring on the sensor housing. Lubricate the O-ring with a light coat of engine oil.
84  Installation is the reverse of removal.

## Throttle angle sensor
### Check
**All Legend models and 1986 and 1987 Integra models**
85  Turn the ignition switch to OFF.
86  Remove the ALTERNATOR SENSE fuse in the engine compartment relay box for 10 seconds on all Legends or the HAZARD fuse in the main fuse box on Integra (see Chapter 12) for ten seconds to reset the ECU.
87  Warm the engine to operating temperature and check to make sure the CHECK ENGINE warning light continues to flash.
88  Turn the ignition Off and disconnect the electrical connector from the throttle angle sensor (the sensor is mounted on the side of the throttle body). Install the test harness into the circuit. **Note:** *The test harness can be purchased at a dealer parts department or you can fabricate your own.*
89  With the ignition switch ON, measure the voltage between the RED (+) terminal and the GRN (-) terminal. There should be about five volts.
90  Measure the voltage between the WHITE (+) terminal and the GREEN (-) terminal. There should be about four volts.
91  If the voltage readings check out okay, replace the throttle angle sensor.
92  If these tests don't pinpoint the problem, have the throttle angle sensor circuit diagnosed by a dealer service department or other repair shop, as special tools are required from this point.

**1988 and 1989 Integra models**
93  Turn the ignition switch to OFF.
94  Remove the HAZARD fuse in the main fuse box on Integra (see Chapter 12) for ten seconds to reset the ECU.
95  Warm the engine to operating temperature and check to make sure the CHECK ENGINE warning light continues to flash.
96  Turn the ignition Off and disconnect the 3P connector from the throttle angle sensor.
97  With the ignition switch ON, measure the voltage between the RED (+) terminal and the GRN (-) terminal. There should be about five volts.
98  If not, measure the voltage between the RED (+) terminal and the body ground. There should NOT be voltage.
99  If there is, repair the open in the wire harness between the ECU and the throttle angle sensor.
100  If the voltage readings check out okay, replace the throttle angle sensor.
101  If these tests don't pinpoint the problem, have the throttle angle sensor circuit diagnosed by a dealer service department or other repair shop, as special tools are required from this point.

### Replacement (all models)
102  Disconnect the electrical connector from the sensor.
103  Remove the screws securing the sensor hold-down plate. If shear-head bolts are installed, it will be necessary to drill a small hole in the center of each screw and remove it with a screw extractor. This is most easily accomplished with the throttle body removed from the vehicle (see Chapter 4).
104  Remove the sensor from the throttle body.
105  Insert the switch (with a new gasket) into the throttle body, aligning the pin on the sensor with the groove in the throttle shaft.
106  Tighten the screws securely. If factory replacement shear-head bolts are being used, tighten them until the heads break off.
107  Reconnect the electrical connector.

## Intake air temperature (TA) sensor
### Check
*Refer to illustration 4.108*
108  With the ignition switch ON, disconnect the electrical connector from the TA sensor, which is located on the intake manifold **(see illustration)**. Using an ohmmeter, measure the resistance between the two terminals on the sensor. It should be between 1 and 4 ohms.
109  If the test results are incorrect, replace the TA sensor.
110  If the sensor checks out okay but there is still a problem, have the vehicle checked at a dealer service department or other qualified repair shop, as the ECU may be malfunctioning.

### Replacement
111  Unplug the electrical connector from the TA sensor.
112  Remove the screws that retain the sensor to the intake manifold and remove the TA sensor.
113  Installation is the reverse of removal.

# Chapter 6  Emissions control systems

## 5  Electronic Air/Idle Control Valve (EACV/EICV) – check and replacement

**Note:** *1988 and later models are equipped with an EACV, while 1986 and 1987 models are equipped with an EICV.*

### Check

*Refer to illustration 5.4*

1  Turn the ignition switch OFF.
2  Remove the HAZARD fuse in the main fuse box for 10 seconds to re-set the ECU.
3  Remove the ALTERNATOR SENSE fuse in the engine compartment relay box for 10 seconds on all Legends or the HAZARD fuse in the main fuse box on Integras (see Chapter 12) for ten seconds to reset the ECU.
4  Locate the EACV/EICV. The EACV or EICV is either a box shaped or cylinder shaped unit attached to the intake manifold near the throttle body **(see illustration)**. Measure the resistance with an ohmmeter between the two terminals. On the EACV, it should read between 8 and 15 ohms. On the EICV, it should read between 6 and 20 ohms.
5  Check for continuity to ground on each terminal. Continuity should not exist. If any of the results are incorrect, replace the EACV/EICV with a new unit.
6  Turn the ignition switch On. Measure the voltage on the wiring harness side of the connector, between the black/yellow (+) wire and the blue/yellow (-) wire (EACV) or the blue/red (-) wire (EICV).
7  If there is battery voltage, the ECU might be faulty. Have the vehicle checked at a dealer service department or other qualified repair shop.
8  If there is no voltage, measure the voltage between the black/yellow terminal and ground.
9  If there is no voltage, repair the open circuit in the black/yellow wire between the EACV and the #4 fuse or main relay (EACV) or between the EICV and main relay (EICV).
10  If there is voltage present, have the vehicle checked at a dealer service department or other qualified repair shop.

### Replacement

11  Disconnect the electrical connector from the EACV.
12  Remove the bolts that retain the valve to the intake manifold and remove the EACV/EICV from the engine.
13  Installation is the reverse of removal.

## 6  Positive Crankcase Ventilation (PCV) system

1  The Positive Crankcase Ventilation (PCV) system reduces hydrocarbon emissions by scavenging crankcase vapors. It does this by circulating fresh air from the air cleaner through the crankcase, where it mixes with blow-by gases and is then rerouted through a PCV valve to the intake manifold.
2  The main components of the PCV system are the PCV valve, a blow-by filter and the vacuum hoses connecting these two components with the engine.
3  To maintain idle quality, the PCV valve restricts the flow when the intake manifold vacuum is high. If abnormal operating conditions (such as piston ring problems) arise, the system is designed to allow excessive amounts of blow-by gases to flow back through the crankcase vent tube into the air cleaner to be consumed by normal combustion.
4  Checking and replacement of the PCV valve and filter is covered in Chapter 1.

## 7  Fuel evaporative control system

### General description

1  The fuel evaporative control system absorbs fuel vapors and, during

5.4  Mounting details of the EACV (arrow) on Legend models

engine operation, releases them into the engine intake where they mix with the incoming air-fuel mixture.
2  Every evaporative system employs a canister filled with activated charcoal to absorb fuel vapors. The means by which these vapors are controlled, however, varies considerably from one system to another. The following descriptions of typical systems for the models covered by this manual should provide you enough information to understand the system on your vehicle. **Note:** *The following descriptions are not intended as a specific description of the evaporative system on your particular vehicle. Rather, they are intended as a general description of a typical system used on fuel injected vehicles. Although the following components are most likely all used on your particular system, there may also be other devices, not included here, which are unique to your system. Check with the VECI label and the Vacuum Hose Routing Diagram under the hood.*

### Legend models

3  The fuel filler cap is fitted with a two-way valve as a safety device. The valve vents fuel vapors to the atmosphere if the evaporative control system fails.
4  Another fuel cut-off valve, mounted on the fuel tank, regulates fuel vapor flow from the fuel tank to the charcoal canister, based on the pressure or vacuum caused by temperature changes.
5  After passing through the two-way valve, fuel vapor is carried by vent hoses to the charcoal canister in the engine compartment. The activated charcoal in the canister absorbs and stores these vapors.
6  When the engine is running and warmed to a pre-set temperature, a thermo valve on top of the canister closes, allowing a purge control diaphragm valve in the charcoal canister to be opened by intake manifold vacuum. Fuel vapors from the canister are then drawn through the purge control diaphragm valve by intake manifold vacuum.

### Integra models

7  When fuel vapor pressure in the fuel tank exceeds a pre-set level, a two-way valve on the fuel tank opens and allows the fuel vapors to flow to the charcoal canister.
8  The charcoal canister temporarily stores fuel vapors until they can be purged from the charcoal canister into the engine and burned.
9  Canister purging is controlled by a purge control diaphragm valve which is opened or closed by a purge cut-off solenoid valve. When the engine coolant temperature is below about 149-degrees F, the purge cut-off solenoid valve provides no manifold vacuum to the purge control diaphragm valve. When the temperature exceeds 149-degrees F, the purge cut-off solenoid valve directs manifold vacuum to the purge control diaphragm valve, which admits ported vacuum to the canister and draws fresh air through the canister into a port on the throttle body. The purge cut-off solenoid valve is controlled by the ECU.

## Checking

**Note:** *Complete checking of the fuel evaporative control system is beyond the scope of the home mechanic. Fortunately, the evaporative control system, like all emission control systems, is protected by a Federally-mandated extended warranty (5 years or 50,000 miles at the time this manual was written). The fuel evaporative system probably won't fail during the service life of the vehicle; however, if it does, the hoses or charcoal canister are usually to blame.*

### Hoses

10  Always check the hoses first. A disconnected, damaged or missing hose is the most likely cause of a malfunctioning evaporative system. Refer to the Vacuum Hose Routing Diagram (attached to the underside of the hood) to determine whether the hoses are correctly routed and attached. Repair any damaged hoses or replace any missing hoses as necessary.

### Charcoal canister

11  Detach the canister from the firewall.
12  Detach the intake tube hose (refer to the Vacuum Hose Routing Diagram attached to the underside of the hood) and put your finger on the end of the canister inlet fitting.
13  Warm the engine to normal operating temperature, then increase the engine speed to 2500 rpm. If the canister is functioning correctly, air will be drawn into the canister through the inlet tube and you will feel suction. If you don't feel suction, replace the canister.

## 8  Air injection system (Legend models only)

### General information

*Refer to illustration 8.4*

1  The air injection system is designed to improve emissions performance by supplying fresh air from the air cleaner into the exhaust manifold by way of the air suction valve. When the air suction control solenoid valve is activated by the ECU, the manifold vacuum raises the diaphragm of the air suction valve and fresh air from the air cleaner is transferred to the exhaust manifold through the reed valve in the air suction valve by the pulsation of the exhaust gas. The air suction control solenoid valve is delayed at the first starting approximately 10 to 60 seconds depending on the coolant temperature.

### Check

2  Check all the vacuum lines for cracks, damage or disconnected hoses.
3  Warm the engine to operating temperature (the cooling fan comes on).
4  Stop the engine and then restart the engine. Within 10 seconds check for a distinct bubbling sound from the air suction valve **(see illustration)**.
5  If the bubbling noise is not heard, disconnect the number 7 vacuum hose from the top of the air suction valve and check for vacuum within 10 seconds of restarting.
6  If there is vacuum, replace the air suction valve and retest. If there is no vacuum, repair the circuit or replace the solenoid valve.

## 9  Catalytic converter

**Note:** *Because of a Federally mandated extended warranty which covers emissions-related components such as the catalytic converter, check with a dealer service department before replacing the converter at your own expense.*

### General description

1  The catalytic converter is an emission control device added to the exhaust system to reduce pollutants from the exhaust gas stream. There are two types of converters. The conventional oxidation catalyst reduces the levels of hydrocarbon (HC) and carbon monoxide (CO). The three-way catalyst lowers the levels of oxides of nitrogen (NOx) as well as hydrocarbons (HC) and carbon monoxide (CO).

### Checking

2  The test equipment for a catalytic converter is expensive and highly sophisticated. If you suspect that the converter on your vehicle is malfunctioning, take it to a dealer or authorized emissions inspection facility for diagnosis and repair.
3  Whenever the vehicle is raised for servicing of underbody components, check the converter for leaks, corrosion, dents and other damage. Check the welds/flange bolts that attach the front and rear ends of the converter to the exhaust system. If damage is discovered, the converter should be replaced.
4  Although catalytic converters don't break too often, they can become plugged. The easiest way to check for a restricted converter is to use a vacuum gauge to diagnose the effect of a blocked exhaust on intake vacuum.
  a) Open the throttle until the engine speed is about 2000 rpm.
  b) Release the throttle quickly.
  c) If there is no restriction, the gauge will quickly drop to not more than 2 in-Hg or more above its normal reading.
  d) If the gauge does not show 5 in-Hg or more above its normal reading, or seems to momentarily hover around its highest reading for a moment before it returns, the exhaust system, or the converter, is plugged (or an exhaust pipe is bent or dented, or the core inside the muffler has shifted).

### Component replacement

5  Refer to the exhaust system removal and installation section in Chapter 4.

## 10  Exhaust Gas Recirculation (EGR) system (Legend models only)

**Note:** *Integra models are not equipped with an EGR system.*

### General description

1  The EGR system reduces oxides of nitrogen by recirculating exhaust gas through the EGR valve and intake manifold into the combustion chambers.
2  The EGR system consists of the EGR valve, the CVC valve, the EGR control solenoid valve, the Electronic Control Unit (ECU) and various sensors. The ECU memory is programmed to produce the ideal EGR valve lift for each operating condition. An EGR valve lift sensor detects the amount of EGR valve lift and sends this information to the ECU. The ECU then compares it with the ideal EGR valve lift, which is determined by data received from the other sensors. If there's any difference between the two, the ECU triggers the EGR control solenoid valve to reduce the amount of vacuum applied to the EGR valve.

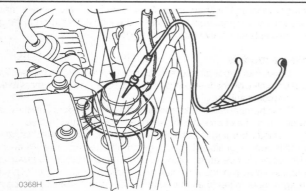

**8.4  Listen carefully for a bubbling sound from the air suction valve (if there is no outside interference, it's possible to hear it without the use of a mechanics stethoscope)**

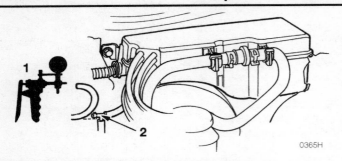

**10.6 Detach the no. 5 vacuum hose from the three-way joint (located near the control box on the firewall) and connect a vacuum gauge**

1  Vacuum pump/gauge    2  3-way joint

## Checking

### EGR valve

3  Start the engine and allow it to idle.
4  Detach the vacuum hose from the EGR valve and attach a hand vacuum pump in its place.
5  Apply vacuum to the EGR valve. Vacuum should remain steady and the engine should run poorly.
   a) If vacuum doesn't remain steady and the engine doesn't run poorly, replace the EGR valve and recheck it.
   b) If vacuum remains steady but the engine doesn't run poorly, remove the EGR valve and check the valve and the intake manifold for blockage. Clean or replace as necessary and recheck.

### EGR control system

*Refer to illustration 10.6*

6  Detach the number 5 vacuum hose from the three way joint and attach a vacuum gauge to the hose **(see illustration)**.
7  Start the engine and warm it to its normal operating temperature (wait for the electric cooling fan to come on).
8  There should be manifold vacuum.

**10.12 EGR valve mounting details**

9  Further checking of the EGR system requires special tools and equipment. Take the vehicle to a dealer service department or other qualified repair shop for checking.

## EGR valve replacement

*Refer to illustration 10.12*

10  If necessary, remove the air intake duct.
11  Unplug the electrical connector for the EGR valve lift sensor.
12  Remove the two nuts that secure the EGR valve **(see illustration)** and detach the EGR valve.
13  Clean the mating surfaces of the EGR valve and adapter.
14  Install the EGR valve, using a new gasket. Tighten the nuts securely.
15  Plug in the electrical connector.

# Chapter 7 Part A  Manual transaxle

## Contents

| | |
|---|---|
| General information ................................... 1 | Manual transaxle – removal and installation .................. 5 |
| Lubricant change ........................... See Chapter 1 | Shift assembly – removal and installation .................... 4 |
| Lubricant level check ....................... See Chapter 1 | Oil seal replacement ..................................... 2 |
| Manual transaxle overhaul – general information ............. 6 | Transaxle mount – check and replacement .................. 3 |

## Specifications

### Torque specifications                                              Ft-lbs
Shift rod fork bolt ........................................ 16
Torque rod front mounting bolt ........................... 7
Transaxle-to-engine bolts
   Integra .............................................. 42
   Legend .............................................. 55

## 1  General information

The vehicles covered by this manual are equipped with either a five-speed manual transaxle or a four-speed automatic transaxle. Information on the manual transaxle is included in this Part of Chapter 7. Service procedures for the automatic transaxle are contained in Chapter 7, Part B.

The manual transaxle is a compact, two-piece, lightweight aluminum alloy housing containing both the transmission and differential assemblies.

## 2  Oil seal replacement

*Refer to illustration 2.4*

1  Oil leaks frequently occur due to wear of the driveaxle oil seals. Replacement of these seals is relatively easy, since the repair can usually be performed without removing the transaxle from the vehicle.

2  The driveaxle oil seals are located at the sides of the transaxle, where the driveaxles are attached. If leakage at the seal is suspected, raise the vehicle and support it securely on jackstands. If the seal is leaking, lubri-

2.4  Insert the tip of a large screwdriver or prybar behind the oil seal and very carefully pry it out

3.1  Pry on the transaxle mount to check for movement

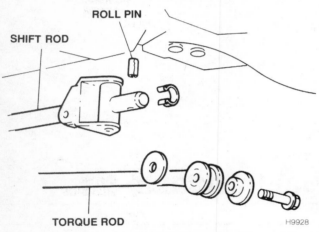

4.1a  On Integra models, remove the clip and drive out the roll pin to disconnect the shift rod from the transaxle

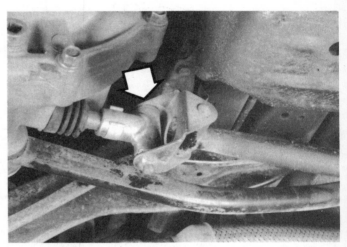

4.1b  On Legend models, remove the bolt (arrow) to disconnect the shift rod

cant will be found on the sides of the transaxle, below the seals.
3  Refer to Chapter 8 and remove the driveaxles.
4  Using screwdriver or prybar, carefully pry the oil seal out of the transaxle bore **(see illustration)**.
5  If the oil seal cannot be removed with a screwdriver or prybar, a special oil seal removal tool (available at auto parts stores) will be required.
6  Using a large section of pipe or a large deep socket (as large as the outside diameter of the seal) as a drift, install the new oil seal. Drive it into the bore squarely and make sure it's completely seated. Coat the seal lip with transaxle lubricant.
7  Install the driveaxle(s). Be careful not to damage the lip of the new seal.

### 3  Transaxle mount – check and replacement

*Refer to illustration 3.1*

1  Insert a large screwdriver or prybar between the mount and the transaxle and pry up **(see illustration)**.
2  The transaxle should not move more than about 1/2 to 3/4-inch away from the mount. If it does, replace the mount.
3  To replace the mount, support the transaxle with a jack, remove the nuts and bolts and remove the mount. It may be necessary to raise the

transaxle slightly to provide enough clearance to remove the mount.
4  Installation is the reverse of removal.

### 4  Shift assembly – removal and installation

*Refer to illustrations 4.1a, 4.1b and 4.1c*

1  From under the vehicle, remove the fasteners from the ends and lower the shift rod and torque rod from the vehicle **(see illustrations)**.
2  Inside the vehicle, remove the shift knob, console and shift lever boot (see Chapter 11).
3  Also from inside the vehicle, remove the screws from the shift lever plate and lift the gear shift lever assembly out.
4  Installation is the reverse of removal. Prior to installation, lubricate the contact surfaces of the assembly with general purpose grease.

### 5  Manual transaxle – removal and installation

*Refer to illustration 5.7*

#### Removal

1  Disconnect the negative cable from the battery. Shift the transaxle to Neutral.

## Chapter 7 Part A  Manual transaxle

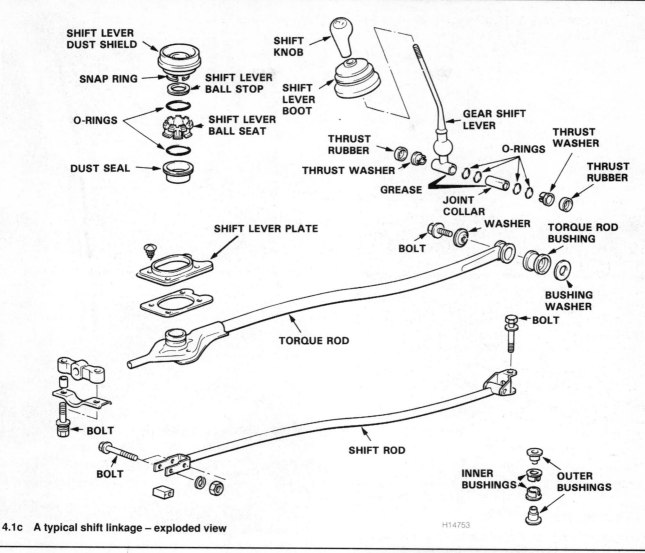

**4.1c  A typical shift linkage – exploded view**

2   Raise the vehicle and support it securely on jackstands.
3   Drain the transaxle lubricant (see Chapter 1).
4   If equipped, remove the belly pan and/or splash shield as necessary for access to the underside of the vehicle.
5   Disconnect the shift and torque rods at the transaxle (see Section 4). On some Legend models, it's also necessary to disconnect the torque rod bracket from the transaxle.
6   Disconnect the clutch linkage from the transaxle. On Legend models, remove the clutch slave cylinder and dampener assemblies (see Chapter 8).
7   Disconnect the speedometer cable and wire harness connectors from the transaxle **(see illustration)**.
8   On Legend models, remove the air cleaner assembly, complete with the air intake hose (see Chapter 4).
9   Remove the exhaust system components as necessary for clearance.
10  Support the engine. This can be done from above with an engine hoist, or by placing a jack (with a block of wood as an insulator) under the engine oil pan. The engine must remain supported at all times while the transaxle is out of the vehicle!
11  Remove the starter motor (see Chapter 5) and any chassis or suspension components that will interfere with transaxle removal (see Chapter 10).
12  Disconnect the driveaxles and remove the intermediate shaft from the transaxle (see Chapter 8).
13  Support the transaxle with a jack, preferably one designed for the purpose, then remove the bolts securing the transaxle to the engine.

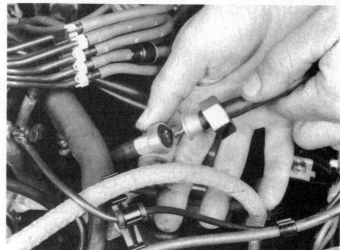

**5.7  On Legend models, disconnect the speedometer cable in the engine compartment**

14  Remove the transaxle mount nuts and bolts.
15  Make a final check that all wires and hoses have been disconnected

# 7A–4  Chapter 7 Part A  Manual transaxle

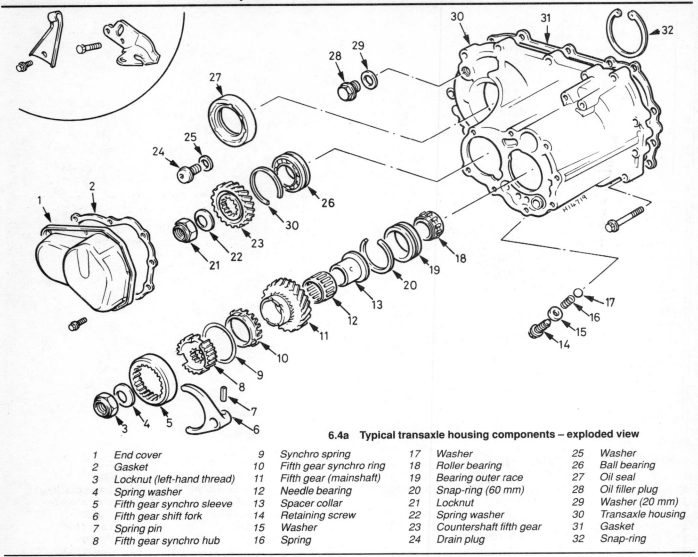

**6.4a  Typical transaxle housing components – exploded view**

| | | | | | |
|---|---|---|---|---|---|
| 1 | End cover | 9 | Synchro spring | 17 | Washer | 25 | Washer |
| 2 | Gasket | 10 | Fifth gear synchro ring | 18 | Roller bearing | 26 | Ball bearing |
| 3 | Locknut (left-hand thread) | 11 | Fifth gear (mainshaft) | 19 | Bearing outer race | 27 | Oil seal |
| 4 | Spring washer | 12 | Needle bearing | 20 | Snap-ring (60 mm) | 28 | Oil filler plug |
| 5 | Fifth gear synchro sleeve | 13 | Spacer collar | 21 | Locknut | 29 | Washer (20 mm) |
| 6 | Fifth gear shift fork | 14 | Retaining screw | 22 | Spring washer | 30 | Transaxle housing |
| 7 | Spring pin | 15 | Washer | 23 | Countershaft fifth gear | 31 | Gasket |
| 8 | Fifth gear synchro hub | 16 | Spring | 24 | Drain plug | 32 | Snap-ring |

from the transaxle, then carefully pull the transaxle and jack away from the engine.
16  Once the input shaft is clear, lower the transaxle and remove it from under the vehicle. **Caution:** *Do not depress the clutch pedal while the transaxle is out of the vehicle.*
17  With the transaxle removed, the clutch components are now accessible and can be inspected. In most cases, new clutch components should be routinely installed when the transaxle is removed (see Chapter 8).

## Installation

18  If removed, install the clutch components (see Chapter 8.)
19  With the transaxle secured to the jack with a chain, raise it into position behind the engine, then carefully slide it forward, engaging the dowel pins on the transaxle with the corresponding holes in the block and the input shaft with the clutch plate hub splines. Do not use excessive force to install the transaxle – if the input shaft does not slide into place, readjust the angle of the transaxle so it is level and/or turn the input shaft so the splines engage properly with the clutch plate hub.
20  Install the transaxle-to-engine bolts. Tighten the bolts to the torque listed in this Chapter's Specifications.
21  Install the transaxle mount nuts and/or bolts.
22  Install the starter and any chassis and suspension components which were removed. Tighten all nuts and bolts securely.
23  Remove the jacks supporting the transaxle and engine.

24  Install the various items removed previously, referring to Chapter 8 for installation of the driveaxles and intermediate shaft and Chapter 4 for information regarding the exhaust system components and air cleaner.
25  Make a final check that all wires, hoses, linkages and the speedometer cable have been connected and that the transaxle has been filled with lubricant to the proper level (Chapter 1).
26  Connect the negative battery cable. Road test the vehicle for proper operation and check for leaks.

## 6  Manual transaxle overhaul – general information

*Refer to illustrations 6.4a, 6.4b, 6.4c, 6.4d and 6.4e*

Overhauling a manual transaxle is a difficult job for the do-it-yourselfer. It involves the disassembly and reassembly of many small parts. Numerous clearances must be precisely measured and, if necessary, changed with select fit spacers and snap-rings. As a result, if transaxle problems arise, it can be removed and installed by a competent do-it-yourselfer, but overhaul should be left to a transmission repair shop. Rebuilt transaxles may be available – check with your dealer parts department and auto parts stores. At any rate, the time and money involved in an overhaul is almost sure to exceed the cost of a rebuilt unit.

Nevertheless, it's not impossible for an inexperienced mechanic to rebuild a transaxle if the special tools are available and the job is done in a

# Chapter 7 Part A  Manual transaxle

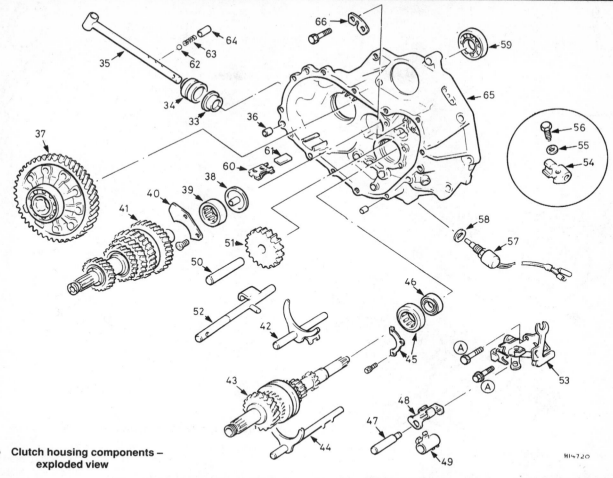

**6.4b  Clutch housing components – exploded view**

| | | | | | | | |
|---|---|---|---|---|---|---|---|
| 33 | Oil seal | 42 | First and second gearshift shaft | 50 | Reverse idler gear shaft | 59 | Differential bearing |
| 34 | Boot | 43 | Mainshaft | 51 | Reverse idler gear | 60 | Magnet holder |
| 35 | Shift rod | 44 | Third and fourth gearshift shaft | 52 | Fifth and reverse gearshift shaft | 61 | Magnet |
| 36 | Dowel pin | 45 | Mainshaft bearing and bearing retainer plate | 53 | Shift arm holder | 62 | Detent ball |
| 37 | Differential assembly | | | 54 | Shift rod guide | 63 | Spring |
| 38 | Oil barrier plate | 46 | Oil seal | 55 | Washer | 64 | Spring collar |
| 39 | Countershaft bearing | 47 | Shift guide shaft | 56 | Bolt | 65 | Clutch housing |
| 40 | Bearing retainer plate | 48 | Interlock | 57 | Backup light switch | 66 | Breather chamber plate |
| 41 | Countershaft assembly | 49 | Shift guide | 58 | Washer | | |

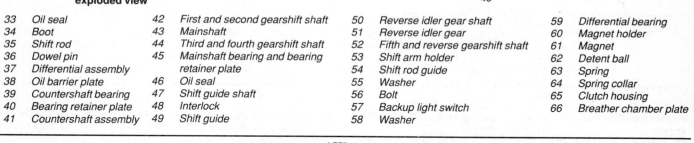

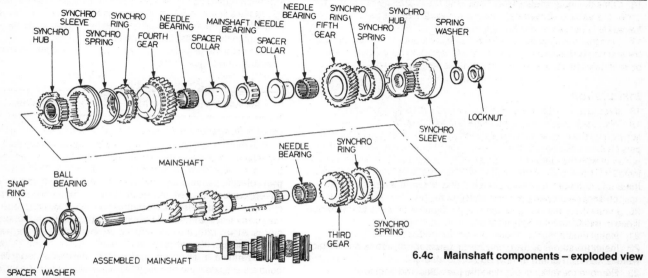

**6.4c  Mainshaft components – exploded view**

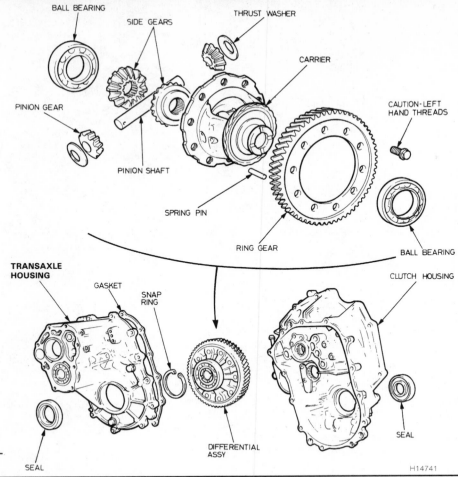

6.4d Differential components – exploded view

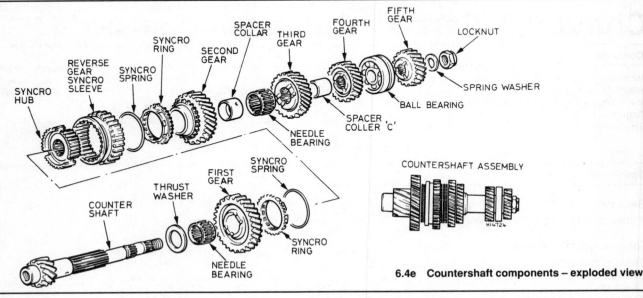

6.4e Countershaft components – exploded view

deliberate step-by-step manner so nothing is overlooked.

The tools necessary for an overhaul include internal and external snap-ring pliers, a bearing puller, a slide hammer, a set of pin punches, a dial indicator and possibly a hydraulic press. In addition, a large, sturdy workbench and a vise or transaxle stand will be required.

During disassembly of the transaxle, make careful notes of how each piece comes off, where it fits in relation to other pieces and what holds it in place. Exploded views are included (see illustrations) to show where the parts go – but actually noting how they are installed when you remove the parts will make it much easier to get the transaxle back together.

Before taking the transaxle apart for repair, it will help if you have some idea what area of the transaxle is malfunctioning. Certain problems can be closely tied to specific areas in the transaxle, which can make component examination and replacement easier. Refer to the Troubleshooting section at the front of this manual for information regarding possible sources of trouble.

# Chapter 7 Part B  Automatic transaxle

## Contents

| | |
|---|---|
| Automatic transaxle fluid change .................. See Chapter 1 | Neutral start switch – check and replacement ................. 4 |
| Automatic transaxle fluid level check ............. See Chapter 1 | Oil seal replacement ........................ See Chapter 7A |
| Automatic transaxle – removal and installation ................ 6 | Shift cable – check and adjustment ....................... 3 |
| Diagnosis – general ....................................... 2 | Throttle Valve (TV) cable – adjustment .................... 5 |
| General information ...................................... 1 | Transaxle mount – check and replacement ....... See Chapter 7A |

## Specifications

### General

Throttle Valve (TV) cable freeplay at transaxle lever ........... 5/32 to 11/64 in (2 to 4 mm)

### Torque specifications                                         Ft-lbs

Transaxle-to-engine bolts
    Integra ............................................... 42
    Legend ............................................... 55
Torque converter-to-driveplate bolts ...................... 9

## 1  General information

All vehicles covered in this manual come equipped with either a 5-speed manual transaxle or a 4-speed automatic transaxle. All information on the automatic transaxle is included in this Part of Chapter 7. Information for the manual transaxle can be found in Part A of this Chapter.

Due to the complexity of the automatic transaxles covered in this manual and the need for specialized equipment to perform most service operations, this Chapter contains only general diagnosis, routine maintenance, adjustment and removal and installation procedures.

# 7B-2   Chapter 7 Part B   Automatic transaxle

If the transaxle requires major repair work, it should be left to a dealer service department or an automotive or transmission repair shop. You can, however, remove and install the transaxle yourself and save the expense, even if the repair work is done by a transmission shop.

## 2   Diagnosis – general

**Note:** *Automatic transaxle malfunctions may be caused by five general conditions: poor engine performance, improper adjustments, hydraulic malfunctions, mechanical malfunctions or malfunctions in the computer or its signal network. Diagnosis of these problems should always begin with a check of the easily repaired items: Fluid level and condition (see Chapter 1), shift linkage adjustment and throttle valve (TV) cable adjustment. Next, perform a road test to determine if the problem has been corrected or if more diagnosis is necessary. If the problem persists after the preliminary tests and corrections are completed, additional diagnosis should be done by a dealer service department or transmission repair shop. Refer to the Troubleshooting section at the front of this manual for transaxle problem diagnosis.*

### *Preliminary checks*

1   Drive the vehicle to warm the transaxle to normal operating temperature.
2   Check the fluid level as described in Chapter 1:
   a) If the fluid level is unusually low, add enough fluid to bring the level within the designated area of the dipstick, then check for external leaks.
   b) If the fluid level is abnormally high, drain off the excess, then check the drained fluid for contamination by coolant. The presence of engine coolant in the automatic transmission fluid indicates that a failure has occurred in the internal radiator walls that separate the coolant from the transmission fluid (see Chapter 3).
   c) If the fluid is foaming, drain it and refill the transaxle, then check for coolant in the fluid or a high fluid level.
3   Check the engine idle speed. **Note:** *If the engine is malfunctioning, do not proceed with the preliminary checks until it has been repaired and runs normally.*
4   Check the throttle valve (TV) cable for freedom of movement. Adjust it if necessary (see Section 5). **Note:** *The throttle valve cable may function properly when the engine is shut off and cold, but it may malfunction once the engine is hot. Check it cold and at normal engine operating temperature.*
5   Inspect the shift linkage (see Section 3). Make sure that it's properly adjusted and that the linkage operates smoothly.

### *Fluid leak diagnosis*

6   Most fluid leaks are easy to locate visually. Repair usually consists of replacing a seal or gasket. If a leak is difficult to find, the following procedure may help.
7   Identify the fluid. Make sure it's transmission fluid and not engine oil or brake fluid (automatic transmission fluid is a deep red color).
8   Try to pinpoint the source of the leak. Drive the vehicle several miles, then park it over a large sheet of cardboard. After a minute or two, you should be able to locate the leak by determining the source of the fluid dripping onto the cardboard.
9   Make a careful visual inspection of the suspected component and the area immediately around it. Pay particular attention to gasket mating surfaces. A mirror is often helpful for finding leaks in areas that are hard to see.
10   If the leak still cannot be found, clean the suspected area thoroughly with a degreaser or solvent, then dry it.
11   Drive the vehicle for several miles at normal operating temperature and varying speeds. After driving the vehicle, visually inspect the suspected component again.
12   Once the leak has been located, the cause must be determined before it can be properly repaired. If a gasket is replaced but the sealing flange is bent, the new gasket will not stop the leak. The bent flange must be straightened.
13   Before attempting to repair a leak, check to make sure that the following conditions are corrected or they may cause another leak. **Note:** *Some of the following conditions cannot be fixed without highly specialized tools and expertise. Such problems must be referred to a transmission shop or a dealer service department.*

**Gasket leaks**

14   Check the end cover periodically. Make sure the bolts are tight, no bolts are missing and the gasket is in good condition.
15   If the end cover gasket is leaking, the fluid level or the fluid pressure may be too high, the vent may be plugged, the end cover bolts may be too tight, the sealing surface of the end cover or transaxle housing may be damaged, the gasket may be damaged or the transaxle casting may be cracked or porous. If sealant instead of gasket material has been used to form a seal between the end cover and the transaxle housing, it may be the wrong sealant.

**Seal leaks**

16   If a transaxle seal is leaking, the fluid level or pressure may be too high, the vent may be plugged, the seal bore may be damaged, the seal itself may be damaged or improperly installed, the surface of the shaft protruding through the seal may be damaged or a loose bearing may be causing excessive shaft movement.
17   Make sure the dipstick guide pipe seal is in good condition and the pipe is properly seated. Periodically check the area around the speedometer gear or sensor for leakage. If transmission fluid is evident, check the O-ring for damage. Also inspect the side gear shaft oil seals for leakage.

**Case leaks**

18   If the case itself appears to be leaking, the casting is porous and will have to be repaired or replaced.
19   Make sure the oil cooler hose fittings are tight and in good condition.

**Fluid comes out vent pipe or guide pipe**

20   If this condition occurs, the transaxle is overfilled, there is coolant in the fluid, the case is porous, the dipstick is incorrect, the vent is plugged or the drain back holes are plugged.

## 3   Shift cable – check and adjustment

*Refer to illustrations 3.3 and 3.4*

### *Check*

1   Try to start the engine in each shift lever position; the starter should operate in Park and Neutral only. If the starter does not operate in Park or Neutral or operates in any position other than Park and Neutral, the shift cable is in need of adjustment or the Neutral start switch is defective (see Section 4).

### *Adjustment*

2   Remove the center console (see Chapter 11).
3   Place the shift lever in Drive (Integra) or Reverse (Legend) and remove the lock pin from the adjuster **(see illustration)**.
4   Make sure the hole in the adjuster is perfectly aligned with hole in the shift cable **(see illustration)**. There are two holes in the end of the shift cable, positioned 90-degrees apart, allowing cable adjustment in 1/4-turn increments.
5   If the cable holes are not aligned perfectly, loosen the locknut on the adjuster **(see illustration 3.3)** and turn the adjuster until the hole in the adjuster is perfectly aligned with the hole in the shift cable.
6   Tighten the locknut and install the lock pin. If the pin binds when you install it, readjust the cable.
7   Try to start the engine in each shift lever position; the starter should operate in the Park and Neutral only. If it doesn't, the neutral start switch is bad (see the next Section) or the shift cable needs readjustment.
8   Reinstall the center console (see Chapter 11).

# Chapter 7 Part B  Automatic transaxle

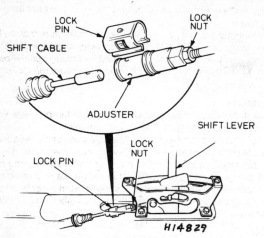

3.3  Shift cable connection details

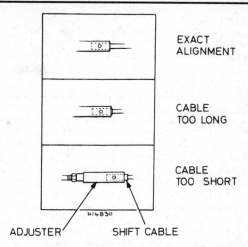

3.4  The shift cable is properly adjusted when the holes in the adjuster and shift cable are lined up

## 4  Neutral start switch – check and replacement

Refer to illustrations 4.2a and 4.2b

1  Try to start the engine in each shift lever position; the starter should operate in Park and Neutral only. If the starter does not operate, or operates in any position other than Park or Neutral, first check the shift cable adjustment (see Section 3). If the shift cable is properly adjusted and the problem persists, check the Neutral start switch which, on these models, incorporates the backup light switch as well.

2  Remove the center console (see Chapter 11) and unplug the connector from the Neutral start switch. Check the connector for continuity with the shift lever in each position (see illustrations).

### Neutral start switch

| SWITCH POSITION | TERMINAL | |
|---|---|---|
| | Bl/w | Bl/w |
| N | ● | ● |
| R | | |
| P | ● | ● |

### Back-up switch

| SWITCH POSITION | TERMINAL | |
|---|---|---|
| | Y | G/Bl |
| N | | |
| R | ● | ● |
| P | | |

### Shift position switch

| SWITCH POSITION | TERMINAL | |
|---|---|---|
| | Bl | P |
| 2 | ● | ● |
| D₃ | ● | ● |
| D4 | ● | ● |
| N | | |
| R | | |
| P | | |

### Shift lever position switch

| SWITCH POSITION | TERMINAL | | | | | | |
|---|---|---|---|---|---|---|---|
| | Bl | G/Y | G/BU | G/Bl | G | G/R | G/W |
| 2 | ● | ● | | | | | |
| D₃ | ● | | ● | | | | |
| D₂ | ● | | | ● | | | |
| N | ● | | | | ● | | |
| R | ● | | | | | ● | |
| P | ● | | | | | | ● |

4.2a  Terminal guide and continuity tables for the Neutral start switch (Integra models)

## 7B–4  Chapter 7 Part B  Automatic transaxle

### Shift position switch (for cruise control)

| SWITCH POSITION | TERMINAL | |
|---|---|---|
| | 1 | 2 |
| 2 | ● | ● |
| S | ● | ● |
| D | ● | ● |

### Back-up light switch

| SWITCH POSITION | TERMINAL | |
|---|---|---|
| | 4 | 5 |
| R | ● | ● |

### Neutral start switch

| SWITCH POSITION | TERMINAL | |
|---|---|---|
| | 11 | 12 |
| 2 | | |
| S | | |
| D | | |
| N | ● | ● |
| R | | |
| P | ● | ● |

### Indicator switch

| SWITCH POSITION | TERMINAL | | | | | | |
|---|---|---|---|---|---|---|---|
| | 1 | 9 | 8 | 3 | 10 | 6 | 7 |
| 2 | ● | ● | | | | | |
| S | ● | | ● | | | | |
| D | ● | | | ● | | | |
| N | ● | | | | ● | | |
| R | ● | | | | | ● | |
| P | ● | | | | | | ● |

**4.2b  Terminal guide and continuity tables for the Neutral start switch (Legend models)**

3  On 1989 and later Legend models, move the shift lever back and forth slightly while making the check. If there is no continuity while moving the lever within it's freeplay range, the position of the switch can be adjusted. Place the shift lever in Park, loosen the bolts and move the switch toward the Drive detent until there is continuity between the number 1 and 7 terminals.

4  On all models, if the continuity is not as specified, replace the switch. Do this by removing the two retaining bolts and detaching the switch from the gearshift selector mounting bracket. On Legends, place the shift lever in Neutral and move the slider on the switch so it will line up with the shift lever pin, then place the switch over the pin and install the bolts. Installation is the reverse of removal.

5  On Legend models, when installing a new switch, place the switch slider and shift lever in Neutral, place the switch in place over the actuator pin and tighten the bolts securely **(see illustration 4.4b)**.

## 5  Throttle Valve (TV) cable – adjustment

*Refer to illustrations 5.2 and 5.5*

1  Before beginning this procedure, the engine must be at normal operating temperature (the cooling fan must come on at least twice), the idle speed must be correct (see Chapter 1) and the throttle cable must be properly adjusted.

2  Loosen locknuts A and B on the TV cable at the bracket **(see illustration)**.

3  Push down on the TV cable lever on the transaxle, hold it down and

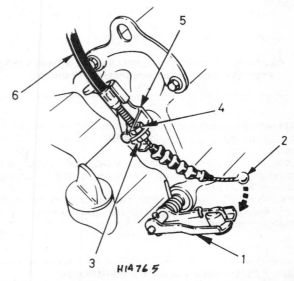

**5.2  TV cable details – to remove the cable, loosen locknut B, slide the TV cable out of the bracket, then slide the TV cable end out of the TV control lever**

1  TV control lever
2  TV cable end
3  Locknut B
4  Locknut A
5  TV cable bracket
6  TV cable

## Chapter 7 Part B  Automatic transaxle

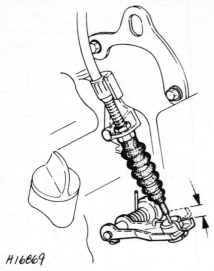

5.5  Measure the TV lever freeplay at the point shown

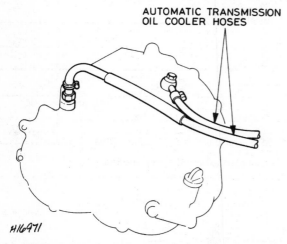

6.8  Detach the oil cooler hoses, plug them to prevent leakage and contamination and wire them out of the way

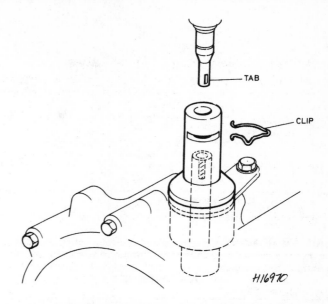

6.6  To disconnect the speedometer cable, remove the retaining clip and pull the cable out of the housing; don't remove the housing itself (Integra model shown)

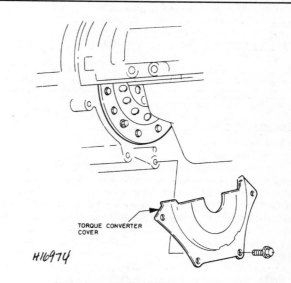

6.14  Remove the torque converter cover

check the TV cable freeplay at the fuel injection unit throttle link. If there is any freeplay, remove it by turning locknut A until none can be felt at the throttle link.
4    Continue to the press down on the TV cable lever and pull on the throttle link at the fuel injection unit. The throttle link and TV cable lever should move at precisely the same time.
5    Have an assistant depress the throttle pedal and hold it to the floor. Check the freeplay at the control lever **(see illustration)**. If it's not as specified, readjust the cable.

6    **Automatic transaxle – removal and installation**

*Refer to illustrations 6.6, 6.8, 6.14, 6.15, 6.17a, 6.17b and 6.17c*

### Removal
1    Disconnect the negative cable from the battery.
2    Raise the vehicle and support it securely on jackstands.
3    Drain the transaxle fluid (see Chapter 1).
4    Remove the starter motor (see Chapter 5).
5    Disconnect the driveaxles from the transaxle (see Chapter 8).
6    Disconnect the speedometer cable **(see illustration)**.
7    Disconnect the wire harness from the transaxle.
8    Disconnect and plug the oil cooler hoses **(see illustration)**.
9    Remove any exhaust components which will interfere with transaxle removal (see Chapter 4).
10   Disconnect the TV cable (see Section 5).
11   Disconnect the shift cable (see Section 3).
12   Support the engine using a hoist from above or a jack and a block of wood under the oil pan to spread the load.
13   Support the transaxle with a jack – preferably a special jack made for this purpose. Safety chains will help steady the transaxle on the jack.
14   Remove the torque converter cover **(see illustration)**.

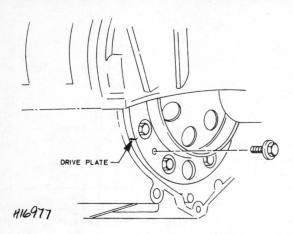

6.15 Before removing the driveplate-to-torque converter bolts, mark the edge of the driveplate and torque converter to ensure that they're reattached in exactly the same relationship when the transaxle is reinstalled

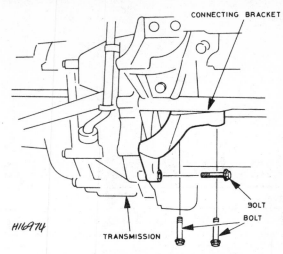

6.17a Typical Integra front mount details

15  Mark the relationship of the torque converter to the driveplate with white paint so they can be installed in the same position **(see illustration)**.
16  Remove the torque converter-to-driveplate bolts **(see illustration 6.15)**. Turn the crankshaft pulley bolt for access to each bolt.
17  Remove the transaxle mount nuts and bolts **(see illustrations)**.
18  Remove the bolts securing the transaxle to the engine.
19  Remove any chassis or suspension components which will interfere with transaxle removal.
20  Move the transaxle back to disengage it from the engine block dowel pins and make sure the torque converter is detached from the driveplate. Secure the torque converter to the transaxle so it will not fall out during removal. Lower the transaxle from the vehicle.

### Installation

21  Prior to installation, make sure that the torque converter hub is securely engaged in the pump.
22  With the transaxle secured to the jack, raise it into position. Be sure to keep it level so the torque converter does not slide out.
23  Turn the torque converter to line it up with the driveplate. The white paint mark on the torque converter and the driveplate in Step 15 must line up.
24  Move the transaxle forward carefully until the dowel pins and the torque converter are engaged.
25  Install the transaxle housing-to-engine bolts. Tighten them to the torque listed in the Specifications Section at the beginning of this Chapter.
26  Install the torque converter-to-driveplate bolts. Tighten the bolts to the torque listed in the Specifications Section at the beginning of this Chapter.
27  Install any suspension and chassis components which were removed. Tighten the bolts and nuts securely.
28  Remove the jacks supporting the transaxle and the engine.
29  Install the starter motor (see Chapter 5).
30  Connect the shift and TV cables.
31  Plug in the transaxle electrical connectors.
32  Install the torque converter cover.
33  Connect the driveaxles (see Chapter 8).
34  Connect the speedometer cable.
35  Connect the fluid cooler lines.
36  Adjust the shift linkage (see Section 3).
37  Install any exhaust system components that were removed or disconnected.
38  Lower the vehicle.
39  Fill the transaxle (see Chapter 1), run the vehicle and check for fluid leaks.

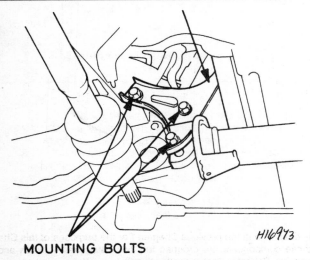

6.17b Typical Integra rear mount details

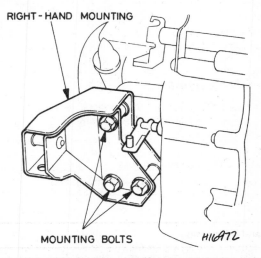

6.17c Typical Integra side mount details

# Chapter 8  Clutch and driveaxles

## Contents

| | |
|---|---|
| Clutch cable – removal, installation and adjustment | 3 |
| Clutch components – removal, inspection and installation | 9 |
| Clutch – description and check | 2 |
| Clutch hydraulic system – bleeding | 6 |
| Clutch master cylinder – removal, overhaul and installation | 4 |
| Clutch pedal – adjustment (Legend models) | 7 |
| Clutch release bearing and fork – removal, inspection and installation | 8 |
| Clutch release cylinder – removal, overhaul and installation | 5 |
| Driveaxle boot replacement and constant velocity (CV) joint overhaul | 13 |
| Driveaxles – removal and installation | 11 |
| Flywheel – removal and installation | See Chapter 2 |
| General information | 1 |
| Intermediate shaft – removal and installation | 12 |
| Starter/clutch interlock switch – check, replacement and adjustment | 10 |
| Transaxle lubricant level check | See Chapter 1 |

## Specifications

### General

Clutch pedal disengagement height
    Integra ................................................ 2-3/16 in
    Legend ................................................ 2-13/16 in
Clutch pedal freeplay
    Integra ................................................ 1/16 to 5/64 in
    Legend ................................................ 3/64 to 9/32 in
Clutch pedal standard height
    Integra ................................................ 5-43/64 in
    Legend ................................................ 7 in
Clutch pedal stroke
    Integra ................................................ 5-5/16 to 5-1/2 in
    Legend ................................................ 5-11/16 to 6 in
Release arm freeplay (Integra models) ................................................ 5/32 to 13/64 in

### Driveaxles

Driveaxle length
    Integra ................................................ 18-51/64 to 18-63/64 in
    Legend ................................................ 21 to 21-13/64 in

### Torque specifications                                 Ft-lbs

Clutch pressure plate bolts ................................................ 19
Release fork bolt ................................................ 22
Driveaxle/hub nut
    Integra ................................................ 134
    Legend ................................................ 180
Intermediate shaft bearing support bolts
    Integra ................................................ 29
    Legend ................................................ 16

## 1  General information

The information in this Chapter deals with the components from the rear of the engine to the the front wheels, except for the transaxle, which is dealt with in the previous Chapter. For the purposes of this Chapter, these components are grouped into two categories – clutch and driveaxles. Separate Sections within this Chapter offer general descriptions and checking procedures for components in each of the two groups.

# Chapter 8  Clutch and driveaxles

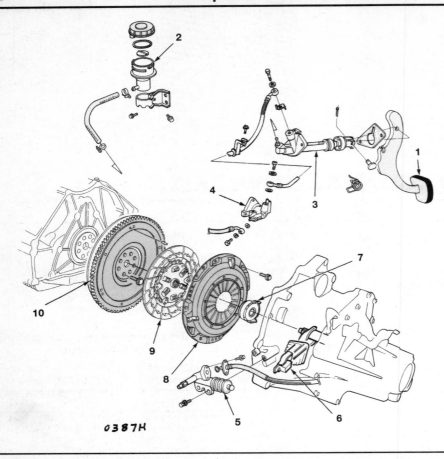

**2.2a  Exploded view of the clutch components (Legend models)**

1. Clutch pedal
2. Reservoir
3. Clutch master cylinder
4. Clutch damper
5. Release cylinder
6. Release fork
7. Release bearing
8. Pressure plate
9. Clutch disc
10. Flywheel

Since nearly all the procedures covered in this Chapter involve working under the vehicle, make sure it's securely supported on sturdy jackstands or on a hoist where the vehicle can be easily raised and lowered.

## 2  Clutch – description and check

*Refer to illustrations 2.2a and 2.2b*

1  All vehicles with a manual transmission use a single dry-plate, diaphragm-spring type clutch. The clutch disc has a splined hub which allows it to slide along the splines of the transmission input shaft. The clutch and pressure plate are held in contact by spring pressure exerted by the diaphragm in the pressure plate.

2  On Legend models, the clutch release system is operated by hydraulic pressure; on Integra models a mechanical system is used **(see illustrations)**. The hydraulic release system consists of the clutch pedal, a master cylinder and fluid reservoir, the hydraulic line, a release (or slave) cylinder which actuates the clutch release lever and the clutch release (or throwout) bearing. The mechanical release system includes the clutch pedal, a clutch cable which actuates the clutch release lever and the release bearing.

3  When pressure is applied to the clutch pedal to release the clutch, hydraulic or mechanical pressure is exerted against the outer end of the clutch release lever. As the lever pivots the shaft fingers push against the release bearing. The bearing pushes against the fingers of the diaphragm spring of the pressure plate assembly, which in turn releases the clutch plate.

4  Terminology can be a problem when discussing the clutch components because common names are in some cases different from those used by the manufacturer. For example, the driven plate is also called the clutch plate or disc, the clutch release bearing is sometimes called a throwout bearing, the release cylinder is sometimes called the operating or slave cylinder.

5  Other than to replace components with obvious damage, some preliminary checks should be performed to diagnose clutch problems. These checks assume that the transaxle is in good working condition.
   a) The first check should be of the fluid level in the clutch master cylinder (vehicles with a hydraulic release system). If the fluid level is low, add fluid as necessary and inspect the hydraulic system for leaks. If the master cylinder reservoir has run dry, bleed the system as described in Section 6 and retest the clutch operation.
   b) To check "clutch spin down time," run the engine at normal idle speed with the transmission in Neutral (clutch pedal up – engaged). Disengage the clutch (pedal down), wait several seconds and shift the transmission into Reverse. No grinding noise should be heard. A grinding noise would most likely indicate a problem in the pressure plate or the clutch disc.
   c) To check for complete clutch release, run the engine (with the parking brake applied to prevent movement) and hold the clutch pedal approximately 1/2-inch from the floor. Shift the transmission between 1st gear and Reverse several times. If the shift is rough, component failure is indicated. On vehicles with a hydraulic release system, check the release cylinder pushrod travel. With the clutch pedal depressed completely, the release cylinder pushrod should extend substantially. If it doesn't, check the fluid level in the clutch master cylinder. On vehicles with mechanical release systems, check the adjustment of the clutch cable.
   d) Visually inspect the pivot bushing at the top of the clutch pedal to make sure there is no binding or excessive play.
   e) On vehicles with mechanical release systems, a clutch pedal that is difficult to operate is most likely caused by a faulty clutch cable. Check the cable where it enters the housing for frayed wires, rust and other signs of corrosion. If it looks good, lubricate the cable with penetrating oil. If pedal operation improves, the cable is worn out and should be replaced.
   f) Crawl under the vehicle and make sure the clutch release lever is solidly mounted on the ball stud (Legend models only).

# Chapter 8 Clutch and driveaxles

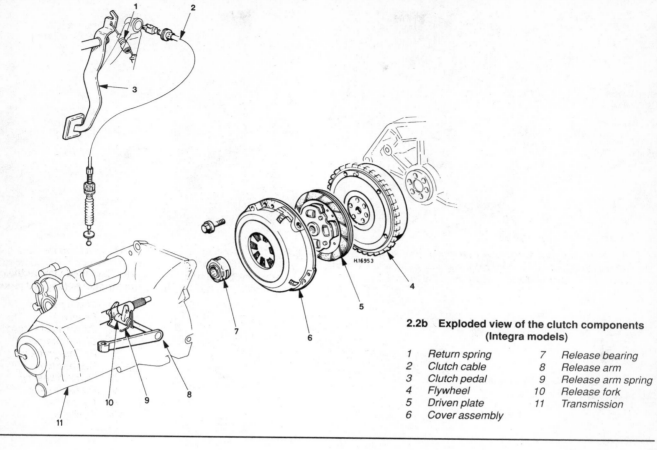

**2.2b Exploded view of the clutch components (Integra models)**

1. Return spring
2. Clutch cable
3. Clutch pedal
4. Flywheel
5. Driven plate
6. Cover assembly
7. Release bearing
8. Release arm
9. Release arm spring
10. Release fork
11. Transmission

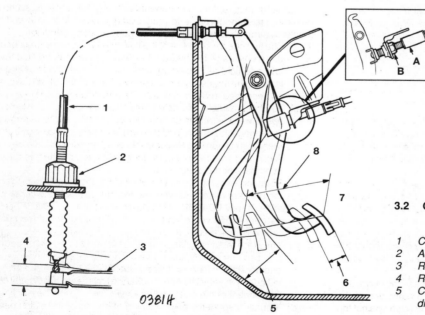

**3.2 Clutch cable and pedal adjustment details (Integra models)**

1. Clutch cable
2. Adjusting nut
3. Release arm
4. Release arm freeplay
5. Clutch pedal disengagement height
6. Clutch pedal freeplay
7. Clutch pedal standard height
8. Clutch pedal stroke

## 3 Clutch cable – removal, installation and adjustment

### Removal and installation

*Refer to illustration 3.2*

1. Detach the cable from the negative battery terminal.
2. Locate the adjusting nut for the clutch cable (**see illustration**) at the release arm on the front of the transaxle bellhousing. Loosen this nut by turning it counterclockwise.
3. Disconnect the cable from the release arm.
4. Detach the cable from the support bracket.
5. Detach any cable clips in the engine compartment.
6. From inside the vehicle, disconnect the cable from the clutch pedal.

# 8-4  Chapter 8  Clutch and driveaxles

4.4a  Pinch off the fluid feed hose with a pair of locking pliers to prevent the fluid in the reservoir from running out the lower end of the hose when you disconnect it from the master cylinder

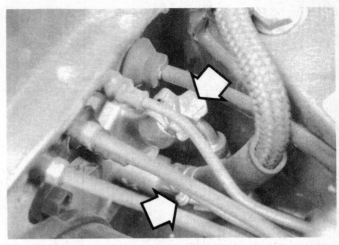

4.4b  To disconnect the clutch fluid hoses from the clutch master cylinder, loosen the hose clamp (arrow) on the feed hose and remove the banjo bolt from the pressure hose, then remove the two mounting nuts (arrows) and remove the master cylinder

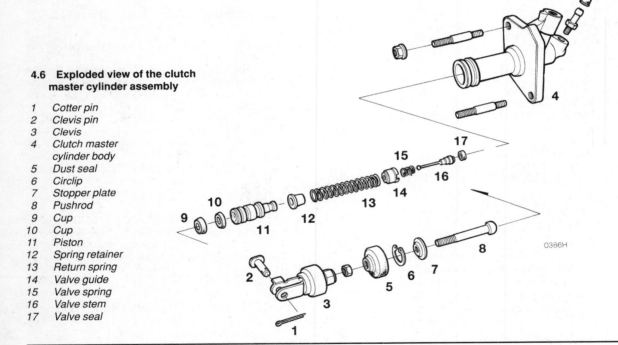

4.6  Exploded view of the clutch master cylinder assembly

1. Cotter pin
2. Clevis pin
3. Clevis
4. Clutch master cylinder body
5. Dust seal
6. Circlip
7. Stopper plate
8. Pushrod
9. Cup
10. Cup
11. Piston
12. Spring retainer
13. Return spring
14. Valve guide
15. Valve spring
16. Valve stem
17. Valve seal

7  Pull the cable assembly through the firewall from the engine compartment side.

8  Installation is the reverse of removal. Be sure to lubricate the clutch pedal end of the cable with moly-based grease. Adjust the cable (see Step 9).

## Adjustment

9  Measure the freeplay of the clutch pedal.

10  Adjust the freeplay by turning the adjusting nut until the freeplay is as listed in this Chapter's Specifications.

11  Verify that there's 5/32 to 13/64-inch freeplay at the tip of the release arm after the adjustment.

12  If the vehicle is equipped with cruise control, turn the adjuster (A) until the pedal stroke is 5-5/16 to 5-1/2 inches, then tighten the locknut (B) **(see illustration 3.2).**

## 4  Clutch master cylinder – removal, overhaul and installation

**Note:** *Before beginning this procedure, contact local parts stores and dealer service departments concerning the purchase of a rebuild kit or a new master cylinder. Availability and cost of the necessary parts may dictate whether the cylinder is rebuilt or replaced with a new one. If it's decided to rebuild the cylinder, inspect the bore as described in Step 9 before purchasing parts.*

### Removal

*Refer to illustrations 4.4a and 4.4b*

1  Disconnect the cable from the negative battery terminal.

2  Working under the dashboard, remove the cotter pin from the master cylinder pushrod clevis. Pull out the clevis pin to disconnect the pushrod from the pedal.

# Chapter 8  Clutch and driveaxles

**5.3  Using a flare-nut wrench, disconnect the clutch fluid line at the release cylinder, then remove the mounting bolts (arrows)**

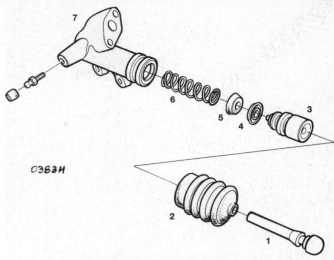

**5.6  An exploded view of the clutch release cylinder**

| 1 | Pushrod | 4 | Piston cup B | 7 | Release |
|---|---|---|---|---|---|
| 2 | Boot | 5 | Piston cup A | | cylinder body |
| 3 | Piston | 6 | Spring | | |

3   Working in the engine compartment, remove the control box from the firewall.
4   Clamp a pair of locking pliers onto the clutch fluid feed hose, a couple of inches downstream of the reservoir (see illustration). The pliers should be just tight enough to prevent fluid flow when the hose is disconnected. Disconnect the hydraulic lines at the cylinder (see illustration). Loosen the fluid feed hose clamp and detach the hose from the cylinder. Have rags handy as some fluid will be lost as the line is removed. Cap or plug the ends of the lines (and/or hose) to prevent fluid leakage and the entry of contaminants. **Caution:** *Don't allow brake fluid to come into contact with the paint as it will damage the finish.*
5   Working under the dash, unscrew the two clutch master cylinder retaining bolts and remove the cylinder.

## Overhaul

*Refer to illustration 4.6*

6   Pull back the dust boot on the pushrod and remove the snap-ring **(see illustration)**.
7   Remove the stopper and pushrod from the cylinder. Invert the cylinder and tap it against a block of wood to eject the piston assembly. If it won't come out, apply compressed air to the fluid inlet port (make sure the open end of the cylinder is not pointed toward anyone and use just enough pressure to push the piston out).
8   Carefully disassemble the piston and one-way valve assembly, noting how the components are arranged. Discard the rubber parts and clean the remaining parts with clean brake fluid or brake system cleaner. DO NOT use petroleum-based solvents. If available, blow the components dry with compressed air.
9   Inspect the bore of the cylinder for scratches, score marks, ridges and corrosion. The surface must be smooth to the touch. If the bore isn't perfectly smooth, the cylinder must be replaced with a new or factory rebuilt unit.
10   If the cylinder will be rebuilt, use all of the new parts contained in the kit and follow any specific instructions which may be included in the kit. Use plenty of clean brake fluid when assembling the components.
11   Place the valve seal on the valve stem, then install the valve spring, the valve guide, the return spring, the spring retainer and the master piston. Make sure all the pieces are aligned as shown in illustration 4.6.
12   Carefully stretch the rubber piston cups into position on the piston. The lips of the cups must be pointing toward the return spring.
13   Lubricate the cylinder bore and piston with brake fluid, then insert the valve and piston assembly into the cylinder bore, being careful not to distort the piston cups.
14   Position the pushrod and stopper plate in the bore, compress the spring and install the snap-ring.
15   Apply a film of rubber grease or equivalent to the inside of the dust boot and push it into position, making sure the lip of the boot seats in the groove on the cylinder body.

## Installation

16   Place the master cylinder in position and install the mounting bolts finger tight.
17   Connect the hydraulic lines to the master cylinder. Move the cylinder slightly as necessary to thread the banjo bolt properly into its threaded bore. Don't cross-thread the bolt. Attach the fluid feed hose to the cylinder and tighten the hose clamp.
18   Tighten the mounting bolts securely, then tighten the hydraulic-line hose clamp and banjo bolt.
19   Connect the pushrod to the clutch pedal. Use a new cotter pin to secure the clevis pin.
20   Remove the locking pliers from the feed hose. Fill the clutch master cylinder reservoir with the with brake fluid conforming to DOT 3 specifications and bleed the clutch system as outlined in Section 6.

## 5   Clutch release cylinder – removal, overhaul and installation

**Note:** *Before beginning this procedure, contact local parts stores and dealer service departments concerning the purchase of a rebuild kit or a new release cylinder. Availability and cost of the necessary parts may dictate whether the cylinder is rebuilt or replaced with a new one. If it's decided to rebuild the cylinder, inspect the bore as described in Step 8 before purchasing parts.*

### Removal

*Refer to illustration 5.3*

1   Disconnect the negative cable from the battery.
2   Raise the vehicle and support it securely on jackstands.
3   Disconnect the fluid hose at the release cylinder. Use a flare nut wrench so you don't strip the corners off the fitting **(see illustration)**. Have a small can and rags handy – some fluid will be spilled as the line is removed.
4   Remove the two release cylinder mounting bolts.
5   Remove the release cylinder.

### Overhaul

*Refer to illustration 5.6*

6   Remove the pushrod and the boot **(see illustration)**.
7   Tap the release cylinder body on a block of wood to eject the piston, the cups and the spring from inside the cylinder.
8   Carefully inspect the bore of the cylinder. Check for scratches, score marks and ridges. The bore must be smooth to the touch. If any imperfections are found, the release cylinder must be replaced with a new one.

# Chapter 8 Clutch and driveaxles

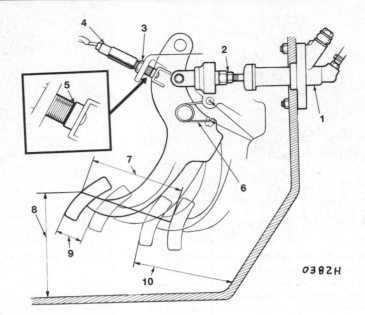

**7.1 Clutch pedal adjustment details (Legend models)**

1. Clutch master cylinder
2. Locknut B
3. Locknut A
4. Clutch pedal switch
5. Pedal in contact with switch
6. Clutch assist spring
7. Stroke at pedal
8. Clutch pedal height
9. Clutch pedal freeplay
10. Clutch pedal disengagement height

9  Wash the components with clean brake fluid or brake cleaner. Do not use petroleum-based solvents.

10  Using the new parts in the rebuild kit, assemble the components using plenty of fresh brake fluid for lubrication. The lip of piston cup A must face away from the piston; the lip of cup B faces toward the piston. On later models, attach the small end of the spring to the piston.

## Installation

11  Install the release cylinder on the clutch housing. Make sure the push-rod is seated in the release fork pocket.
12  Connect the hydraulic line to the release cylinder. Using a flare-nut wrench, tighten the fitting securely.
13  Fill the clutch master cylinder with brake fluid conforming to DOT 3 specifications.
14  Bleed the system as described in Section 6.
15  Lower the vehicle and connect the negative battery cable.

## 6  Clutch hydraulic system – bleeding

1  Bleed the hydraulic system whenever any part of the system has been removed or the fluid level has fallen so low that air has been drawn into the master cylinder. The bleeding procedure is very similar to bleeding a brake system.
2  Fill the master cylinder with new brake fluid conforming to DOT 3 specifications. **Caution:** *Do not re-use any of the fluid coming from the system during the bleeding operation or use fluid which has been inside an open container for an extended period of time.*
3  Raise the vehicle and place it securely on jackstands to gain access to the release cylinder, which is located on the front of the transaxle.
4  Remove the dust cap which fits over the bleeder valve and push a length of plastic hose over the valve. Place the other end of the hose into a clear container with about two inches of brake fluid. The hose end must be in the fluid at the bottom of the container.
5  Have an assistant depress the clutch pedal and hold it. Open the bleeder valve on the release cylinder, allowing fluid to flow through the hose. Close the bleeder valve when the flow of fluid (and bubbles) ceases. Once closed, have your assistant release the pedal.
6  Continue this process until all air is evacuated from the system, indicated by a solid stream of fluid being ejected from the bleeder valve each time with no air bubbles in the hose or container. Keep a close watch on the fluid level inside the clutch master cylinder reservoir – if the level drops too far, air will get into the system and you'll have to start all over again.
7  Install the dust cap and lower the vehicle. Check carefully for proper operation before placing the vehicle into normal service.

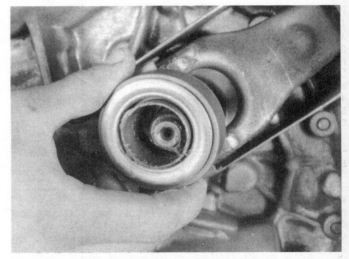

**8.3a  On Legend models, slide the bearing off the input shaft, . . .**

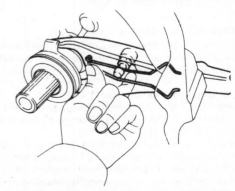

**8.3b  . . . then disengage the fork from the ball stud by pulling on the retention spring**

# Chapter 8  Clutch and driveaxles

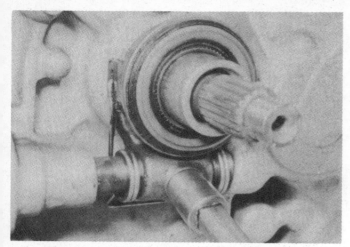

8.4a  On Integra models, remove the release bearing fork-to-release shaft bolt, . . .

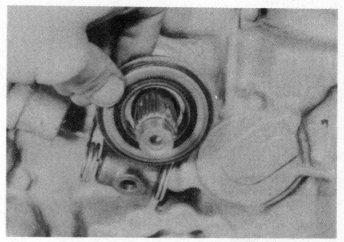

8.4b  . . . slide the shaft out of the transaxle housing to free the release bearing . . .

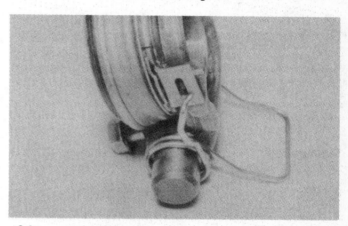

8.4c  . . . and pull the spring ends outward, disengaging them from the slots in the bearing locating tabs

## 7  Clutch pedal – adjustment (Legend models)

*Refer to illustration 7.1*

**Note:** *The clutch pedal adjustment procedure for Integra models is in Section 3.*

1  Loosen locknut A **(see illustration)**, then turn the pedal switch.
2  Loosen locknut B, then turn the pushrod in or out until the stroke and height of the pedal are within the range listed in the Specifications at the beginning of this Chapter.
3  Tighten locknut B.
4  Screw in the clutch pedal switch until it contacts the clutch pedal.
5  Turn the switch another 1/4 to 1/2-turn.
6  Tighten locknut A.

## 8  Clutch release bearing and fork – removal, inspection and installation

**Warning:** *Dust produced by clutch wear and deposited on clutch components may contain asbestos, which is hazardous to your health. DO NOT blow it out with compressed air and DO NOT inhale it. DO NOT use gasoline or petroleum-based solvents to remove the dust. Brake system cleaner should be used to flush the dust into a drain pan. After the clutch components are wiped clean with a rag, dispose of the contaminated rags and cleaner in a labeled, covered container.*

### Removal

*Refer to illustrations 8.3a, 8.3b, 8.4a, 8.4b and 8.4c*

1  On Integra models, disconnect the clutch cable (see Section 3). On Legend models, unbolt the clutch release cylinder, but don't disconnect the fluid line between the master cylinder and the release cylinder (see Section 5).
2  Remove the transaxle (see Chapter 7).
3  On Legend models, slide the release bearing off the input shaft, lift the clutch release fork off the ball pivot stud and remove the fork **(see illustrations)**.
4  On Integra models, remove the release bearing fork-to-release shaft bolt and slide the shaft out of the transaxle housing to free the release bearing **(see illustrations)**. To separate the bearing from the fork, pull the spring ends outward, disengaging them from the slots in the bearing locating tabs **(see illustration)**.

### Inspection

5  Hold the bearing by the outer race and rotate the inner race while applying pressure. If the bearing doesn't turn smoothly or if it's noisy, replace the bearing/hub assembly with a new one. Wipe the bearing with a clean rag and inspect it for damage, wear and cracks. It's common practice to

8.7  Fill the inner groove of the release bearing with high temperature grease and apply a light coat of the same grease to the transaxle input shaft splines and the front bearing retainer

replace the bearing with a new one whenever a clutch job is performed, to decrease the possibility of a bearing failure in the future. Don't immerse the bearing in solvent – it's sealed for life and to do so would ruin it. Also check the release lever for cracks and bends.
6  If the new bearing is not equipped with a bearing holder (hub), drive the holder from the old bearing and install it to the new one. A seal/bushing driver or an appropriately sized socket can be used to accomplish this.

### Installation

*Refer to illustrations 8.7 and 8.8*

7  Fill the inner groove of the release bearing with high temperature grease. Also apply a light coat of the same grease to the transaxle input

# Chapter 8  Clutch and driveaxles

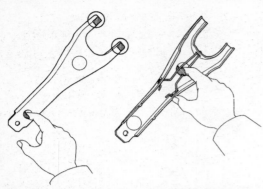

**8.8  On Legend models, lubricate the release fork ball socket, fork ends and release cylinder pushrod socket with high temperature grease**

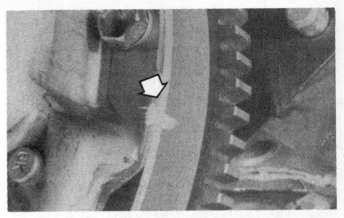

**9.5  Index the pressure plate to the flywheel (just in case you're going to reuse the same pressure plate)**

**9.6a  Remove the pressure plate bolts (arrows) gradually and evenly in a criss-cross pattern**

**9.6b  Once the pressure plate bolts are removed and the pressure plate is pulled away from the flywheel, the disc will fall if unsupported, so grasp it firmly, and make sure you don't breathe the dust deposited on the clutch components – it may contain asbestos, which is carcinogenic**

shaft splines and the front bearing retainer **(see illustration)**.
8   On Legend models, lubricate the release fork ball socket, fork ends and release cylinder pushrod socket with high temperature grease **(see illustration)**.
9   On Legend models, attach the release bearing to the release fork.
10  On Legend models, slide the release bearing onto the transaxle input shaft front bearing retainer while passing the end of the release fork through the opening in the clutch housing. Push the clutch release fork onto the ball stud until it's firmly seated.
11  On Integra models, position the bearing and fork assembly on the transaxle input shaft and slide the release shaft through the release fork. Align the holes and install the bolt, tightening it to the torque listed in this Chapter's Specifications.
12  Apply a light coat of high temperature grease to the face of the release bearing where it contacts the pressure plate diaphragm fingers.
13  The remainder of the installation is the reverse of the removal procedure, tightening all bolts to the specified torque.

## 9  Clutch components – removal, inspection and installation

**Warning:** *Dust produced by clutch wear and deposited on clutch components may contain asbestos, which is hazardous to your health. DO NOT blow it out with compressed air and DO NOT inhale it. DO NOT use gasoline or petroleum-based solvents to remove the dust. Brake system cleaner should be used to flush the dust into a drain pan. After the clutch components are wiped clean with a rag, dispose of the contaminated rags and cleaner in a covered, marked container.*

### Removal

*Refer to illustrations 9.5, 9.6a and 9.6b*
1   Access to the clutch components is normally accomplished by removing the transaxle, leaving the engine in the vehicle. If the engine is being removed for major overhaul, check the clutch for wear and replace worn components as necessary. However, the relatively low cost of the clutch components compared to the time and trouble spent gaining access to them warrants their replacement anytime the engine or transaxle is removed, unless they are new or in near perfect condition. The following procedures are based on the assumption the engine will stay in place.
2   Remove the transaxle from the vehicle (see Chapter 7, Part A). Support the engine while the transaxle is out. Preferably, an engine hoist should be used to support it from above. However, if a jack is used underneath the engine, make sure a piece of wood is positioned between the jack and oil pan to spread the load. **Caution:** *The pickup for the oil pump is very close to the bottom of the oil pan. If the pan is bent or distorted in any way, engine oil starvation could occur.*
3   The clutch fork and release bearing can remain attached to the transaxle housing for the time being.
4   To support the clutch disc during removal, install a clutch alignment tool through the clutch disc hub.
5   Carefully inspect the flywheel and pressure plate for indexing marks. The marks are usually an X, an O or a white letter. If they cannot be found,

# Chapter 8 Clutch and driveaxles

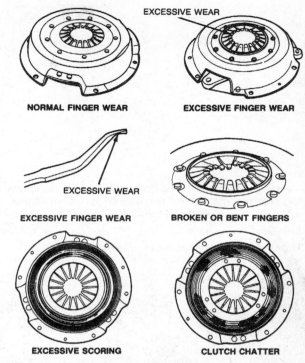

**9.9 The clutch disc**

1. Lining – this will wear down in use
2. Springs or dampers – check for cracking and deformation
3. Splined hub – the splines must not be worn and should slide smoothly on the transaxle input shaft splines
4. Rivets – these secure the lining and will damage the flywheel or pressure plate if allowed to contact the surfaces

**9.11 Replace the pressure plate if any of these conditions are noted**

**9.13 Center the clutch disc in the pressure plate with a clutch alignment tool**

scribe or paint marks yourself so the pressure plate and the flywheel will be in the same alignment during installation **(see illustration)**.

6  Turning each bolt a little at a time, loosen the pressure plate-to-flywheel bolts **(see illustration)**. Work in a criss-cross pattern until all spring pressure is relieved. Then hold the pressure plate securely and completely remove the bolts, followed by the pressure plate and clutch disc **(see illustration)**.

## Inspection

*Refer to illustrations 9.9 and 9.11*

7  Ordinarily, when a problem occurs in the clutch, it can be attributed to wear of the clutch driven plate assembly (clutch disc). However, all components should be inspected at this time.

8  Inspect the flywheel for cracks, heat checking, grooves and other obvious defects. If the imperfections are slight, a machine shop can machine the surface flat and smooth, which is highly recommended regardless of the surface appearance. Refer to Chapter 2 for the flywheel removal and installation procedure.

9  Inspect the lining on the clutch disc. There should be at least 1/16-inch of lining above the rivet heads. Check for loose rivets, distortion, cracks, broken springs and other obvious damage **(see illustration)**. As mentioned above, ordinarily the clutch disc is routinely replaced, so if in doubt about the condition, replace it with a new one.

10  The release bearing should also be replaced along with the clutch disc (see Section 8).

11  Check the machined surfaces and the diaphragm spring fingers of the pressure plate **(see illustration)**. If the surface is grooved or otherwise damaged, replace the pressure plate. Also check for obvious damage, distortion, cracking, etc. Light glazing can be removed with emery cloth or sandpaper. If a new pressure plate is required, new and factory-rebuilt units are available.

## Installation

*Refer to illustration 9.13*

12  Before installation, clean the flywheel and pressure plate machined surfaces with brake cleaner, lacquer thinner or acetone. It's important that no oil or grease is on these surfaces or the lining of the clutch disc. Handle the parts only with clean hands.

13  Position the clutch disc and pressure plate against the flywheel with the clutch held in place with an alignment tool **(see illustration)**. Make sure it's installed properly (most replacement clutch plates will be marked "flywheel side" or something similar – if not marked, install the clutch disc with the damper springs toward the transaxle).

14  Tighten the pressure plate-to-flywheel bolts only finger tight, working around the pressure plate.

15  Center the clutch disc by ensuring the alignment tool extends through the splined hub and into the pocket in the crankshaft. Wiggle the tool up, down or side-to-side as needed to center the disc. Tighten the pressure plate-to-flywheel bolts a little at a time, working in a criss-cross pattern to prevent distorting the cover. After all of the bolts are snug, tighten them to the torque listed in this Chapter's Specifications. Remove the alignment tool.

# Chapter 8 Clutch and driveaxles

11.1 If the driveaxle hub nut is "staked," use a chisel to unstake it

11.2 To prevent the hub from turning while you're breaking the hub nut loose, place a pry bar between two of the wheel studs

11.6 Swing the hub/knuckle assembly out (away from the vehicle) and pull the driveaxle from the hub

11.7 Use a large screwdriver or a pry bar to pop the inner end of the driveaxle loose from the differential

16 Using high-temperature grease, lubricate the inner groove of the release bearing (see Section 8). Also place grease on the release lever contact areas and the transaxle input shaft bearing retainer.
17 Install the clutch release bearing (see Section 8).
18 Install the transaxle and all components removed previously. Tighten all fasteners to the proper torque specifications.

## 10 Starter/clutch interlock switch – check, replacement and adjustment

### Check

1 The starter/clutch interlock switch is located near the upper end of the clutch pedal on some models. It has two wires – one coming from the starter relay and one going to ground. When the ignition switch key is turned to the Start position and the clutch pedal is depressed, the starter relay's path to ground is closed by the starter/clutch interlock switch and the starter motor is activated.
2 If the engine won't start when the clutch pedal is depressed, adjust the switch (see Step 6) and try again. If it still won't start, check the switch (see Step 3) and, if necessary, replace it (see Step 5). If the engine starts when the clutch pedal isn't depressed, adjust the switch and try again.
3 If the engine won't start when the clutch pedal is depressed, either there's no voltage from the starter relay to the switch, there's no continuity from the switch to ground, or the switch itself is faulty.

4 Using the wiring diagrams at the end of Chapter 12, check the voltage to the switch with a voltmeter or test light. When you turn the ignition key to the Start position and depress the clutch pedal, there should be voltage in the wire from the starter relay. If there isn't, look for an open or short circuit condition somewhere between the starter relay and the switch. If there is voltage in this wire, check the ground wire side of the switch for voltage. If there's voltage on the ground side as well, the switch should be operating correctly. Try adjusting it (see Step 6). If there's no voltage on the ground side, the switch is bad.

### Replacement

5 Unplug the electrical connector, loosen the adjustment nut and unscrew the switch from its mounting bracket. Installation is the reverse of removal.

### Adjustment

6 Loosen the locknut and turn the switch in or out, as necessary, to provide continuity through the switch when the clutch pedal is depressed.

## 11 Driveaxles – removal and installation

### Removal

*Refer to illustrations 11.1, 11.2, 11.6 and 11.7*

1 If the vehicle has steel wheels, remove the hub cap; if it has mag wheels, remove the wheel lug nut cover. If the wheel hub nut is staked,

# Chapter 8 Clutch and driveaxles

11.8a Pry the old spring clip from the inner end of the driveaxle with a small screwdriver or awl

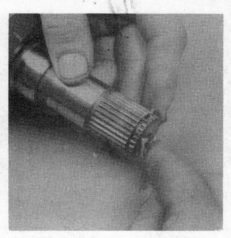

11.8b To install the new spring clip, start one end in the groove and work the clip over the shaft end, into the groove

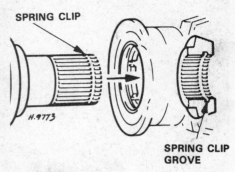

11.9 When installing the driveaxle, make sure the spring clip pops into place in its groove – if it's seated properly, you shouldn't be able to pull it out by hand

unstake it with a chisel (see illustration); if it's secured by locking tabs, bend the tabs out. Break the hub nut loose with a socket and large breaker bar. Loosen the wheel lug nuts, raise the front of the vehicle and support it securely on jackstands. Remove the wheel lug nuts and the front wheel.
2   Remove the driveaxle hub nut. To prevent the hub from turning, place a pry bar between two of the wheel studs, then loosen the nut (see illustration).
3   Drain the transaxle lubricant (see Chapter 1).
4   On Legend models, disconnect the damper fork from the shock absorber assembly and the lower arm (see Chapter 10).
5   Separate the lower control arm and the tie-rod end from the steering knuckle (see Chapter 10).
6   Swing the knuckle/hub assembly out (away from the vehicle) until the end of the driveaxle is free of the hub (see illustration). Support the outer end of the driveaxle with a piece of wire to avoid unnecessary strain on the inner CV joint.
7   Carefully pry the inner end of the driveaxle from the transaxle or intermediate shaft, using a large screwdriver or pry bar positioned between the transaxle or bearing support and the CV joint housing (see illustration). Support the CV joints and carefully remove the driveaxle from the vehicle. To prevent damage to the intermediate shaft seal or the differential seal, hold the inner CV joint horizontal until the driveaxle is clear of the intermediate shaft.

## Installation

*Refer to illustrations 11.8a, 11.8b and 11.9*

8   Pry the old spring clip from the inner end of the driveaxle and install a new one (see illustrations). Lubricate the differential seal with multi-purpose grease and raise the driveaxle into position while supporting the CV joints.
9   Insert the splined end of the inner CV joint into the differential side gear or the intermediate shaft and make sure the spring clip locks in its groove (see illustration).
10   Apply a light coat of multi-purpose grease to the outer CV joint splines, pull out on the strut/steering knuckle assembly and install the stub axle into the hub.
11   Insert the studs of the lower control arm balljoint and the tie-rod end into the steering knuckle and tighten the nuts (see Chapter 10). Be sure to use new cotter pins. On Legend models, install the damper fork (see Chapter 10).
12   Install the hub nut (and, if applicable, a new locking tab washer). Lock the disc as described in Step 2 so it can't turn, then tighten the hub nut securely. Don't try to tighten it to the actual torque specification until you've lowered the vehicle to the ground.
13   Grasp the inner CV joint housing (not the driveaxle) and pull out to make sure the driveaxle has seated securely in the transaxle.

12.5 Bearing support bolts (arrows) – Integra models

14   Install the wheel and lug nuts, then lower the vehicle.
15   Tighten the lug nuts to the torque listed in the Chapter 1 Specifications. Tighten the hub nut to the torque listed in this Chapter's Specifications. Using a hammer and punch, stake the nut to the groove in the driveaxle. If the hub nut uses a locking tab, be sure to bend the tabs up against the nut. Install the wheel cover (if applicable).
16   Refill the transaxle with the recommended type and amount of lubricant (see Chapter 1).

## 12   Intermediate shaft – removal and installation

### Removal

*Refer to illustration 12.5*

1   Loosen the wheel lug nuts, raise the front of the vehicle and support it securely on jackstands. Remove the wheel.
2   Drain the transaxle lubricant or automatic transmission fluid (see Chapter 1).
3   Separate the suspension lower arm from the steering knuckle (see Chapter 10).
4   Pry the inner CV joint housing from the intermediate shaft. Position the driveaxle out of the way and hang it with a piece of wire. Do not allow it to hang unsupported, as the outer CV joint may be damaged.
5   Unscrew the bearing support-to-engine block bolts (see illustration) and slide the intermediate shaft out of the transaxle. Be careful not to damage the differential oil seal when pulling the shaft out.

# Chapter 8 Clutch and driveaxles

13.3a Cut off the boot clamps and discard them – don't try to reuse old clamps

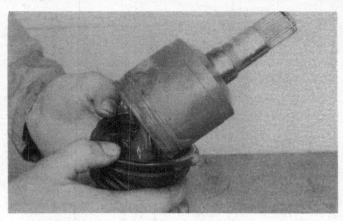

13.3b Slide the boot down the driveaxle, out of the way

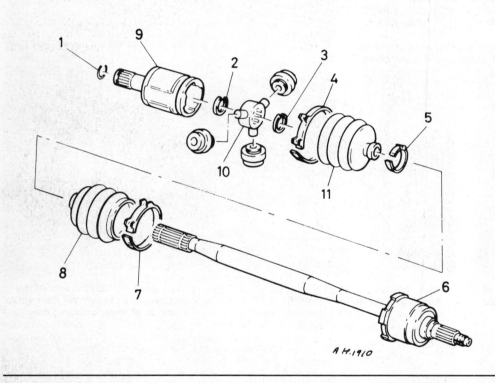

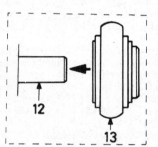

13.3c An exploded view of a typical driveaxle assembly with a tri-pot type inner CV joint

1. Spring clip
2. Snap-ring
3. Stop-ring
4. Boot clamp
5. Boot clamp
6. Outer CV joint assembly
7. Boot clamp
8. Boot
9. Inner CV joint housing/outer race
10. Tri-pot assembly
11. Boot
12. Tri-pot post
13. Roller bearing

6  Check the support bearing for smooth operation by turning the shaft while holding the bearing. If you feel any roughness, have a new bearing installed by a dealer service department or other repair shop. To do the job at home, you'd need specialized tools.

## Installation

7  Lubricate the lips of the differential oil seal with multi-purpose grease. Carefully guide the intermediate shaft into the differential side gear then install the mounting bolts through the bearing support. Tighten the bolts to the torque listed in this Chapter's Specifications.

8  Install a new spring clip on the inner CV joint (**see illustration 11.9**) and seat the driveaxle into the intermediate shaft splines.

9  Connect the lower arm to the steering knuckle and tighten the balljoint stud nut to the torque listed in the Chapter 10 Specifications.

10  Install the wheel and lug nuts, lower the vehicle and tighten the lug nuts to the torque listed in the Chapter 1 Specifications.

11  Refill the transaxle with the proper type and amount of lubricant (see Chapter 1).

## 13 Driveaxle boot replacement and constant velocity (CV) joint overhaul

**Note:** *If the CV joints are worn, indicating the need for an overhaul (usually due to torn boots), explore all options before beginning the job. Complete rebuilt driveaxles are available on an exchange basis, which eliminates much time and work. If you decide to rebuild a CV joint, check on the cost and availability of parts before disassembling the driveaxle.*

1  Remove the driveaxle from the vehicle (see Section 11).
2  Mount the driveaxle in a vise. The jaws of the vise should be lined with wood or rags to prevent damage to the driveaxle.

### Inner CV joint and boot
#### Disassembly
*Refer to illustrations 13.3a, 13.3b, 13.3c, 13.4, 13.5 and 13.6*

3  Cut off both boot clamps and slide the boot towards the center of the driveaxle (**see illustrations**).

# Chapter 8  Clutch and driveaxles

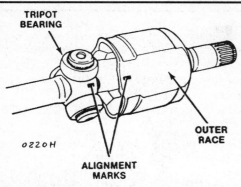

**13.4  Scribe or paint alignment marks on the tri-pot assembly and the outer race, then slide the outer race off**

4  Mark or paint alignment marks on the outer race and the tri-pot bearing assembly **(see illustration)** so they can be returned to their original position, then slide the outer race off the tri-pot bearing assembly.
5  Remove the snap-ring from the end of the axleshaft, then mark the relationship of the tri-pot bearing assembly to the axleshaft **(see illustration)**.
6  Secure the bearing rollers with tape, then remove the tri-pot bearing assembly from the axleshaft with a brass drift and a hammer **(see illustration)**. Remove the tape, but don't let the rollers fall off and get mixed up.
7  Remove the stop-ring, slide the old boot off the driveaxle and discard it.

### Inspection

8  Clean the old grease from the outer race and the tri-pot bearing assembly. Paint or scribe marks on each bearing roller and its respective shaft to ensure proper reassembly, then carefully disassemble each section of the tri-pot assembly, one at a time, and clean the needle bearings with solvent.
9  Inspect the rollers, tri-pot, bearings and outer race for scoring, pitting or other signs of abnormal wear, which will warrant the replacement of the inner CV joint.

### Reassembly

*Refer to illustrations 13.10a, 13.10b, 13.12, 13.13, 13.14, 13.15, 13.16, 13.17a and 13.17b*

10  Wrap the splines of the axleshaft with tape to avoid damaging the new boot, then slide the boot onto the axleshaft **(see illustration)**. Remove the tape and slide the inner snap-ring into place **(see illustration)**.
11  Align the match marks you made before disassembly and tap the tri-pot assembly onto the axleshaft with a hammer and brass drift.
12  Install the outer snap-ring **(see illustration)**.
13  Apply a coat of CV joint grease to the inner bearing surfaces to hold the needle bearings in place when reassembling the tri-pot assembly **(see illustration)**. Make sure each roller is installed on the same post as before.

**13.5  Remove the snap-ring from end of the axleshaft, then mark the relationship of the tri-pot to the axleshaft**

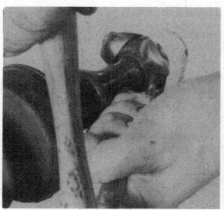

**13.6  Secure the bearing rollers with tape and drive the tri-pot off the shaft with a hammer and a brass drift, then remove the stop-ring**

**13.10a  Wrap the splined area of the axleshaft with tape to prevent damage to the boot when installing it**

**13.10b  Install the stop-ring on the axleshaft, making sure it seats in its groove**

**13.12  Install the tri-pot assembly on the axleshaft, making sure the punch marks are lined up, then install the snap-ring**

**13.13  Use plenty of CV joint grease to hold the needle bearings in place when you install the roller assemblies on the tri-pot and make sure you put each roller in its original position**

14  Pack the outer race with half of the grease furnished with the new boot and place the remainder in the boot. Install the outer race **(see illustration)**. Make sure the marks you made on the tri-pot assembly and the outer race are aligned.
15  Seat the boot in the grooves in the outer race and the axleshaft, then adjust the driveaxle to the proper length **(see illustration)**.
16  With the driveaxle set to the proper length, equalize the pressure in the boot by inserting a blunt screwdriver between the boot and the outer race **(see illustration)**. Don't damage the boot with the tool.
17  Install and tighten the new boot clamps **(see illustrations)**.
18  Install the driveaxle assembly (see Section 11).

### Outer CV joint and boot
#### Disassembly
19  Following Steps 3 through 7, remove the inner CV joint from the driveaxle and disassemble it.
20  If the driveaxle is equipped with a dynamic damper, scribe or paint an alignment mark on the axleshaft along the outer edge of the damper (the side facing the outer CV joint), cut the retaining clamp and slide the damper off.
21  Cut the boot clamps from the outer CV joint. Slide the boot off the shaft.

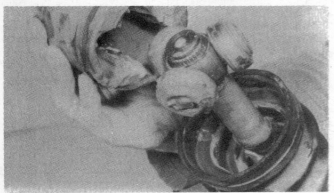

13.14  Pack the outer race with grease and slide it over the tri-pot assembly – make sure the match marks on the outer race and tri-pot line up

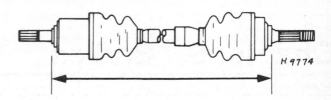

13.15  Before tightening the boot clamps, adjust the driveaxle length to the dimension listed in this Chapter's Specifications

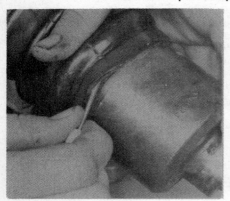

13.16  Equalize the pressure inside the boot by inserting a small, dull screwdriver between the boot and the outer race

13.17a  To install the new clamps, bend the tang down . . .

13.17b  . . . and flatten the tabs to hold it in place

#### Inspection
*Refer to illustration 13.23*
22  Thoroughly wash the inner and outer CV joints in clean solvent and blow them dry with compressed air, if available. **Note:** *Because the outer joint can't be disassembled, it is difficult to wash away all the old grease and to rid the bearing of solvent once it's clean. But it is imperative that the job be done thoroughly, so take your time and do it right.*
23  Bend the outer CV joint housing at an angle to the axleshaft to expose the bearings, inner race and cage **(see illustration)**. Inspect the bearing surfaces for signs of wear. If the bearings are damaged or worn, replace the driveaxle.

#### Reassembly
24  Slide the new outer boot onto the axleshaft. It's a good idea to wrap tape around the splines of the shaft to prevent damage to the boot **(see illustration 13.10a)**. When the boot is in position, add the specified amount of grease (included in the boot replacement kit) to the outer joint and the boot (pack the joint with as much grease as it will hold and put the rest into the boot). Slide the boot on the rest of the way and install the new clamps **(see illustrations 13.17a and 13.17b)**.
25  Slide the dynamic damper, if equipped, onto the shaft. Make sure its outer edge is aligned with the previously applied mark. Install a new retaining clamp.
26  Clean and reassemble the inner CV joint by following Steps 8 through 17, then install the driveaxle as outlined in Section 11.

13.23  After the old grease has been rinsed away and the solvent has been blown out with compressed air, rotate the outer joint assembly through its full range of motion and inspect the bearing surfaces for wear and damage – if any of the ball bearings, the race or the cage look damaged, replace the driveaxle and outer joint assembly

# Chapter 9  Brakes

## Contents

| | |
|---|---|
| Anti-Lock Brake (ALB) system – general information | 17 |
| Brake check | See Chapter 1 |
| Brake disc – inspection, removal and installation | 6 |
| Brake hoses and lines – inspection and replacement | 11 |
| Brake hydraulic system – bleeding | 12 |
| Brake light switch – check, replacement and adjustment | 16 |
| Disc brake caliper (front) – removal, overhaul and installation | 3 |
| Disc brake caliper (rear) – removal and installation | 5 |
| Disc brake pads (front) – replacement | 2 |
| Disc brake pads (rear) – replacement | 4 |
| Drum brake shoes – replacement | 7 |
| Fluid level checks | See Chapter 1 |
| General information | 1 |
| Master cylinder – removal, overhaul and installation | 9 |
| Parking brake – adjustment | 14 |
| Parking brake cable(s) – replacement | 15 |
| Power brake booster – check, removal and installation | 13 |
| Proportioning valve – general information, removal and installation | 10 |
| Wheel cylinder – removal, overhaul and installation | 8 |

## Specifications

### General

Brake pedal
 Height
  Integra .................................................. 7-3/64 in
  Legend .................................................. 6-45/64 in
 Freeplay
  Integra .................................................. 3/64 to 13/64 in
  Legend .................................................. 1/16 to 13/64 in
Parking brake lever travel
 Integra ................................................... 4 to 8 clicks
 Legend ................................................... 7 to 11 clicks
Power brake booster pushrod-to-master cylinder piston clearance (with a vacuum of 20 in-Hg applied to booster) .............................. 0.0 to 0.016 in (0.0 to 0.4 mm)

### Disc brakes

Brake pad minimum thickness ................................. See Chapter 1
Disc minimum thickness ...................................... Refer to minimum thickness cast into disc
Thickness variation (parallelism) ............................. No more than 0.0006 in (0.015 mm)
Runout limit ................................................. 0.004 in (0.10 mm)

### Drum brakes

Brake lining minimum thickness .............................. See Chapter 1
Drum diameter .............................................. Refer to maximum diameter cast into drum

### Torque specifications                                   Ft-lbs (unless otherwise indicated)

#### General

Brake hose-to-caliper banjo bolt (front or rear) .............. 25
Master cylinder mounting nuts ............................... 132 in-lbs
Brake booster mounting nuts ................................ 108 in-lbs

### Front disc brake
Caliper guide pin bolts
    Integra ............................................. 33
    Legend ............................................. 24
Caliper mounting bracket bolts
    Integra ............................................. 53
    Legend ............................................. 56
Caliper mounting bolts
    Integra ............................................. 33
    Legend ............................................. 24

### Rear disc brake
Caliper flange (guide pin) bolts
    Integra ............................................. 17
    Legend ............................................. 20
    Caliper mounting bracket bolts ........................... 28

### Rear drum brake
Wheel cylinder nuts ...................................... 72 in-lbs
Brake backing plate nuts (some Integra models) .............. 132 in-lbs

## 1 General information

### General

All vehicles covered by this manual are equipped with hydraulically operated power assisted brake systems. All front and most rear brake systems are disc type. Some Integra models use drum type brakes at the rear.

All brakes are self-adjusting. The front and rear disc brakes automatically compensate for pad wear, while the rear drum brakes incorporate an adjustment mechanism which is activated as the brakes are applied, either through the pedal or the parking brake lever.

The hydraulic system is a diagonally-split design, meaning there are separate circuits for the left front/right rear and the right front/left rear brakes. If one circuit fails, the other circuit will remain functional and a warning indicator will light up on the dashboard when a substantial amount of brake fluid is lost, showing that a failure has occurred.

### Master cylinder

The master cylinder is bolted to the power brake booster, which is mounted on the driver's side of the firewall on all Integra models and most Legend models; on some Legend coupes equipped with the Anti-Lock Brake (ALB) system, the master cylinder and power brake booster are mounted on the passenger's side of the firewall. To locate the master cylinder, look for the large fluid reservoir on top. The fluid reservoir is a removable plastic cup, secured to the master cylinder by a clamp.

The master cylinder is designed for the "split system" mentioned earlier and has separate primary and secondary piston assemblies, the piston nearest the firewall being the secondary piston.

### Proportioning valve

The proportioning valve assembly is bolted to the right strut tower on Integra models; it's located near the front left wheel well on Legend models. On vehicles equipped with the Anti-Lock Brake (ALB) system, it's an integral part of the modulator/solenoid unit, which is located just in front of the left front wheel well. The proportioning valve is really two separate valves – one valve for each circuit.

The proportioning valve regulates the hydraulic pressure to the rear brakes during heavy braking to eliminate rear wheel lock-up. Under normal braking conditions, it allows full pressure to the rear brake system until a predetermined pedal pressure is reached. Above that point, the pressure to the rear brakes is limited.

The proportioning valve is not serviceable – if a problem develops with the valve, it must be replaced as an assembly.

### Power brake booster

The power brake booster, which uses engine manifold vacuum and atmospheric pressure to provide assistance to the hydraulically operated brakes, is mounted on the firewall in the engine compartment (see *Master cylinder* above for typical locations).

### Parking brake

A parking brake lever inside the vehicle operates a single front cable attached to a pair of rear cables, each of which is connected to its respective rear brake. When the parking brake lever is pulled up on drum brake models, each rear cable pulls on a lever attached to the brake shoe assembly, causing the shoes to expand against the drum. When the lever is pulled on models with rear disc brakes, the rear cables pull on levers that are attached to screw-type actuators in the caliper housings, which apply force to the caliper pistons, clamping the brake pads against the brake disc.

### Precautions

There are some general cautions and warnings involving the brake system on these vehicles:

a) Use only brake fluid conforming to DOT 3 specifications.
b) The brake pads and linings may contain asbestos fibers which are hazardous to your health if inhaled. Whenever you work on brake system components, clean all parts with brake system cleaner or denatured alcohol. Do not allow the fine dust to become airborne.
c) Safety should be paramount whenever any servicing of the brake components is performed. Do not use parts or fasteners which are not in perfect condition, and be sure that all clearances and torque specifications are adhered to. If you are at all unsure about a certain procedure, seek professional advice. Upon completion of any brake system work, test the brakes carefully in a controlled area before putting the vehicle into normal service.
d) No part of the brake hydraulic system on a vehicle equipped with an ALB (Anti-Lock Brake) system should be disconnected, as special tools are required to properly bleed the system. Take the vehicle to a dealer service department or other qualified repair shop for repairs which require opening of the system.

If a problem is suspected in the brake system, don't drive the vehicle until it's fixed.

## 2 Disc brake pads (front) – replacement

*Refer to illustrations 2.5, 2.6a through 2.6o and 2.7*

**Warning:** *Disc brake pads must be replaced on both front wheels at the same time – never replace the pads on only one wheel. Also, the dust created by the brake system may contain asbestos, which is harmful to your health. Never blow it out with compressed air and don't inhale any of it. An approved filtering mask should be worn when working on the brakes.*

# Chapter 9  Brakes

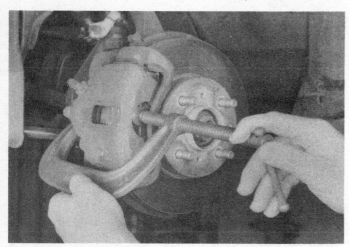

2.5  Using a large C-clamp, push the piston back into the caliper – note that one end of the clamp is on the flat area on the back side of the caliper and the other end (screw end) is pressing on the outer brake pad

2.6a  Before removing anything, spray the caliper and brake pads with brake cleaner to remove the dust produced by brake pad wear – DO NOT blow the dust off with compressed air!

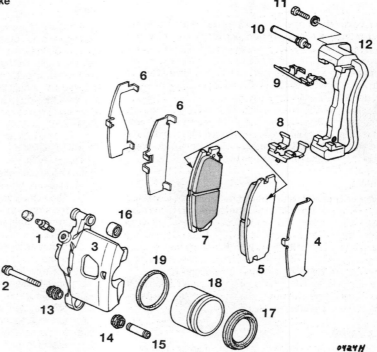

2.6b  An exploded view of a typical front brake caliper assembly (Integra models)

1. Bleeder screw
2. Caliper mounting bolt
3. Caliper body
4. Outer pad shim
5. Outer brake pad
6. Inner pad shims
7. Inner brake pad
8. Lower pad retainer
9. Upper pad retainer
10. Caliper pin
11. Caliper bracket bolt
12. Caliper bracket
13. Inner sleeve boot
14. Outer sleeve boot
15. Sleeve
16. Pin boot
17. Piston boot
18. Piston
19. Piston seal

Do not, under any circumstances, use petroleum-based solvents to clean brake parts. Use brake cleaner or denatured alcohol only! When servicing the disc brakes, use only high-quality, nationally recognized name-brand pads.

**Note:** *Most models have disc brakes at the front and rear wheels; some Integra models have rear drum brakes. This Section covers the replacement procedure for front brake pads; rear brake pads are covered in Section 4; rear drum brake shoes are covered in Section 7.*

1. Remove the cap from the brake fluid reservoir.
2. Loosen the front wheel lug nuts, raise the front of the vehicle and support it securely on jackstands.
3. Remove the front wheels. Work on one brake assembly at a time, using the assembled brake for reference if necessary.
4. Inspect the brake disc carefully as outlined in Section 6. If machining is necessary, follow the information in that Section to remove the disc, at which time the calipers and pads can be removed as well.
5. Push the piston back to the bore to provide room for the new brake pads. A C-clamp can be used to accomplish this **(see illustration).** As the piston is depressed to the bottom of the caliper bore, the fluid in the master cylinder will rise. Make sure it doesn't overflow. If necessary, siphon off some of the fluid.
6. Follow the accompanying illustrations, beginning with **2.6a**, then proceed to **2.6b through 2.6h** (Integra models) or **2.6i through 2.6o** (Legend models) for the actual pad replacement procedure. Be sure to stay in order and read the caption under each illustration. When you're done, proceed to the next Step.

2.6c  On Integra models, if you're going to overhaul the caliper, disconnect the brake hose banjo fitting (arrow). To remove the brake pads, remove the lower caliper bolt as shown . . .

2.6d  . . . swing the caliper up . . .

2.6e  . . . remove the outer pad shim . . .

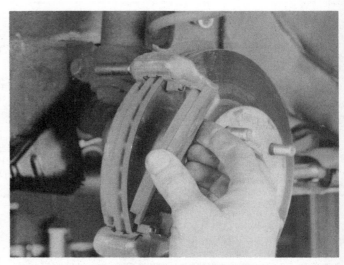

2.6f  . . . remove the outer brake pad . . .

2.6g  . . . remove the inner pad shims (there are two, so note the order in which they're removed, because they're different) . . .

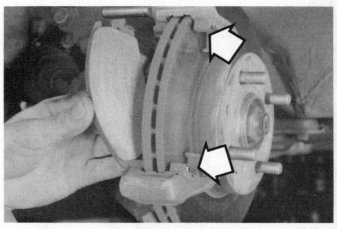

2.6h  . . . and remove the inner brake pad – if you're also removing the caliper mounting bracket and/or the brake disc, note how the upper and lower pad retainers (arrows) are installed in the bracket, then remove them

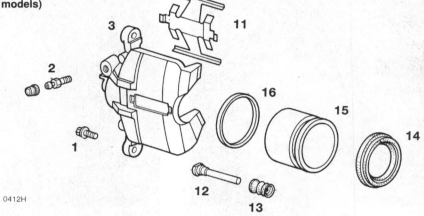

**2.6i  An exploded view of a typical front brake caliper assembly (Legend models)**

1. Caliper mounting bolt
2. Bleeder screw
3. Caliper body
4. Outer pad shim
5. Outer brake pad
6. Inner pad shims
7. Inner brake pad
8. Pad retainers
9. Caliper bracket bolt
10. Caliper bracket
11. Pad spring
12. Caliper pin
13. Pin boot
14. Piston boot
15. Piston
16. Piston seal

2.6j  On Legend models, if you're going to overhaul the caliper, remove the brake hose banjo bolt (lower arrow). To remove the brake pads, unscrew the caliper mounting bolts (wrench on lower bolt, upper arrow indicates upper bolt) . . .

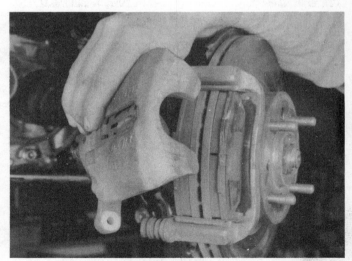

2.6k  . . . lift the caliper off the caliper mounting bracket and hang it out of the way with a piece of rope or wire . . .

2.6l  . . . remove the outer pad shim . . .

2.6m  . . . remove the outer brake pad . . .

2.6n  . . . remove the inner pad shims (there are two, so note the order in which they're removed, because they're different) . . .

2.6o  . . . and remove the inner brake pad – if you're also removing the caliper mounting bracket and/or the brake disc, note how the upper and lower pad retainers (arrows) are installed in the bracket, then remove them

7  Apply a thin coat of disc brake anti-squeal compound, in accordance with the manufacturer's recommendations, on the backing plates of the new pads **(see illustration)**.
8  Install the shims onto their respective pads.
9  Install the pad retainers in the caliper mounting bracket. Lubricate the retainers with a thin film of silicone grease.
10  Install the new pads and shims to the caliper mounting bracket.
11  Install the caliper and caliper mounting bolt(s) and tighten them to the torque listed in this Chapter's Specifications, then proceed to the next Step.
12  Install the wheel and lug nuts, lower the vehicle and tighten the lug nuts to the torque specified in Chapter 1.
13  Check the brake fluid level and add fluid, if necessary (see Chapter 1).
14  Apply and release the brake pedal and, if you replaced the rear pads, the hand brake lever several times to bring the pads into contact with the brake discs. Check the operation of the brakes in an isolated area before driving the vehicle in traffic.

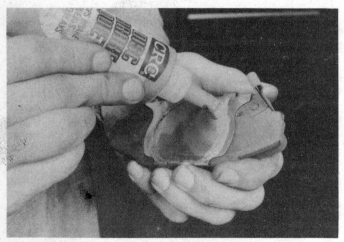

2.7  Before installing the brake pads, apply a coat of disc brake anti-squeal compound to the backing plates of the pads – follow the manufacturer's instructions on the label

**3  Disc brake caliper (front) – removal, overhaul and installation**

**Warning:** *Dust created by the brake system may contain asbestos, which*

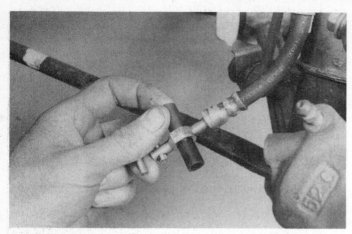

3.2  Using a short piece of rubber hose of the appropriate diameter, plug the brake line banjo fitting like this to prevent brake fluid from leaking

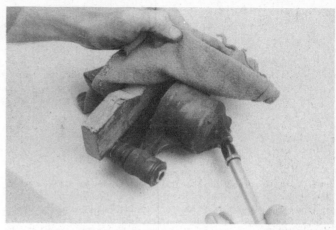

3.5  With the caliper padded to catch the piston, use compressed air to force the piston out of its bore – make sure your hands or fingers are not between the piston and the caliper

3.7  The piston seal should be removed with a plastic or wooden tool to avoid damage to the bore and seal groove – a pencil will do the job

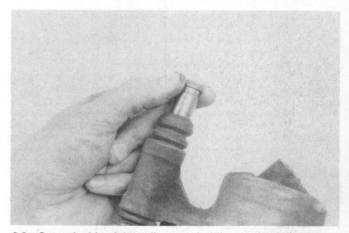

3.8  On each side of the caliper, push the mounting bolt sleeves through the boot and pull them free, then remove the dust boots (not applicable to all calipers)

is harmful to your health. Never blow it out with compressed air and don't inhale any of it. An approved filtering mask should be worn when working on the brakes. Do not, under any circumstances, use petroleum-based solvents to clean brake parts. Use brake cleaner or denatured alcohol only!

**Warning:** This procedure should not be undertaken on a vehicle equipped with an ALB (Anti-Lock Brake) system, since special tools are needed to properly bleed the brakes. Take the vehicle to a dealer service department or other repair shop that has the proper tools.

**Note:** If an overhaul is indicated (usually because of fluid leakage), explore all options before beginning the job. New and factory rebuilt calipers are available on an exchange basis, which makes this job quite easy. If you decide to rebuild the calipers, make sure a rebuild kit is available before proceeding. Always rebuild the calipers in pairs – never rebuild just one of them.

## Removal

*Refer to illustration 3.2*

1   Loosen – but don't remove – the lug nuts on the front wheels. Raise the front of the vehicle and place it securely on jackstands. Remove the front wheels.
2   Disconnect the brake line **(see illustrations 2.6c and 2.6j)** from the caliper and plug it **(see illustration)** to keep contaminants out of the brake system and to prevent losing any more brake fluid than is necessary.
3   Remove the caliper (see Section 2).

## Overhaul

*Refer to illustrations 3.5, 3.7, 3.8, 3.13 and 3.14*

**Note:** In addition to the illustrations accompanying this Section, refer to the exploded views accompanying Section 2 in this Chapter. The models covered by this book include a number of different caliper assemblies. They're all similar in design, but when you buy a caliper rebuild kit, be sure to tell your dealer or auto parts store the year and model of your vehicle so you don't get the wrong kit.

4   Place the caliper on a clean workbench. If there are any pad retainers in the caliper, note how they're installed, then remove them. Pry out the dust boot.
5   Before you remove the piston, place a wood block between the piston and caliper to prevent damage as it is removed. To remove the piston from the caliper, apply compressed air to the brake fluid hose connection on the caliper body **(see illustration)**. Use only enough pressure to ease the piston out of its bore. **Warning:** *Be careful not to place your fingers between the piston and the caliper, as the piston may come out with some force.*
6   Inspect the mating surfaces of the piston and caliper bore wall. If there is any scoring, rust, pitting or bright areas, replace the complete caliper unit with a new one.
7   If these components are in good condition, remove the piston seal from the caliper bore using a wooden or plastic tool **(see illustration)**. Metal tools may damage the cylinder bore.
8   Remove all mounting bolt(s), dust boot(s) and sleeve(s) from the caliper ears **(see illustration)**.

# Chapter 9 Brakes

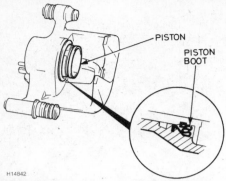

**3.13** With the piston boot positioned in the caliper bore, stretch the boot over the bottom of the piston and push the piston into the bore – the folds of the boot should be even, with no distortion or twist

9  Wash all the components in brake cleaner, clean brake fluid or alcohol.
10  To reassemble the caliper, you should already have the correct rebuild kit for the vehicle.
11  Submerge the new piston seal in brake fluid and install it in the lower groove in the caliper bore.
12  Install the boot in the upper groove in the caliper bore.
13  Lubricate the piston with clean brake fluid, carefully slide it through the new boot, position it squarely in the caliper bore and apply firm (but not excessive) pressure to install it. Make sure the piston boot seats in the groove in the piston **(see illustration)**.
14  If equipped, lubricate the sleeves with silicone-based grease (supplied in the kit) and push them into the caliper ears. Install the boots. Also lubricate the caliper pin(s) with silicone grease **(see illustration)**.

## Installation

15  Install the caliper by reversing the removal procedure. Remember to replace the copper sealing washer on either side of the brake line fitting (they should be included with the rebuild kit).
16  Bleed the brake system (see Section 12).
17  Install the wheels, hand tighten the wheel lug nuts, remove the jackstands and lower the vehicle. Tighten the wheel lug nuts to the torque listed in the Chapter 1 Specifications.

## 4  Disc brake pads (rear) – replacement

*Refer to illustrations 4.6a through 4.6t and 4.11*

**Warning:** *Disc brake pads must be replaced on both rear wheels at the same time – never replace the pads on only one wheel. Also, the dust created by the brake system may contain asbestos, which is harmful to your health. Never blow it out with compressed air and don't inhale any of it. An approved filtering mask should be worn when working on the brakes. Do not, under any circumstances, use petroleum-based solvents to clean brake parts. Use brake cleaner or denatured alcohol only! When servicing the disc brakes, use only high-quality, nationally recognized name brand pads.*

**Note:** *Most models have disc brakes at the front and rear. Some Integra models have rear drum brakes. This Section covers the replacement procedure for rear brake pads; front brake pads are covered in Section 2; rear drum brake shoes are covered in Section 7.*

1  Remove the cap from the brake fluid reservoir.
2  Loosen the rear wheel lug nuts, raise the rear of the vehicle and support it securely on jackstands.
3  Remove the rear wheels. Work on one brake assembly at a time, using the assembled brake for reference if necessary.
4  Inspect the brake disc carefully as outlined in Section 6. If machining is necessary, follow the information in that Section to remove the disc, at which time the calipers and pads can be removed as well.

**3.14  Apply a thin film of silicone grease to the pin(s)**

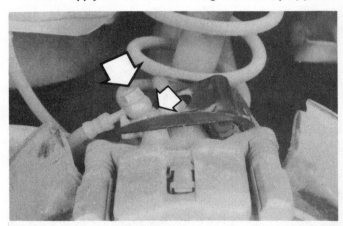

**4.6a  On Integra models, if you're going to replace the caliper, disconnect the brake hose banjo fitting (left arrow) and discard the two sealing washers; use new ones on reassembly. Remove the upper bolt (arrow) . . .**

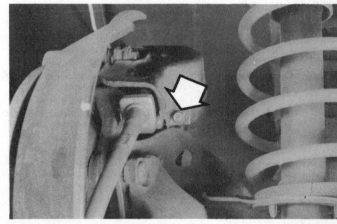

**4.6b  . . . and the lower bolt (arrow) for the caliper protector and remove the protector, . . .**

5  Before you remove anything, spray the caliper and brake pads with brake cleaner to remove the dust produced by brake pad wear **(see illustration 2.6a)**.
6  Follow illustrations **4.6a** through **4.6j** (Integra models) or **4.6k** through **4.6t** (Legend models), for the actual pad replacement procedure. Be sure to stay in order and read the caption under each illustration. When those Steps have been completed, proceed to Step 7.

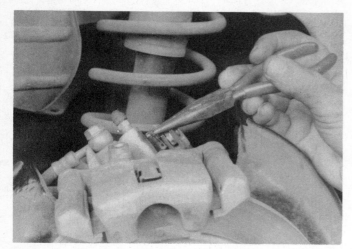
4.6c ... remove the cotter pin from the clevis pin for the parking brake cable, ...

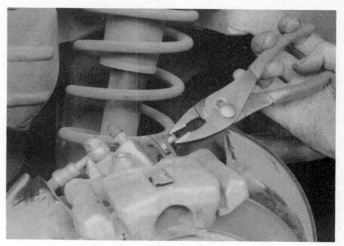
4.6d ... remove the clevis pin, ...

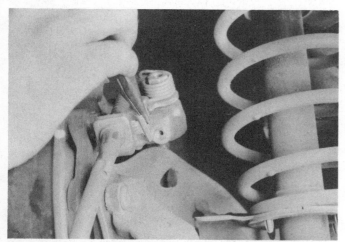
4.6e ... remove the clip from the bracket for the parking brake cable, pull the parking brake cable through the bracket, set it aside, ...

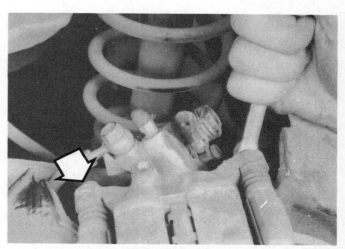
4.6f ... remove the two caliper mounting bolts, ...

4.6g ... lift the caliper off and hang it out of the way with a piece of rope or wire, ...

4.6h ... remove the outer pad shim, ...

4.6i  ... remove the outer brake pad, ...

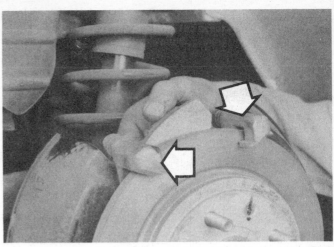

4.6j  ... and remove the inner pad and shim – if you're planning to remove the caliper mounting bracket and/or the brake disc, note how the pad retainers (arrows) are installed in the bracket, then remove them

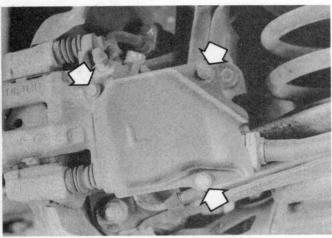

4.6k  On Legend models, remove the caliper protector bolts and the protector ...

4.6l  ... remove the cotter pin from the clevis pin for the parking brake cable ...

4.6m  ... remove the clevis pin, ...

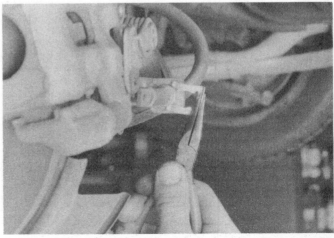

4.6n  ... remove the clip from the bracket for the parking brake cable, ...

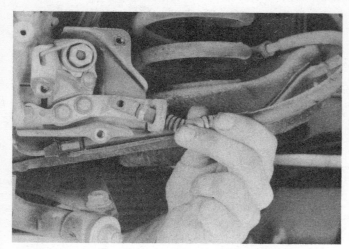

4.6o ... pull the cable through the bracket, set it aside, ...

4.6p ... remove the caliper mounting bolts (wrench on lower bolt, upper bolt shown by left arrow – right arrow is the brake hose fitting bolt), ...

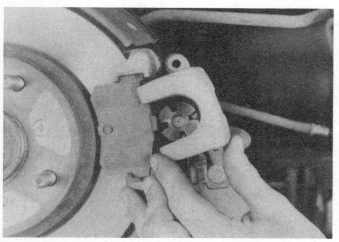

4.6q ... lift the caliper off the pads, ...

4.6r ... remove the outer pad shim, then the pad, ...

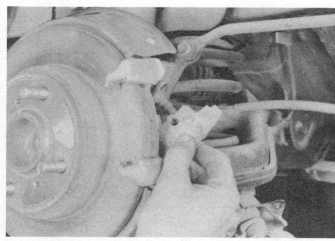

4.6s ... remove the inner pad shim, ...

4.6t ... and remove the inner brake pad – if you're planning to remove the caliper mounting bracket and/or the brake disc, note how the pad retainers are installed in the bracket, then remove them

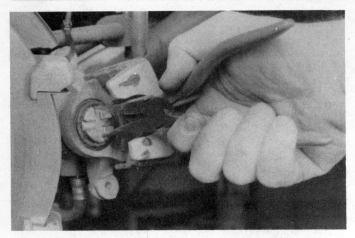

**4.11 To provide enough clearance between the caliper piston and the disc, back the piston into the caliper bore by rotating it with a pair of needle-nose pliers**

**6.2a Before you can remove the front disc, you'll have to remove these caliper mounting bracket-to-steering knuckle bolts (arrows) and the bracket**

7  Apply a thin coat of disc brake anti-squeal compound, in accordance with the manufacturer's recommendations, on the backing plates of the new pads **(see illustration 2.7)**.
8  Install the shims onto their respective pads.
9  If you removed them, install the pad retainers in the caliper mounting bracket. Lubricate the retainers with a thin film of silicone grease.
10  Install the new pads and shims into the caliper mounting bracket.
11  Retract the piston by engaging the tips of a pair of needle-nose pliers with two of the grooves in the top of the piston and turning it clockwise until it bottoms out **(see illustration)**. Now, rotate the piston out until one of its grooves is aligned with the tab on the inner brake pad when you install the caliper. You may have to adjust the piston position by turning it back and forth until the tab fits. If the piston dust boot becomes distorted when the piston is turned, turn the piston in the opposite direction to restore the shape of the boot, but make sure the cut-out still lines up.
12  Install the caliper protector.
13  Install the wheel and lug nuts, lower the vehicle and tighten the lug nuts to the torque specified in Chapter 1.
14  Check the brake fluid level and add fluid, if necessary (see Chapter 1).
15  Apply and release the brake pedal, and the hand brake lever, several times to bring the pads into contact with the brake discs. Check the operation of the brakes in an isolated area before driving the vehicle in traffic.

**5  Disc brake caliper (rear) – removal and installation**

**Warning:** *Dust created by the brake system may contain asbestos, which is harmful to your health. Never blow it out with compressed air and don't inhale any of it. An approved filtering mask should be worn when working on the brakes. Do not, under any circumstances, use petroleum-based solvents to clean brake parts. Use brake cleaner or denatured alcohol only!*
**Warning:** *This procedure should not be undertaken on a vehicle equipped with an ALB (Anti-Lock Brake) system, since special tools are needed to properly bleed the brakes. Take the vehicle to a dealer service department or other repair shop that has the proper tools.*

### Removal

1  Loosen – but don't remove – the lug nuts on the rear wheels. Raise the rear of the vehicle and place it securely on jackstands. Remove the rear wheels.
2  Remove the caliper protector, remove the cotter pin from the clevis pin that connects the parking brake cable to the parking brake lever, pull out the pin and detach the cable (see Section 4).
3  Disconnect the brake hose fitting from the caliper **(see illustrations 4.6c or 4.6p)** and plug it to keep contaminants out of the brake system and to prevent losing any more brake fluid than is necessary **(see illustration 3.2)**.

**6.2b To remove the rear disc, remove these caliper mounting bracket-to-spindle bolts (arrows) and the bracket**

4  Remove the caliper mounting bolts and lift the caliper off its mounting bracket (see Section 4).
5  Disassembly of the rear caliper requires special tools not generally available to the home mechanic. If the rear caliper needs to be overhauled, take it to a dealer service department or other repair shop or obtain a rebuilt unit from an auto parts store or dealer service department.

### Installation

6  Install the caliper by reversing the removal procedure. Remember to replace the copper sealing washers on either side of the brake line fitting.
7  Bleed the brake system (see Section 12).
8  Install the wheels, hand tighten the wheel lug nuts, remove the jackstands and lower the vehicle. Tighten the wheel lug nuts to the torque listed in the Chapter 1 Specifications.

**6  Brake disc – inspection, removal and installation**

**Note:** *This procedure applies to both the front and rear brake discs (on vehicles so equipped).*

### Inspection

*Refer to illustrations 6.2a, 6.2b, 6.3, 6.4a, 6.4b, 6.5a and 6.5b*

1  Loosen the wheel lug nuts, raise the vehicle and support it securely on jackstands. Remove the wheel and install two lug nuts with 3 mm thick

6.3  The brake pads on this vehicle were obviously neglected, as they wore down to the rivets; the rivets then cut deep grooves into the disc, and now the disc must be replaced

6.4a  Make sure the disc retaining screws or lug nuts are tight, then rotate the disc and check the runout with a dial indicator – if the reading exceeds the maximum allowable runout limit, the disc will have to be machined or replaced

6.4b  Using a swirling motion, remove the glaze from the disc with emery cloth or sandpaper

6.5a  The minimum allowable thickness is stamped into the disc

washers under them to hold the disc in place (if the two disc retaining screws are still in place, this will be unnecessary). If you're removing the rear disc, release the parking brake.

2  Remove the front or rear brake caliper (see Section 3 or 5). It's not necessary to disconnect the brake hose. After removing the caliper bolts, suspend the caliper out of the way with a piece of wire. Remove the two caliper mounting bracket-to-steering knuckle bolts **(see illustration)** or, on rear calipers, the bracket-to-spindle bolts **(see illustration)**, and remove the mounting bracket.

3  Visually inspect the disc surface for scoring or damage. Light scratches and shallow grooves are normal after use and may not always be detrimental to brake operation, but deep scoring (over 0.015 inch) requires refinishing by an automotive machine shop. Be sure to check both sides of the disc **(see illustration)**.

4  If you've noted pulsation during braking, suspect disc runout. To check disc runout, place a dial indicator at a point about 1/2-inch from the outer edge of the disc **(see illustration)**. Set the indicator to zero and turn the disc. The indicator reading should not exceed the specified allowable runout limit. If it does, have the disc refinished by an automotive machine shop. **Note:** *You should resurface the discs regardless of the dial indicator reading, as this will impart a smooth finish and ensure a perfectly flat surface, eliminating any brake pedal pulsation or other undesirable symptoms related to questionable discs. At the very least, if you elect not to have the discs resurfaced, remove the glazing from the surface with*

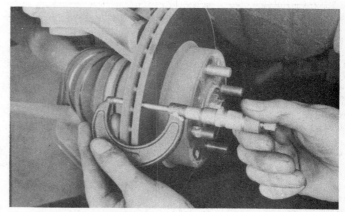

6.5b  A micrometer is used to measure disc thickness

emery cloth or sandpaper using a swirling motion **(see illustration)**.

5  It is absolutely critical that the disc not be machined to a thickness less than the minimum allowable thickness. The minimum wear (or discard) thickness is stamped on the disc **(see illustration)**. The disc thickness can be checked with a micrometer **(see illustration)**.

6.6a  If the disc retaining screws are stuck, use an impact screwdriver to loosen them

6.6b  If the disc is stuck, thread two bolts into the holes in the disc and tighten them

6.6c  As you remove the disc, make sure you don't damage the threads on the studs for the wheel lug nuts

7.2  If the drum is hard to pull off, thread a pair of 8 mm bolts into the holes provided and press the drum off

### Removal

*Refer to illustration 6.6a, 6.6b and 6.6c*

6  Remove the two lug nuts which were put on to hold the disc in place, or the two disc retaining screws, if present **(see illustration)** and remove the disc from the hub. If the disc is stuck to the hub and won't come off, thread two bolts into the holes provided **(see illustration)** and tighten them. Alternate between the bolts, turning them a couple of turns at a time, until the disc is free **(see illustration)**.

### Installation

7  Place the disc in position over the threaded studs.
8  Install the caliper mounting bracket, brake pads and caliper over the disc. Tighten the mounting bracket and caliper bolts to the specified torque.
9  Install the wheel, then lower the vehicle to the ground. Depress the brake pedal a few times to bring the brake pads into contact with the disc. Bleeding of the system will not be necessary unless the fluid hose was disconnected from the caliper. Check the operation of the brakes carefully before placing the vehicle into normal service.

### 7  Drum brake shoes – replacement

*Refer to illustrations 7.2, 7.4a through 7.4s and 7.5*

**Warning:** *Drum brake shoes must be replaced on both wheels at the same time – never replace the shoes on only one wheel. Also, the dust created by the brake system may contain asbestos, which is harmful to your health. Never blow it out with compressed air and don't inhale any of it. An approved filtering mask should be worn when working on the brakes. Do not, under any circumstances, use petroleum-based solvents to clean brake parts. Use brake cleaner or denatured alcohol only!*

**Caution:** *Whenever the brake shoes are replaced, the return and holddown springs should also be replaced. Due to the continuous heating/cooling cycle that the springs are subjected to, they lose their tension over a period of time and may allow the shoes to drag on the drum and wear at a much faster rate than normal. When replacing the rear brake shoes, use only high-quality, nationally recognized brand-name parts.*

1  Loosen the wheel lug nuts, raise the rear of the vehicle and support it securely on jackstands. Block the front wheels to keep the vehicle from rolling. Remove the rear wheels. Release the parking brake.
2  Remove the brake drum. It should simply pull straight off the hub. If the drums won't come off, tap them carefully with a soft-faced mallet, or screw a couple of 8.0 mm bolts into the tapped holes **(see illustration)**. If it still won't budge, the shoes have probably carved wear grooves into the drum. To get the drum off, you'll have to retract them. Remove the rubber plug in the backing plate. Use one screwdriver inserted through the hole in the backing plate to hold the self-adjuster lever away from the adjuster bolt **(see illustration 7.4a)**, then use another screwdriver to rotate the adjuster bolt until the drum can be removed.
3  Replacing the shoes is a lot easier if you remove the rear wheel bearing cap, spindle nut and washer, and slide off the hub unit.

# Chapter 9 Brakes

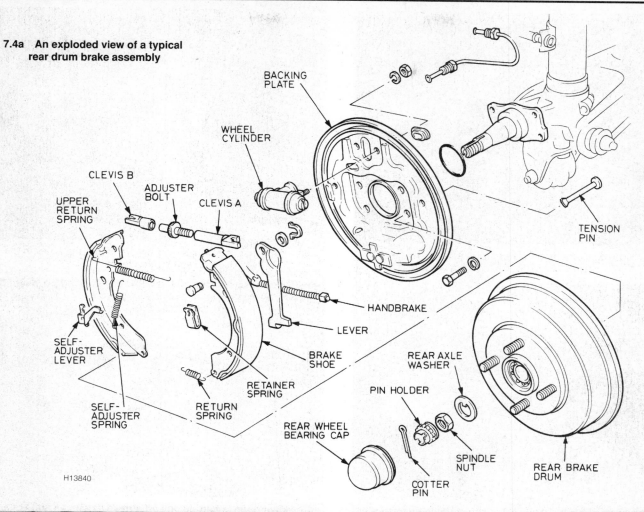

7.4a An exploded view of a typical rear drum brake assembly

7.4b Before removing anything, clean the brake assembly with brake cleaner and allow it to dry – position a drain pan under the brake to catch the residue – DO NOT USE COMPRESSED AIR TO BLOW THE DUST FROM THE PARTS!

7.4c Push down on the retainer spring with a screwdriver, then turn the tension pin to align its blade with the slot in the retainer spring – the spring should pop off (repeat this on the other spring)

4  Follow the accompanying illustrations (**7.4a through 7.4s**) for the inspection and replacement of the brake shoes. Be sure to stay in order and read the caption under each illustration. All four rear brake shoes must be replaced at the same time, but to avoid mixing up parts, work on only one brake assembly at a time.

5  Before reinstalling the drum it should be checked for cracks, score marks, deep scratches and hard spots, which will appear as small discolored areas. If the hard spots cannot be removed with fine emery cloth or if any of the other conditions listed above exist, the drum must be taken to an automotive machine shop to have it turned. **Note:** *Professionals recom-*

7.4d  Pull the shoe assembly down and over the spindle

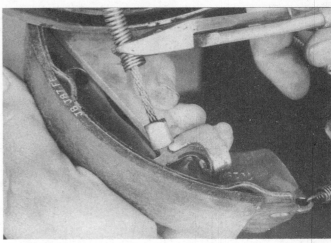

7.4e  Using a pair of diagonal cutting pliers, pull back on the parking brake cable spring and squeeze the pliers just enough to grip the cable, holding the spring in the compressed position (be careful not to cut the cable); unhook the cable end from the parking brake lever

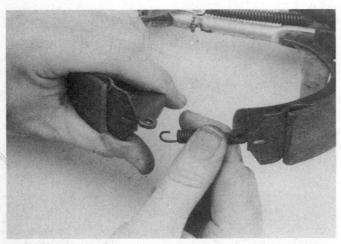

7.4f  With the brake shoe assembly on a clean working surface, unhook the lower return spring from the shoes

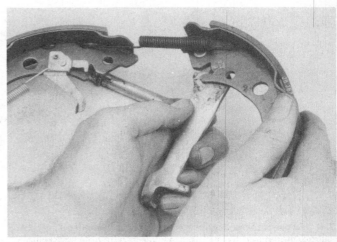

7.4g  Swing the parking brake lever away from the trailing shoe, which will force the adjuster bolt clevis out of its groove in the shoe; the two shoes can now be separated

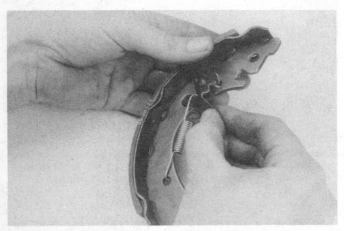

7.4h  Remove the self adjuster lever and spring from the leading shoe

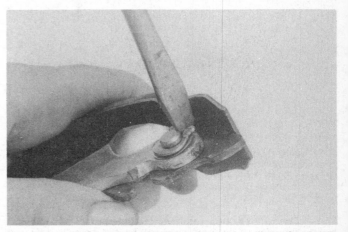

7.4i  Pry open the parking brake lever retaining clip and separate the lever from the shoe; be careful not to lose the wave washer that is under the clip

7.4j  Put the new trailing shoe on the lever, place the wave washer over the pin, then install the retaining clip; crimp the ends of the clip together with a pair of needle-nose pliers

7.4k  Clean the adjuster bolt and clevis, then lubricate the threads and ends with high-temperature grease

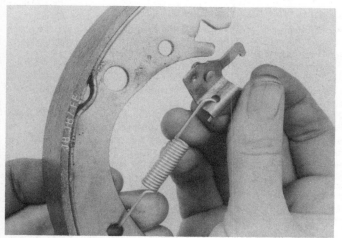

7.4l  Connect the self adjuster lever spring to the leading brake shoe, then insert the pin on the lever into its hole in the shoe

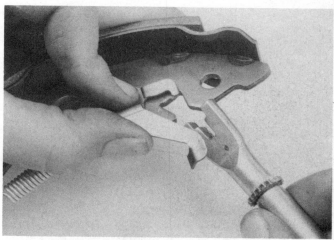

7.4m  Insert the short clevis of the adjuster bolt into its slot in the leading shoe, making sure it catches the self adjuster lever

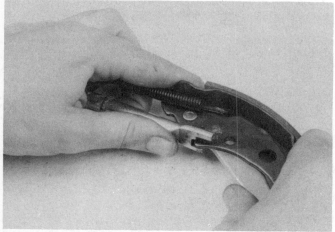

7.4n  Connect the upper return spring between the two shoes, pry the lower ends of the shoes apart and insert the clevis at the other end of the adjuster bolt into the slot in the shoe; notice the position of the stepped portion of the clevis opening

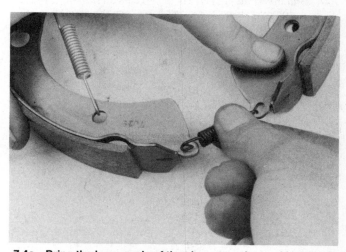

7.4o  Bring the lower ends of the shoes together and install the lower return spring

# 9-18  Chapter 9  Brakes

7.4p  Lubricate the brake shoe contact areas on the backing plate with high-temperature grease

7.4r  Place the brake shoe assembly against the backing plate and slide it up, engaging the upper ends of the shoes in the slots in the wheel cylinder pistons

7.4s  With the brake shoes in position on the backing plate, pass the tension pins through the holes in the backing plate and brake shoes, then install the retainer springs (see illustration 7.4c) – make sure the parking brake cable spring and the lower return spring are seated behind the anchor plate, as shown here

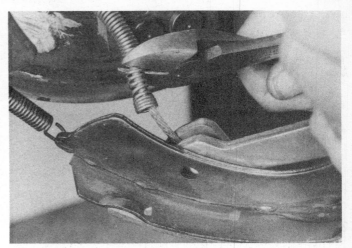

7.4q  Compress the parking brake cable spring, hold it in position and connect the cable end to the parking brake lever

mend resurfacing the drums whenever a brake job is done. Resurfacing will eliminate the possibility of out-of-round drums. If the drums are worn so much that they can't be resurfaced without exceeding the maximum allowable diameter (stamped into the drum) **(see illustration)**, then new ones will be required. At the very least, if you elect not to have the drums resurfaced, remove the glazing from the surface with sandpaper or emery cloth using a swirling motion.

6  Install the hub and bearing unit, the washer and a new spindle nut (if removed previously). Tighten the nut to the torque listed in the Chapter 10 Specifications. Install the brake drum.

7  Mount the wheel, install the lug nuts, then lower the vehicle. Tighten the lug nuts to the torque listed in the Chapter 1 Specifications.

8  Make a number of forward and reverse stops to adjust the brakes until satisfactory pedal action is obtained.

9  Check brake operation before driving the vehicle in traffic.

## 8  Wheel cylinder – removal, overhaul and installation

**Note 1:** *This procedure applies to models with rear drum brakes only.*

**Note 2:** *If an overhaul is indicated (usually because of fluid leakage or sticky operation), explore all options before beginning the job. New wheel cylinders are available, which makes this job quite easy. If you decide to rebuild the wheel cylinder, make sure that a rebuild kit is available before proceeding. Never overhaul one wheel cylinder – always rebuild both of them at the same time.*

7.5  The maximum diameter is cast into the drum

# Chapter 9  Brakes

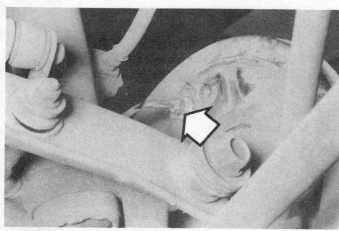

**8.4 Unscrew the brake line fitting (arrow), then remove the two mounting nuts or bolts**

## Removal
*Refer to illustration 8.4*

1  Raise the rear of the vehicle and support it securely on jackstands. Block the front wheels to keep the vehicle from rolling.
2  Remove the brake shoe assembly (see Section 7).
3  Remove all dirt and foreign material from around the wheel cylinder.
4  Unscrew the brake line fitting **(see illustration)**. Don't pull the brake line away from the wheel cylinder.
5  Remove the wheel cylinder mounting fasteners.
6  Detach the wheel cylinder from the brake backing plate and place it on a clean workbench. Immediately plug the brake line to prevent fluid loss and contamination. **Note:** *If the brake shoe linings are contaminated with brake fluid, install new brake shoes and clean the drums with brake cleaner.*

## Overhaul
*Refer to illustration 8.7*

7  Remove the bleed screw, dust covers, pistons, piston cups and spring assembly from the wheel cylinder body **(see illustration)**.
8  Clean the wheel cylinder with brake fluid, denatured alcohol or brake system cleaner. **Warning:** *Do not, under any circumstances, use petroleum-based solvents to clean brake parts!*
9  Use compressed air to remove excess fluid from the wheel cylinder and to blow out the passages.
10  Check the cylinder bore for corrosion and score marks. Crocus cloth can be used to remove light corrosion and stains, but the cylinder must be replaced with a new one if the defects cannot be removed easily, or if the bore is scored.
11  Lubricate the new cups with brake fluid.
12  Assemble the wheel cylinder components. Make sure the cup lips face in.

## Installation

13  Apply silicone sealant to the mating surface of the wheel cylinder and the brake backing plate, place the cylinder in position and connect the brake line. Don't tighten the fitting completely yet.
14  Install the mounting bolts, tightening them securely. Tighten the brake line fitting. Install the brake shoe assembly.
15  Bleed the brakes (see Section 12).
16  Check brake operation before driving the vehicle in traffic.

## 9  Master cylinder – removal, overhaul and installation

**Warning:** *This procedure should not be undertaken on a vehicle equipped with an ALB (Anti-Lock Brake) system, since special tools are needed to properly bleed the brakes. Take the vehicle to a dealer service department or other repair shop that has the proper tools.*
**Note:** *Before you decide whether to overhaul the master cylinder, compare the cost of a new or factory rebuilt unit versus the cost of a rebuild kit. If you've got a 1990 Legend, the master cylinder can't be rebuilt.*

### Removal
*Refer to illustrations 9.4 and 9.6*

1  The master cylinder is located in the engine compartment, mounted to the power brake booster.
2  Remove as much fluid as you can from the reservoir with a syringe.
3  Place rags under the fluid fittings and prepare caps or plastic bags to cover the ends of the lines once they are disconnected. **Caution:** *Brake fluid will damage paint. Cover all body parts and be careful not to spill fluid during this procedure.*

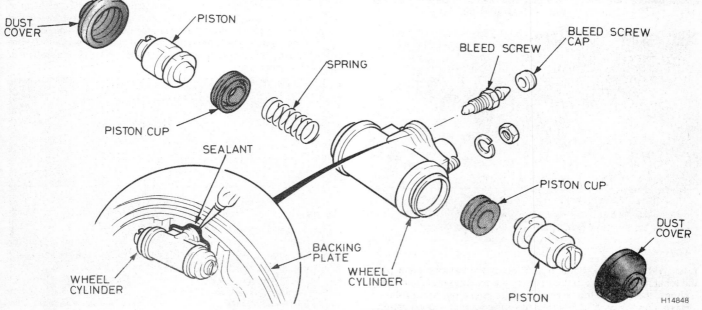

**8.7 An exploded view of a typical wheel cylinder assembly**

## Chapter 9  Brakes

9.4 Use a flare-nut wrench to remove the threaded fittings at the master cylinder – a regular wrench can strip the corners off the soft metal of these fittings

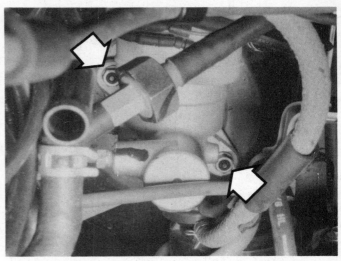

9.6 To detach the master cylinder from the brake booster, remove these two nuts (arrows)

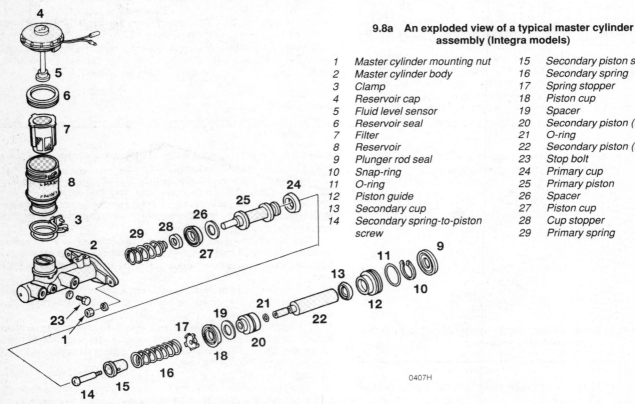

9.8a An exploded view of a typical master cylinder assembly (Integra models)

1  Master cylinder mounting nut
2  Master cylinder body
3  Clamp
4  Reservoir cap
5  Fluid level sensor
6  Reservoir seal
7  Filter
8  Reservoir
9  Plunger rod seal
10  Snap-ring
11  O-ring
12  Piston guide
13  Secondary cup
14  Secondary spring-to-piston screw
15  Secondary piston stopper
16  Secondary spring
17  Spring stopper
18  Piston cup
19  Spacer
20  Secondary piston (front half)
21  O-ring
22  Secondary piston (rear half)
23  Stop bolt
24  Primary cup
25  Primary piston
26  Spacer
27  Piston cup
28  Cup stopper
29  Primary spring

4  Loosen the tube nuts at the ends of the brake lines where they enter the master cylinder (see illustration). To prevent rounding off the flats on these nuts, the use of a flare-nut wrench, which wraps around the nut, is preferred.
5  Pull the brake lines slightly away from the master cylinder and plug the ends to prevent contamination.
6  Disconnect the electrical connector at the master cylinder, then remove the nuts attaching the master cylinder to the power booster (see illustration). Pull the master cylinder off the studs and out of the engine compartment. Again, be careful not to spill the fluid as this is done.

### Overhaul

Refer to illustrations 9.8a, 9.8b and 9.10

7  Before attempting the overhaul of the master cylinder, obtain the proper rebuild kit, which will contain the necessary replacement parts and also any instructions which may be specific to your model.
8  Remove the filter from the reservoir and clean it thoroughly. Pour out any fluid still in the reservoir. On Integra models, loosen the reservoir clamp and pull the reservoir off the master cylinder body (see illustration); on Legend models, remove the reservoir retaining screw (see illus-

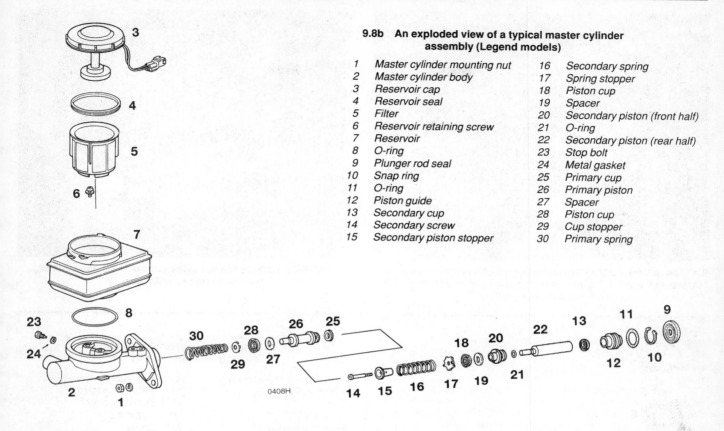

**9.8b  An exploded view of a typical master cylinder assembly (Legend models)**

1  Master cylinder mounting nut
2  Master cylinder body
3  Reservoir cap
4  Reservoir seal
5  Filter
6  Reservoir retaining screw
7  Reservoir
8  O-ring
9  Plunger rod seal
10  Snap ring
11  O-ring
12  Piston guide
13  Secondary cup
14  Secondary screw
15  Secondary piston stopper
16  Secondary spring
17  Spring stopper
18  Piston cup
19  Spacer
20  Secondary piston (front half)
21  O-ring
22  Secondary piston (rear half)
23  Stop bolt
24  Metal gasket
25  Primary cup
26  Primary piston
27  Spacer
28  Piston cup
29  Cup stopper
30  Primary spring

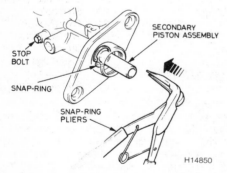

**9.10  Push in on the secondary piston assembly and remove the stop bolt and snap-ring**

tration) and carefully pry the reservoir loose from the master cylinder body.
9  Remove the plunger rod seal.
10  Push in the secondary piston assembly and remove the snap-ring with a pair of snap-ring pliers **(see illustration)**.
11  Place the cylinder in a vise and use a punch or Phillips screwdriver to depress the secondary piston assembly until the internal components bottom against the other end of the master cylinder. Hold the pistons in this position and remove the stop bolt on the side of the master cylinder.
12  Remove the internal components from the cylinder bore in the exact order shown in **illustration 9.8a or 9.8b**. If the number or shape of any of the parts in the master cylinder you're overhauling are different in any way from the typical Integra and Legend units shown, make sure you lay them out in the exact order in which you remove them to ensure that they're reinstalled in the correct sequence. **Note:** *The two springs are of different tension, so pay particular attention to their order.*
13  To disassemble the secondary piston subassembly, remove the Phillips screw that attaches the secondary spring to the piston.
14  Carefully inspect the bore of the master cylinder. Any deep scoring or other damage will mean a new master cylinder is required. DO NOT attempt to hone the master cylinder.
15  Replace all parts included in the rebuild kit, following any instructions in the kit. Clean all reused parts with brake system cleaner or clean brake fluid only. During assembly, lubricate all parts liberally with clean brake fluid.
16  After you've reassembled the secondary piston subassembly, reattach the secondary spring to the piston and secure it with the Phillips screw.
17  Push the assembled components into the bore, bottoming them against the end of the master cylinder, then install the stop bolt and a new copper washer. A screwdriver can be used to bottom the components, but be careful not to scratch the bore.
18  Install the snap-ring and a new plunger rod seal.
19  Install the reservoir. On Integra models, tighten the clamp securely; on Legend models, make sure the reservoir is pushed all the way down into the master cylinder, then install the reservoir retaining screw (don't overtighten it).
20  Before installing the new master cylinder it should be bench bled. Because it will be necessary to apply pressure to the master cylinder piston and, at the same time, control flow from the brake line outlets, it is recommended that the master cylinder be mounted in a vise, with the jaws of the vise clamping on the mounting flange.
21  Insert threaded plugs into the brake line outlet holes and snug them down so that there will be no air leakage past them, but not so tight that they cannot be easily loosened.
22  Fill the reservoir with brake fluid of the recommended type (see Chapter 1).

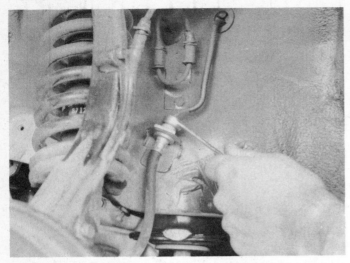

11.4a Use a flare-nut wrench to break loose the brake line-to-hose fitting . . .

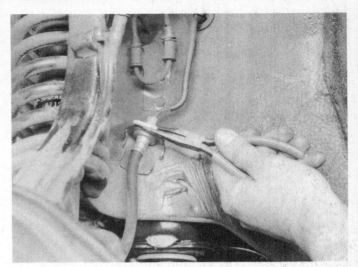

11.4b . . . then remove the clip and slide the hose out of the bracket

23 Remove one plug and push the piston assembly into the master cylinder bore to expel the air from the master cylinder. A large Phillips screwdriver can be used to push on the piston assembly.

24 To prevent air from being drawn back into the master cylinder, the plug must be replaced and snugged down before releasing the pressure on the piston assembly.

25 Repeat the procedure until only brake fluid is expelled from the brake line outlet hole. When only brake fluid is expelled, repeat the procedure with the other outlet hole and plug. Be sure to keep the master cylinder reservoir filled with brake fluid to prevent the introduction of air into the system.

26 Since high pressure is not involved in the bench bleeding procedure, an alternative to the removal and replacement of the plugs with each stroke of the piston assembly is available. Before pushing in on the piston assembly, remove the plug as described in Step 23. Before releasing the piston, however, instead of replacing the plug, simply put your finger tightly over the hole to keep air from being drawn back into the master cylinder. Wait several seconds for brake fluid to be drawn from the reservoir into the piston bore, then depress the piston again, removing your finger as brake fluid is expelled. Be sure to put your finger back over the hole each time before releasing the piston, and when the bleeding procedure is complete for that outlet, replace the plug and snug it before going on to the other port.

## Installation

27 Install the master cylinder over the studs on the power brake booster and tighten the attaching nuts only finger tight at this time.

28 Thread the brake line fittings into the master cylinder. Since the master cylinder is still a bit loose, it can be moved slightly in order for the fittings to thread in easily. Do not strip the threads as the fittings are tightened.

29 Fully tighten the mounting nuts and the brake fittings.

30 Fill the master cylinder reservoir with fluid, then bleed the master cylinder and the brake system as described in Section 12. To bleed the cylinder on the vehicle, have an assistant pump the brake pedal several times and then hold the pedal to the floor. Loosen the fitting nut to allow air and fluid to escape. Repeat this procedure on both fittings until the fluid is clear of air bubbles. Test the operation of the brake system carefully before placing the vehicle in normal service.

## 10 Proportioning valve – general information, removal and installation

1 On Integra models, the proportioning valve is mounted on the right side strut tower; on Legend models, it's mounted in front of the master cylinder, near the left wheel well. Its purpose is to limit hydraulic pressure to the rear brakes under heavy braking conditions to prevent rear wheel lock-up.

2 The valve is not serviceable; if you suspect it's malfunctioning, have it checked by a dealer service department or repair shop equipped with the necessary pressure gauges.

3 If the valve is defective, replace it by unscrewing the brake lines (using a flare-nut wrench, if available) and unbolting the valve from the strut tower or its mounting bracket (on Integra models, note that the bolt underneath which secures the lower bracket to the valve must not be removed – you'll have to remove the entire bracket from the vehicle). After the new valve is installed, bleed the complete brake system as described in Section 12.

## 11 Brake hoses and lines – inspection and replacement

*Refer to illustrations 11.4a and 11.4b*

**Warning:** *The brake hoses on vehicles equipped with an ALB (Anti-Lock Brake) system should not be disconnected or replaced by the home mechanic, since special tools are required to properly bleed the brakes. Take the vehicle to a dealer service department or other qualified repair shop.*

1 About every six months the flexible hoses which connect the steel brake lines with the rear brakes and front calipers should be inspected for cracks, chafing of the outer cover, leaks, blisters, and other damage.

2 Replacement steel and flexible brake lines are commonly available from dealer parts departments and auto parts stores. Do not, under any circumstances, use anything other than genuine steel lines or approved flexible brake hoses as replacement items.

3 When installing the brake line, leave at least 0.75 in (19 mm) clearance between the line and any moving or vibrating parts.

4 To disconnect a hose and line, use a flare-nut wrench (**see illustration**). Then remove the clip and slide the hose out of the bracket (**see illustration**).

5 When disconnecting two hoses, use normal wrenches on the hose fittings. When connecting two hoses, make sure they are not twisted or strained.

6 Steel brake lines are usually retained along their span with clips. Always remove these clips completely before removing a fixed brake line. Always reinstall these clips, or new ones if the old ones are damaged, when replacing a brake line, as they provide support and keep the lines from vibrating, which can eventually break them.

7 When replacing brake lines be sure to use the correct parts. NEVER use copper tubing! Purchase steel brake lines from a dealer or auto parts store.

8 When installing a steel line, make sure it's securely supported in the

brackets and has plenty of clearance between moving or hot components.
9   After installation, check the fluid level in the master cylinder and add fluid as necessary. Bleed the brake system as described in Section 12 and test the brakes carefully before driving the vehicle in traffic.

## 12   Brake hydraulic system – bleeding

*Refer to illustration 12.8*

**Warning:** *This procedure should not be undertaken on a vehicle equipped with an ALB (Anti-Lock Brake) system, since special tools are needed to properly bleed the brakes. Take the vehicle to a dealer service department or other repair shop that has the proper tools.*

**Warning:** *Wear eye protection when bleeding the brake system. If the fluid comes in contact with your eyes, immediately rinse them with water and seek medical attention.*

1   Bleeding the hydraulic system is necessary to remove any air that manages to find its way into the system when it's been opened during removal and installation of a hose, line, caliper or master cylinder. It will probably be necessary to bleed the system at all four brakes if air has entered the system due to low fluid level, or if the brake lines have been disconnected at the master cylinder.
2   If a brake line was disconnected only at a wheel, then only that caliper or wheel cylinder must be bled.
3   If a brake line is disconnected at a fitting located between the master cylinder and any of the brakes, that part of the system served by the disconnected line must be bled.
4   Remove any residual vacuum from the brake power booster by applying the brake several times with the engine off.
5   Remove the master cylinder reservoir cover and fill the reservoir with brake fluid. Reinstall the cover. **Note:** *Check the fluid level often during the bleeding operation and add fluid as necessary to prevent the fluid level from falling low enough to allow air bubbles into the master cylinder.*
6   Have an assistant on hand, as well as a supply of new brake fluid, a clear container partially filled with clean brake fluid, a length of 3/16-inch plastic, rubber or vinyl tubing to fit over the bleed screw and a wrench to open and close the bleed screw.
7   Beginning at the left front wheel, loosen the bleed screw slightly, then tighten it to a point where it is snug but can still be loosened quickly and easily.
8   Place one end of the tubing over the bleed screw and submerge the other end in brake fluid in the container **(see illustration)**.
9   Have the assistant pump the brakes slowly a few times to get pressure in the system, then hold the pedal firmly depressed.
10   While the pedal is held depressed, open the bleed screw just enough to allow a flow of fluid to leave the screw. Watch for air bubbles to exit the submerged end of the tube. When the fluid flow slows after a couple of seconds, close the screw and have your assistant release the pedal.
11   Repeat Steps 9 and 10 until no more air is seen leaving the tube, then tighten the bleed screw and proceed to the right rear wheel, the right front wheel and the left rear wheel, in that order, and perform the same procedure. Be sure to check the fluid in the master cylinder reservoir frequently.
12   Never use old brake fluid. It contains moisture which will deteriorate the brake system components.
13   Refill the master cylinder with fluid at the end of the operation.
14   Check the operation of the brakes. The pedal should feel solid when depressed, with no sponginess. If necessary, repeat the entire process.

**Warning:** *Do not operate the vehicle if you are in doubt about the effectiveness of the brake system.*

## 13   Power brake booster – check, removal and installation

### *Operating check*

1   Depress the brake pedal several times with the engine off and make sure there is no change in the pedal reserve distance.
2   Depress the pedal and start the engine. If the pedal goes down slightly, operation is normal.

### *Airtightness check*

3   Start the engine and turn it off after one or two minutes. Depress the brake pedal several times slowly. If the pedal goes down farther the first time but gradually rises after the second or third depression, the booster is airtight.
4   Depress the brake pedal while the engine is running, then stop the engine with the pedal depressed. If there is no change in the pedal reserve travel after holding the pedal for 30 seconds, the booster is airtight.

### *Removal*

5   Power brake booster units should not be disassembled. They require special tools not normally found in most automotive repair stations or shops. They are fairly complex and because of their critical relationship to brake performance it is best to replace a defective booster unit with a new or rebuilt one.
6   To remove the booster, first remove the brake master cylinder as described in Section 9.
7   Locate the pushrod clevis pin connecting the booster to the brake pedal. This is accessible from under the dash panel in front of the driver's seat.
8   Remove the cotter pin with pliers and pull out the clevis pin.
9   Disconnect the hose leading from the engine to the booster. Be careful not to damage the hose when removing it from the booster fitting.
10   Remove the four nuts and washers holding the brake booster to the firewall. You may need a light to see these, as they are up under the dash area.
11   Slide the booster straight out from the firewall until the studs clear the holes and pull the booster, brackets and gaskets from the engine compartment area.

### *Installation*

*Refer to illustrations 13.13a and 13.13b*

12   Installation procedures are basically the reverse of those for removal. Tighten the booster mounting nuts to the specified torque. Also, be sure to use a new cotter pin on the clevis pin.

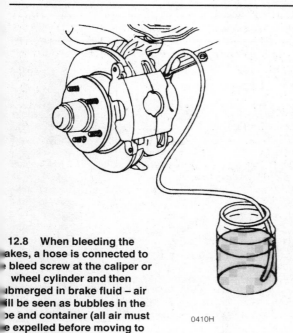

**12.8   When bleeding the brakes, a hose is connected to the bleed screw at the caliper or wheel cylinder and then submerged in brake fluid – air will be seen as bubbles in the tube and container (all air must be expelled before moving to the next wheel)**

# Chapter 9  Brakes

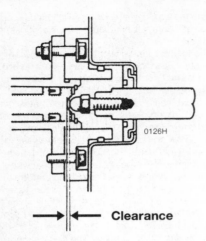

**13.13a  The booster pushrod-to-master cylinder clearance must be as specified – if there is interference between the two, the brakes may drag; if there is too much clearance, there will be excessive brake pedal travel**

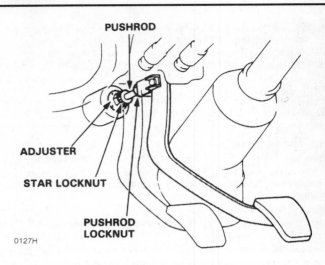

**13.13b  To adjust the length of the booster pushrod, loosen the star locknut and turn the adjuster in or out, as necessary, to achieve the desired setting**

13. If the power booster unit is being replaced, the clearance between the master cylinder piston and the pushrod in the vacuum booster must be measured. Using a depth micrometer or vernier calipers, measure the distance from the seat (recessed area) in the master cylinder piston to the master cylinder mounting flange. Next, apply a vacuum of 20 in-Hg to the booster (using a hand vacuum pump) and measure the distance from the end of the vacuum booster pushrod to the mounting face of the booster (including gasket, if used) where the master cylinder mounting flange seats. Subtract the two measurements to get the clearance **(see illustration)**. If the clearance is more or less than specified, loosen the star locknut and turn the adjuster on the power booster pushrod until the clearance is within the specified limit **(see illustration)**. After adjustment, tighten the locknut.

14. After the final installation of the master cylinder and brake hoses and lines, bleed the brakes as described in Section 12.

## 14  Parking brake – adjustment

*Refer to illustrations 14.4a and 14.4b*

1. Refer to Chapter 11 and remove the console trim around the parking brake lever.
2. Remove the center console (Legend) or the equalizer cover located behind the console (Integra).
3. Block the front wheels, raise the rear of the vehicle and support it securely on jackstands. Apply the parking brake lever until you hear one click.
4. Turn the adjusting nut on the equalizer **(see illustrations)** clockwise while rotating the rear wheels. Stop turning the nut when the brakes just start to drag on the rear wheels.
5. Release the parking brake lever and check to see that the brakes don't drag when the rear wheels are turned. The travel on the parking brake lever should be as listed in the Chapter 1 Specifications when properly adjusted.
6. Lower the vehicle and reinstall the console or cover.

## 15  Parking brake cable(s) – replacement

*Refer to illustrations 15.3, 15.4a and 15.4b*

1. Block the front wheels and loosen the rear wheel lug nuts. Raise the rear of the vehicle and support it securely on jackstands.

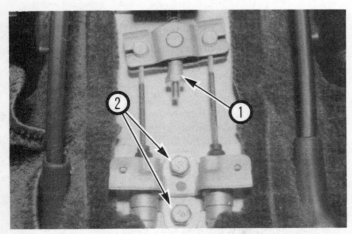

**14.4a  On Integra models, the adjusting nut (1) is on the equalizer assembly – when removing a cable, unscrew the clamp bolts (2) and lift off the clamp**

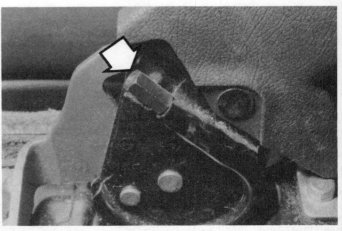

**14.4b  On Legend models, the adjusting nut (arrow) is at the base of the parking brake lever**

# Chapter 9  Brakes

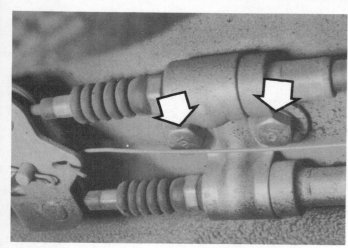

**15.3  On Legend models, after the exhaust system is out of the way, remove the two bolts (arrows) from the clamp**

**15.4a  To detach the parking brake cable from a drum brake, remove the brake shoe and grab the cable end with a pair of needle-nose pliers and pull it out of its slot in the parking brake lever . . .**

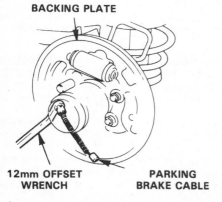

**15.4b  . . . then compress the tangs on the retainer by sliding an offset box wrench over the end of the cable onto the retainer, and pull the cable out the inner side of the backing plate**

8  Installation is the reverse of the removal procedure. After the cable(s) are installed, be sure to adjust them according to the procedure described in Section 12.

## 16  Brake light switch – check, replacement and adjustment

### Check

*Refer to illustration 16.3a and 16.3b*

1  To check the brake light switch, push on the brake pedal and verify that the brake lights come on.
2  If they don't, check the brake light fuse (in the under-hood fusebox on most models – see Chapter 12 or check your owner's manual for fuse locations). Also check the brake light bulbs in both taillight assemblies (don't forget to check the high-mount brake light).
3  Locate the brake light switch at the top of the brake pedal **(see illustrations)**.
4  Unplug the switch connector.

2  On vehicles with rear drum brakes, remove the brake drum(s) (see Section 7).
3  Following the procedure in the previous Section, loosen the cable adjusting nut. Remove the cable clamp from the cable housing (see the accompanying illustration or illustration 14.4a). On Legend models, you must first detach the rear section of the exhaust system and remove a heat shield, since the cables attach to the equalizer at the center of the underside of the floorpan (see Chapter 4). Unhook the cable from the equalizer.
4  On models with rear drum brakes, remove the brake shoes (see Section 7) and disconnect the cable end from the lever on the trailing brake shoe **(see illustration)**. Depress the tangs on the cable housing retainer and pass the cable through the backing plate. You can do this by passing an offset 12mm box end wrench over the end of the cable and onto the retainer **(see illustration)**. This compresses all the tangs simultaneously.
5  On models with rear disc brakes, remove the clip and clevis to disconnect the cable end from the actuator lever on the caliper (see Section 4), then remove the spring clip to free the cable housing from the support bracket.
6  Unbolt the cable housing clamps from the underbody, noting how the cable is routed, then remove the cable from the vehicle. It may be necessary to remove the exhaust pipe heat shield bolts at the rear to allow cable removal.
7  If both cables are to be removed, repeat the above steps to remove the remaining cable.

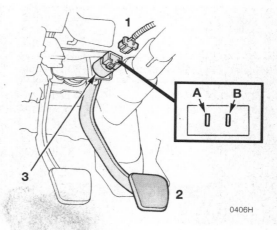

**16.3a  A typical brake light switch assembly (Integra models)**

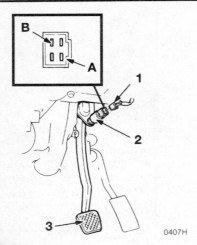

16.3b  A typical brake light switch assembly (Legend models)

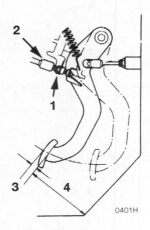

16.9  Typical brake light switch assembly

5  Check for continuity across switch terminals A and B with an ohmmeter. When the brake pedal is depressed, there should be continuity; when it's released, there should be no continuity. If the switch doesn't operate as described, replace it.

### Replacement

6  Disconnect the electrical connector from the switch, if you haven't already done so.
7  Remove the locknut on the pedal side of the switch and unscrew the switch from the bracket.
8  Installation of the brake light switch is the reverse of the removal procedure.

### Adjustment

*Refer to illustration 16.9*

9  Loosen the brake light switch locknut and back off the brake light switch until it's not touching the brake pedal **(see illustration)**.
10  Loosen the pushrod locknut and screw the pushrod in or out with pliers until the pedal height from the floor is correct (listed in this Chapter's Specifications).
11  Tighten the locknut securely.
12  Screw in the brake light switch until its plunger is fully depressed (threaded end touching the pad on the pedal arm), then back off the switch 1/2 turn and tighten the locknut securely.
13  Depress the pedal with your hand and measure the pedal freeplay. It should be within the dimensions listed in this Chapter's Specifications. Make sure the brake lights operate when the pedal is depressed and go off when the pedal is released.

## 17  Anti-Lock Brake (ALB) system – general information

**Warning:** *No part of the hydraulic brake system on a vehicle equipped with an ALB (Anti-Lock Brake) system should be disconnected, as special tools are needed to properly bleed the system. Take the vehicle to a dealer service department or other repair shop for repairs which require opening of the system.*

In a conventional braking system, if you press the brake pedal too hard, the wheels can "lock up" (stop turning) and the vehicle can go into a skid. If the wheels lock up, you can lose control of the vehicle. The Anti-Lock Brake (ALB) system prevents the wheels from locking up by modulating (pulsing on and off) the pressure of the brake fluid at each caliper.

The ALB system has two basic subsystems: One is an electrical system and the other is hydraulic. The electrical half has four "gear pulsers," four wheel sensors, a computer and an electrical circuit connecting all the components. The hydraulic part of the system consists of a solenoid/modulator, the disc brake calipers and the hydraulic fluid lines between the solenoid/modulator and the calipers.

In principle, the system is pretty simple: Each wheel has a wheel sensor monitoring a gear pulser (a ring with evenly-spaced raised ridges cast into its circumference). The wheel sensor "counts" the ridges of the gear pulser as they pass by, converts this information into an electrical output and transmits it back to the computer. The computer constantly "samples" the voltage inputs from all four wheel sensors and compares them to each other. As long as the gear pulsers at all four wheels are rotating at the same speed, the ALB system is inactive. But when a wheel locks up, the voltage signal from that wheel sensor deviates from the signals coming from the other wheels. So the computer "knows" the wheel is locking up. It sends an electrical signal to the solenoid/modulator assembly, which releases the brake fluid pressure to the brake caliper at that wheel. As soon as the wheel unlocks and resumes turning at the same rate of speed as the other wheels, its wheel sensor voltage output once again matches the output of the other wheels and the computer deactivates the signal to the solenoid/modulator.

In reality, the ALB system is far more complex than it sounds, so we don't recommend that you attempt to diagnose or service it. If the ALB system on your vehicle develops problems, take it to a dealer service department or other qualified shop.

# Chapter 10
# Suspension and steering systems

**Contents**

| | |
|---|---|
| Balljoints – replacement | 14 |
| Front shock absorber assembly (Legend models) – removal and installation | 3 |
| Front stabilizer bar and bushings – removal and installation | 10 |
| Front strut assembly (Integra models) – removal and installation | 2 |
| General information | 1 |
| Lower arm (Integra models) – removal and installation | 7 |
| Lower arm (Legend models) – removal and installation | 12 |
| Panhard rod (Integra models) – removal and installation | 18 |
| Power steering fluid level check | See Chapter 1 |
| Power steering pump – removal and installation | 28 |
| Power steering system – bleeding | 29 |
| Radius arm (Integra models) – removal and installation | 8 |
| Radius rod (Legend models) – removal and installation | 11 |
| Rear axle beam (Integra models) – removal and installation | 19 |
| Rear hub and bearing assembly – removal and installation | 16 |
| Rear shock absorber/strut assembly – removal and installation | 15 |
| Rear stabilizer assembly – removal and installation | 21 |
| Rear suspension arms (1986 through 1988 Legend sedans) – removal and installation | 22 |
| Rear suspension arms (1989 and 1990 Legend sedans and all Legend coupes) – removal and installation | 23 |
| Spindle/swing bearing assembly (Integra models) – removal and installation | 20 |
| Steering and suspension check | See Chapter 1 |
| Steering gear boots – replacement | 26 |
| Steering gear – removal and installation | 27 |
| Steering knuckle and hub assembly – removal and installation | 5 |
| Steering/suspension crossmember (Integra models) – removal and installation | 9 |
| Steering wheel – removal and installation | 24 |
| Strut/shock absorber assembly – replacement | 4 |
| Tie-rod ends – removal and installation | 25 |
| Tire and tire pressure check | See Chapter 1 |
| Tire rotation | See Chapter 1 |
| Torsion bar (Integra models) – removal, installation and adjustment | 6 |
| Trailing arm (Integra models) – removal and installation | 17 |
| Upper arm (Legend models) – removal and installation | 13 |
| Wheel alignment – general information | 31 |
| Wheels and tires – general information | 30 |

## Specifications

### General

| | |
|---|---|
| Power steering fluid type | See Chapter 1 |
| Vehicle ride height (Integra) | 25-1/8 to 26-5/16 in |

### Torque specifications

Ft-lbs (unless otherwise indicated)

### Front suspension
**Integra**

| | |
|---|---|
| Lower arm balljoint nut | 29 |
| Lower arm-to-radius arm bolts | 29 |
| Radius arm bushing nut | 60 |
| Stabilizer bar clamp bolts | 16 |
| Stabilizer bar link bolt and nut | 16 |
| Steering knuckle-to-strut pinch bolt | 47 |
| Steering/suspension crossmember bolts | 47 |
| Strut brake hose bracket | 16 |
| Strut damper rod self-locking nut | 33 |
| Strut upper mounting nuts | |
|     Front nut | 29 |
|     Rear nut | 16 |
| Tie-rod end balljoint-to-steering knuckle nut | 29 |
| Torque tube holder bolts | 16 |

**Legend**

| | |
|---|---|
| Damper fork pinch bolt | 39 |
| Damper fork-to-lower arm through bolt | 47 |
| Lower arm inner pivot bolt | 39 |
| Radius rod-to-crossmember nut | 39 |
| Radius rod-to-lower arm bolts | 61 |
| Shock absorber-to-body mounting nuts | 28 |
| Shock absorber damper rod upper nut | 36 |
| Stabilizer bar clamp bolts | 16 |

| Stabilizer bar link bolt and nut | 16 |
| Steering knuckle (lower) balljoint nut | 72 |
| Tie-rod end balljoint-to-steering knuckle nut | 32 |
| Upper arm assembly-to-body mounting nuts | 47 |
| Upper arm balljoint nut | 32 |

## Rear suspension
### Integra
| | |
|---|---|
| Hub retaining nut | 134 |
| Panhard rod-to-axle beam | 54 |
| Panhard rod-to-body bracket bolt | 40 |
| Spindle retaining nuts | 33 |
| Stabilizer assembly-to-axle beam bolts | 54 |
| Stabilizer control plate nuts | 29 |
| Shock absorber lower mounting nut | 40 |
| Shock absorber upper mounting nut | 16 |
| Swing bearing nuts | 33 |
| Trailing arm-to-body pivot bolt | 47 |

### Legend (1986 through 1988 sedans)
| | |
|---|---|
| Hub carrier-to-strut pinch bolt | 47 |
| Lower arm inner pivot bolt | 40 |
| Lower arm-to-hub carrier through bolt | 40 |
| Stabilizer bar clamp bolts | 16 |
| Stabilizer bar link-to-lower arm nut | 16 |
| Stabilizer bar-to-link nut | 108 in-lbs |
| Strut damper rod nut | 54 |
| Strut upper mounting nuts | 16 |
| Trailing link bracket-to-hub carrier through bolt | 54 |
| Trailing link-to-body nut | 33 |

### Legend (1989 and 1990 sedans and all coupes)
| | |
|---|---|
| Front lower arm inner pivot bolt | 40 |
| Front/rear lower arm-to-knuckle through bolt | 47 |
| Rear lower arm inner pivot bolt | 61 |
| Stabilizer bar clamp bolts | 28 |
| Stabilizer bar link-to-trailing arm stud nut | 108 in-lbs |
| Stabilizer link stud nut (on trailing arm) | 26 |
| Stabilizer bar-to-link bolt | 108 in-lbs |
| Shock absorber damper rod nut | 22 |
| Shock absorber-to-knuckle bolt | 40 |
| Trailing arm bracket retaining bolts | 47 |
| Trailing arm-to-body pivot bolt | 47 |
| Trailing arm-to-knuckle bolts | 40 |
| Upper arm balljoint-to-knuckle nut | 32 |
| Upper arm inner mounting bolts | 28 |
| Shock absorber upper mounting nuts | 28 |

## Steering system
| | |
|---|---|
| Steering wheel nut | 36 |
| Steering gear mounting bolts | |
|   Integra | |
|     Clamp bolts | 29 |
|     Gearbox bolts | 32 |
|   Legend | 28 |
| Tie-rod end-to-steering knuckle nut | 32 |

## 1 General information

*Refer to illustrations 1.2, 1.3, 1.4, 1.5a and 1.5b*

The vehicles covered by this manual, while similar in many respects, have two different front and three different rear suspension designs.

The front suspension on Integra models **(see illustration)** is a fully independent design which uses lower arms, radius arms, struts, torsion bar springs and a stabilizer bar.

The front suspension on Legend models **(see illustration)** is also a fully independent design with lower arms and a stabilizer bar, but it uses shock absorber/coil spring units instead of struts, and the upper end of the steering knuckle is supported by an upper control arm.

The rear suspension on Integra models **(see illustration)** is a trailing arm design. It consists of an axle beam laterally located by a Panhard rod and longitudinally located by trailing arms. A pair of shock absorber/coil spring units attached to the body and the axle beam handle suspension and damping. A stabilizer bar is installed inside the right end of the beam axle.

The rear suspension on earlier Legend sedans (1986 through 1988 models) consists of lower arms, trailing links, struts and separate coil springs **(see illustration)**; later sedans (1989 and 1990 models) and all coupes use trailing arms, two unequal length lower arms, an upper arm, shock absorber/coil spring units and a stabilizer bar **(see illustration)**.

All models use either a manual or power-assisted rack-and-pinion steering gear. The power steering system employs an engine-driven pump connected by hoses to the steering gear.

# Chapter 10 Suspension and steering systems

1.2 Front suspension on Integra models

1. Strut assembly
2. Steering knuckle
3. Lower arm
4. Radius arm
5. Torsion bar

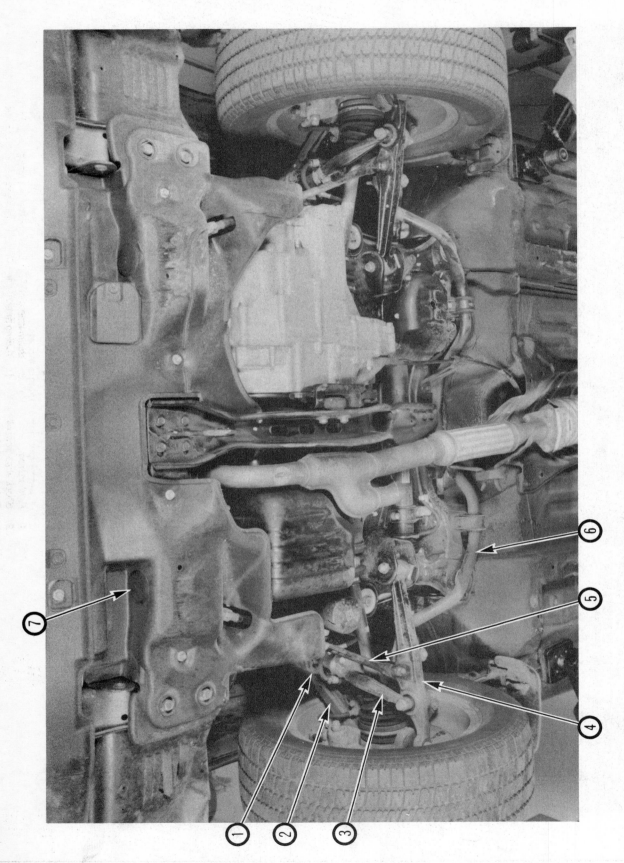

1.3 Front suspension on Legend models

1. Shock absorber/coil spring assembly
2. Steering knuckle
3. Damper fork
4. Lower arm
5. Radius rod
6. Stabilizer bar
7. Removable plug for radius rod nut

## Chapter 10 Suspension and steering systems

1.4 Rear suspension on Integra models

1. Panhard rod
2. Shock absorber/coil spring assembly
3. Beam axle
4. Trailing arm

1.5a Rear suspension on 1986 through 1988 Legend sedans

1. Trailing link
2. Lower arm
3. Coil spring
4. Stabilizer bar
5. Strut unit

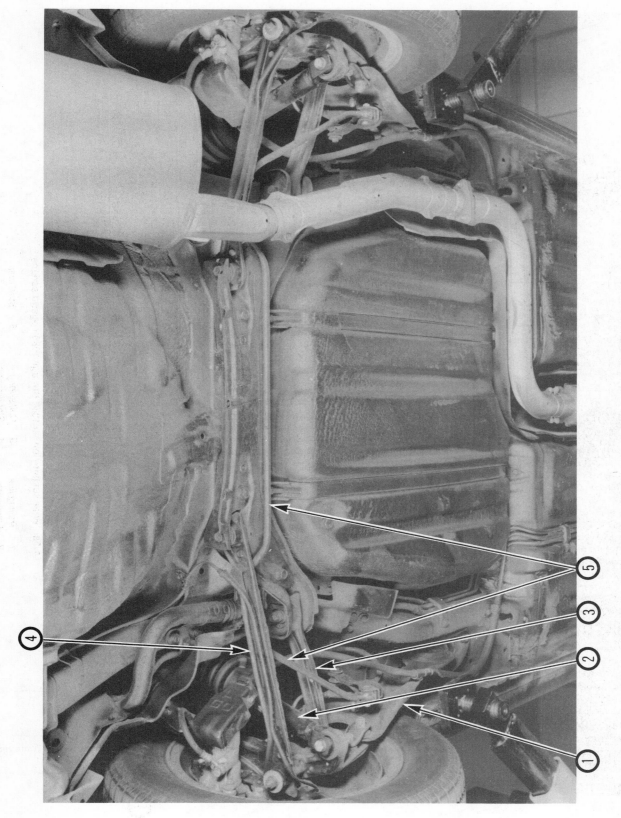

1.5b Rear suspension on 1989 and 1990 Legend sedans and all coupes

1. Trailing arm
2. Shock absorber/coil spring assembly
3. Front lower arm
4. Rear lower arm
5. Stabilizer bar

# Chapter 10 Suspension and steering systems

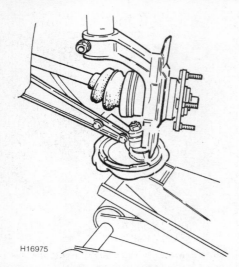

**2.2 Position a floor jack directly beneath the outer end of the lower arm and raise the steering knuckle and lower arm enough to remove the load from the upper mounting nuts**

Frequently, when working on the suspension or steering system components, you may come across fasteners which seem impossible to loosen. These fasteners on the underside of the vehicle are continually subjected to water, road grime, mud, etc., and can become rusted or "frozen," making them extremely difficult to remove. In order to unscrew these stubborn fasteners without damaging them (or other components), be sure to use lots of penetrating oil and allow it to soak in for a while. Using a wire brush to clean exposed threads will also ease removal of the nut or bolt and prevent damage to the threads. Sometimes a sharp blow with a hammer and punch is effective in breaking the bond between a nut and bolt threads, but care must be taken to prevent the punch from slipping off the fastener and ruining the threads. Heating the stuck fastener and surrounding area with a torch sometimes helps too, but isn't recommended because of the obvious dangers associated with fire. Long breaker bars and extension, or "cheater," pipes will increase leverage, but never use an extension pipe on a ratchet – the ratcheting mechanism could be damaged. Sometimes, turning the nut or bolt in the tightening (clockwise) direction first will help to break it loose. Fasteners that require drastic measures to unscrew should always be replaced with new ones.

Since most of the procedures that are dealt with in this chapter involve jacking up the vehicle and working underneath it, a good pair of jackstands will be needed. A hydraulic floor jack is the preferred type of jack to lift the vehicle, and it can also be used to support certain components during various operations. **Warning:** *Never, under any circumstances, rely on a jack to support the vehicle while working on it.* Whenever any of the suspension or steering fasteners are loosened or removed they must be inspected and, if necessary, be replaced with new ones of the same part number or of original equipment quality and design. Torque specifications must be followed for proper reassembly and component retention. Never attempt to heat or straighten any suspension or steering components. Instead, replace any bent or damaged part with a new one.

## 2 Front strut assembly (Integra models) – removal and installation

### Removal

Refer to illustrations 2.2 and 2.3

1  Loosen the wheel lug nuts, raise the front of the vehicle, place it securely on jackstands and remove the front wheel.
2  Place a floor jack under the lower arm **(see illustration)**. Raise the jack just enough to remove the load from the upper mounting nuts.

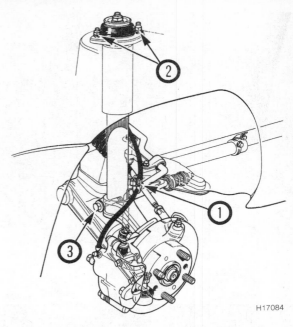

**2.3 Strut assembly mounting details**

| 1 | Brake hose bracket | 3 | Steering knuckle-to-strut pinch bolt |
| 2 | Strut upper mounting nuts | | |

3  Detach the brake hose from its bracket on the strut **(see illustration)**.
4  Remove the steering knuckle pinch bolt at the base of the strut, lower the jack slightly and detach the strut from the steering knuckle. If necessary, tap on the steering knuckle with a hammer to separate the two components. **Warning:** *Make sure the lower arm is properly supported by the jack. It's under a load from the torsion bar, which will react just like a spring if the jack should suddenly slip out from under the lower arm.*
5  Remove the strut mounting nuts and washers from the top of the strut tower inside the engine compartment.
6  Guide the strut out from the fenderwell.

### Installation

7  Place the strut in position, raise it up until the studs protrude through the holes in the top of the strut tower and install the washers and mounting nuts hand tight. **Note:** *These fasteners should be tightened after the vehicle is on the ground, not while it's raised.*
8  Attach the lower end of the strut to the steering knuckle. If necessary, raise the knuckle slightly with the floor jack.
9  Make sure the alignment tab on the strut is aligned with the slot in the steering knuckle. Install the pinch bolt and tighten it to the torque listed in this Chapter's Specifications.
10  Attach the brake hose brackets to the strut.
11  Remove the floor jack.
12  Install the wheel and lug nuts, lower the vehicle and tighten the lug nuts to the torque listed in the Chapter 1 Specifications.
13  Tighten the upper mounting nuts to the torque listed in this Chapter's Specifications.

## 3 Front shock absorber assembly (Legend models) – removal and installation

### Removal

Refer to illustrations 3.4, 3.5a, 3.5b and 3.6

1  Loosen the wheel lug nuts, raise the vehicle and support it securely on jackstands. Remove the wheel.

# Chapter 10  Suspension and steering systems

**3.4  Remove the damper fork pinch bolt**

**3.5a  Remove the through-bolt that connects the damper fork to the lower arm**

**3.5b  Detach the damper fork from the shock absorber**

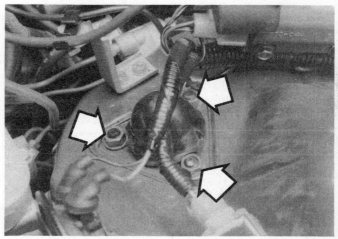

**3.6  Remove these three nuts (arrows) from the shock absorber mounting studs**

2   Unbolt the flexible brake hose from the strut assembly.
3   Disconnect the stabilizer bar from the lower arm (see Section 10).
4   Place a floor jack under the lower arm to support it when the shock absorber assembly is removed. Remove the damper fork pinch bolt **(see illustration)**.
5   Remove the damper fork-to-lower arm bolt and remove the fork **(see illustrations)**. It may be necessary to tap the fork from the shock absorber.
6   Support the shock absorber and coil spring assembly and remove the three upper mounting nuts **(see illustrations)**. Remove the unit from the fenderwell.

## Installation

*Refer to illustration 3.8*

7   Guide the shock absorber assembly up into the fenderwell and insert the three upper mounting studs through the holes in the body. Once the studs protrude from the holes, install the nuts so the assembly won't fall back through, but don't tighten the nuts completely yet. The strut is heavy and awkward, so get an assistant to help you, if possible.
8   Insert the lower end of the shock absorber into the damper fork. Make sure the aligning tab on the back of the strut/shock body enters the slot in the steering knuckle or damper fork **(see illustration)**.

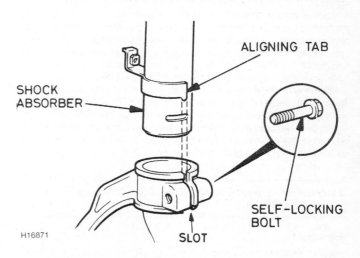

**3.8  Make sure the alignment tab on the shock absorber enters the slot in the damper fork**

# Chapter 10  Suspension and steering systems

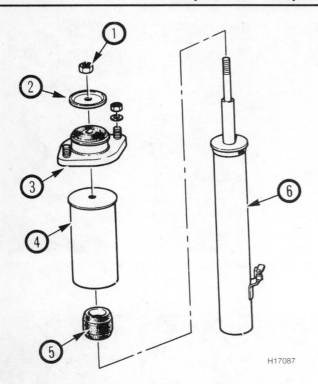

**4.5  Exploded view of a typical Integra front strut assembly**

| | | |
|---|---|---|
| 1  Damper rod nut | 3  Collar | 5  Rubber bump stop |
| 2  Washer | 4  Shield | 6  Strut |

9  Connect the damper fork to the lower arm, tightening the self-locking nut to the torque listed in this Chapter's Specifications. Now tighten the damper fork pinch bolt to the torque listed in this Chapter's Specifications.
10  Attach the brake hose to its bracket and tighten the bolt securely.
11  Install the wheel and lug nuts, lower the vehicle and tighten the lug nuts to the torque listed in the Chapter 1 Specifications.
12  Tighten the upper mounting nuts to the torque listed in this Chapter's Specifications.

## 4  Strut/shock absorber assembly – replacement

1  On Integra models, remove the front strut assembly (see Section 2). On Legend models, remove the front shock absorber assembly (see Section 3).

### Integra models

*Refer to illustration 4.5*

2  Check the strut for leaking fluid, dents, cracks or other obvious damage. Complete strut assemblies are often available on an exchange basis. This eliminates much time and work. So, before disassembling the strut assembly to replace individual components, check on the availability of parts and the price of a complete rebuilt unit.
3  Mount the strut assembly in a vise. Line the vise jaws with wood or rags to prevent damage to the unit and don't tighten the vise excessively.
4  Place an offset box wrench on the damper rod nut, insert an Allen wrench in the recessed hex in the damper rod and remove the nut.
5  Disassemble the strut **(see illustration)**. Pay close attention to the order in which you remove the parts. It's a good idea to lay the parts out in their exact relationship to each other on the work bench because everything must be reassembled exactly the same way it came off.
6  Inspect the piston rod seal for leakage and the piston rod for cracks. Extend and retract the piston rod slowly, then quickly, through its full stroke. It should be smooth, quiet and offer resistance. If it's jerky, noisy or

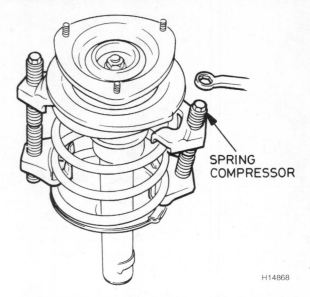

**4.11  Install the spring compressor according to the tool manufacturer's instructions and compress the spring until all pressure is relieved from the mounting base**

offers little or no resistance, replace the strut. It's a sealed unit and can't be rebuilt.
7  Reassembly is the reverse of disassembly. Be sure to use a new damper rod nut. Refer to the exploded view illustration to make sure all of the components are assembled in their proper positions.

### Legend models

*Refer to illustrations 4.11 and 4.12*

8  Check the shock absorber for leaking fluid, dents, cracks or other obvious damage. Check the coil spring for chips or cracks which could cause premature failure and inspect the spring seats for hardness or general deterioration. The shock absorber assemblies, complete with the coil springs, are available on an exchange basis which eliminates much time and work. So, before disassembling your shock to replace individual components, check on the availability of parts and the price of a complete rebuilt unit. **Warning:** *Disassembling a shock absorber/coil spring assembly is a potentially dangerous undertaking and utmost attention must be directed to the job at hand, or serious injury may result. Use only a high quality spring compressor and carefully follow the manufacturer's instructions furnished with the tool. After removing the coil spring from the strut or shock, set it aside in a safe, isolated area (a steel cabinet is preferred).*
9  Mount the shock absorber assembly in a vise. Line the vise jaws with wood or rags to prevent damage to the unit and don't tighten the vise excessively.
10  Mark the relationship of the damper mounting base to the spring (or if the spring is being replaced, put the mark on the damper unit). This will ensure correct positioning of the mounting base when the unit is reassembled.
11  Following the tool manufacturer's instructions, install the spring compressor (which can be obtained at most auto parts stores or equipment yards on a daily rental basis) on the spring and compress it sufficiently to relieve all pressure from the damper mounting base **(see illustration)**.
12  Remove the damper cap **(see illustration)**. Unscrew the self-locking nut while holding the damper shaft with an Allen wrench to prevent it from turning. Remove the parts from the upper part of the shock and lay them out in the exact order in which they're removed.
13  Carefully lift the compressed spring from the assembly and set it in a safe place, such as inside a steel cabinet. **Warning:** *Keep the ends of the spring facing away from your body!*
14  Slide the rest of the parts off of the damper shaft and lay them out in the exact order in which they're removed.

# Chapter 10  Suspension and steering systems

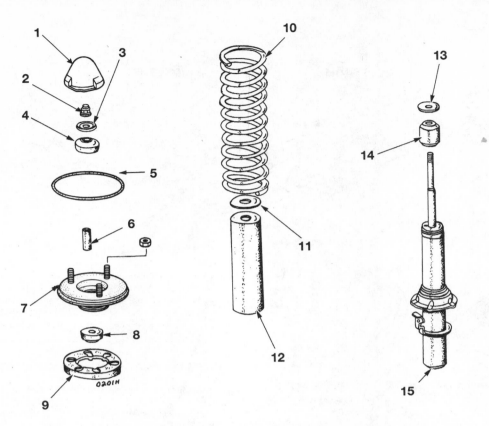

**4.12  Exploded view of a typical Legend front shock absorber/coil spring assembly**

| | | |
|---|---|---|
| 1  Damper cap | 5  Seal | 10  Spring |
| 2  Self-locking nut | 6  Damper mounting collar | 11  Dust cover plate (if equipped) |
| 3  Damper mounting washer | 7  Damper mounting base | 12  Dust cover |
| 4  Upper damper mounting rubber | 8  Lower damper mounting rubber | 13  Bump stop plate |
| | 9  Spring mounting rubber | 14  Bump stop |
| | | 15  Damper unit |

15  Install the bump stop, bump stop plate (if equipped), dust cover and dust cover plate onto the new damper unit. Extend the damper shaft as far as it will go and slide the components down to the damper body.

16  Carefully place the coil spring onto the shock absorber body, with the end of the spring resting in the lowest part of the seat.

17  Install the spring mounting rubber, lower damper mounting rubber, damper mounting collar, damper mounting base, seal, upper damper mounting rubber, damper mounting washer and a new self-locking nut. Before tightening the nut, align the previously applied marks on the mounting base and the spring (or damper body).

18  Tighten the self-locking nut securely, again using the Allen wrench to prevent the shaft from turning. Remove the spring compressor. Install the damper cap.

## All models

19  Install the strut or the shock absorber assembly (see Section 2 or 3, respectively).

### 5  Steering knuckle and hub assembly – removal and installation

## Removal

1  Remove the wheel cover, if equipped. Loosen the driveaxle nut (see Chapter 8). Loosen the wheel lug nuts, raise the front of the vehicle and support it securely on jackstands. Remove the wheel and the driveaxle hub nut.

2  Unbolt the brake hose bracket(s) from the strut and/or steering knuckle. Unbolt the brake caliper, hang it out of the way with a piece of wire, remove the caliper mounting bracket and remove the brake disc (see Chapter 9).

3  Disconnect the tie-rod end from the steering knuckle (see Section 25).

4  On Integra models, separate the radius arm balljoint stud from the bottom of the steering knuckle (see Section 8). On Legend models, separate the lower arm from the balljoint in the bottom of the steering knuckle (see Section 12).

5  On Integra models, remove the pinch bolt from the upper end of the knuckle **(see illustration 2.3)** and tap the knuckle and hub assembly off the lower end of the strut.

6  On Legend models, separate the upper end of the knuckle from the upper arm balljoint (see Section 13).

7  Carefully pull the knuckle and hub assembly off of the driveaxle. Support the driveaxle with a piece of wire to prevent damage to the inner CV joint.

8  Due to the special tools and expertise required to press the hub and bearing from the steering knuckle, the assembly should be taken to a dealer service department or other repair shop to have the bearings replaced, if they're worn.

# Chapter 10 Suspension and steering systems

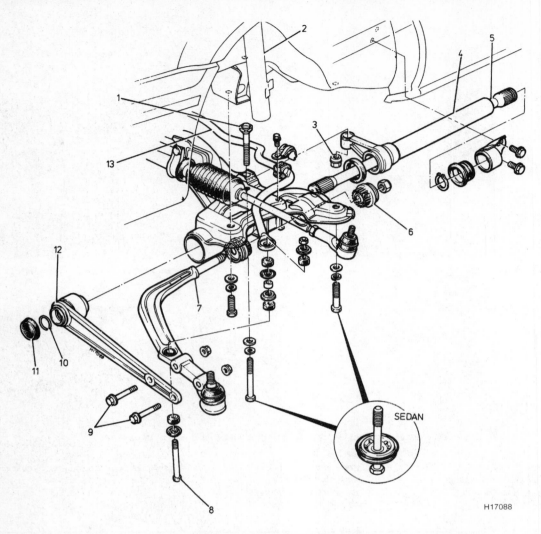

**6.3 An exploded view of the front suspension on Integra models**

| | | | | | |
|---|---|---|---|---|---|
| 1 | Height adjusting bolt | 6 | Radius arm bushing | 10 | Torsion bar snap-ring |
| 2 | Strut unit | 7 | Radius arm | 11 | Torsion bar cap |
| 3 | Height adjusting nut | 8 | Stabilizer bar mounting bolt | 12 | Lower arm |
| 4 | Torque tube | 9 | Lower arm bolt | 13 | Stabilizer bar |
| 5 | Torsion bar | | | | |

## Installation

9  Apply a light coat of wheel bearing grease to the driveaxle splines. Insert the driveaxle through the splined bore of the hub while guiding the steering knuckle into position.
10  On Integra models, attach the upper end of the knuckle to the lower end of the strut, making sure the tab on the strut engages with the split in the pinch joint (see Section 2). Install the pinch bolt and tighten it to the torque listed in this Chapter's Specifications.
11  On Legend models, connect the upper end of the knuckle to the upper arm balljoint (see Section 13). Tighten the balljoint stud nut to the torque listed in this Chapter's Specifications.
12  On Integra models, connect the lower end of the knuckle to the radius arm balljoint stud (see Section 8). On Legend models, connect the balljoint on the bottom of the knuckle to the lower arm (see Section 12).
13  Install the brake disc, caliper mount and caliper (see Chapter 9). Attach the brake hose bracket(s).
14  Install the driveaxle nut and tighten it securely.
15  Install the wheel and lug nuts, lower the vehicle and tighten the lug nuts to the torque listed in the Chapter 1 Specifications.
16  Tighten the driveaxle nut to the torque specified in Chapter 8.
17  Drive the vehicle to an alignment shop and have the front end alignment checked and, if necessary, adjusted.

## 6  Torsion bar (Integra models) – removal, installation and adjustment

### Removal

*Refer to illustrations 6.3, 6.4a, 6.4b, 6.5, 6.6a and 6.6b*

1  Raise the vehicle and place it securely on jackstands.
2  Place a floor jack directly under the lower arm. **Warning:** *The jack will absorb the reaction of the torsion bar during the following removal procedure, so make sure the jack head is centered on the end of the lower arm.*
3  Remove the height adjusting nut **(see illustration)**.
4  Remove the mounting bolts from the torque tube holder **(see illustration)** and remove the torque tube holder and cap bushing **(see illustration)**.

## Chapter 10 Suspension and steering systems

6.4a Remove the torque tube holder mounting bolts ...

6.4b ... and remove the torque tube holder

6.5 Remove the snap-ring from the rear end of the torsion bar with a pair of snap-ring pliers

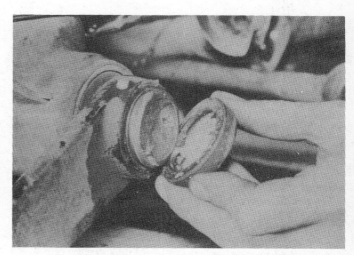

6.6a Remove the front cap from the torsion bar ...

5 Using snap-ring pliers, remove the snap-ring from the rear of the torsion bar **(see illustration)**.
6 Remove the cap from the front end of the torsion bar **(see illustration)**, then tap the bar forward slightly until the snap-ring is accessible. Moving the lower arm up and down slightly will make the torsion bar bar easier to slide forward. When you can get at the snap-ring, remove it **(see illustration)**.
7 There should already be a punch mark or paint mark on the torque tube splines and a cutout or paint mark on the torsion bar splines. If there aren't, paint or scribe alignment marks on the front edges of the tube and the bar to ensure proper reassembly.
8 Using a soft-face hammer, tap the torsion bar out of the lower arm (toward the rear) and remove it from the the torque tube.
9 Remove the torque tube and the torque tube seal.
10 Inspect the torsion bar for cracks and worn splines. If any damage or wear is evident, replace it.

### Installation
11 Liberally lubricate the torque tube sliding surfaces, and a new seal, with multi-purpose grease, then position the seal and torque tube on the steering/suspension crossmember.
12 Lubricate the torsion bar splines with grease and insert the bar into the torque tube from the rear.

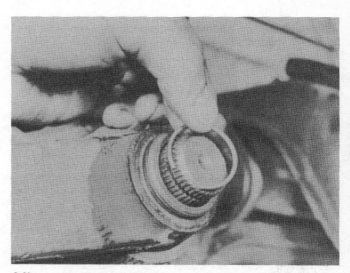

6.6b ... tap the torsion bar forward slightly (until the snap-ring is exposed) and remove the snap-ring from its groove in the front end of the torsion bar

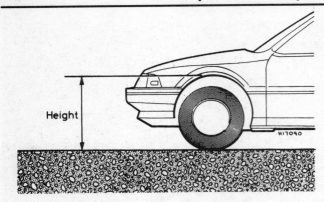

**6.24 Ride height is measured between the highest point of the wheelwell arch and the ground — compare your measurements to the ride height specified at the beginning of the Chapter and adjust it if necessary**

**6.27 To adjust ride height, turn the torsion bar height adjusting nut clockwise to increase the height or counterclockwise to decrease it**

13  Align the mark on the torque tube splines with the mark on the torsion bar splines.
14  Push the torsion bar through the splined bore of the lower arm until the front snap-ring groove is protruding far enough to install the snap-ring. Note that the cut-out in the torsion bar splines must mate with the projection in the splined bore of the lower arm.
15  Install the snap-ring and the cap on the front end of the torsion bar.
16  Install the other snap-ring on the rear end of the torsion bar.
17  Lubricate the cap bushing with multi-purpose grease and slip it over the rear end of the torsion bar.
18  Install the torque tube holder and mounting bolts. Tighten the bolts securely.
19  Lubricate the height adjusting nut and screw it on finger tight.
20  Install the wheel and lug nuts, lower the vehicle and tighten the lug nuts to the torque listed in the Chapter 1 Specifications. Be sure to adjust the vehicle ride height when you're done (see next Step).

### Adjustment

*Refer to illustrations 6.24 and 6.27*

21  Fill the fuel tank. Check the tire pressures (see your owner's manual). Make sure they're correct.
22  Place the vehicle on a level surface with the wheels pointed straight ahead. No one must be inside the vehicle.
23  Bounce the vehicle up and down several times to "settle" the suspension.
24  Measure the vertical distance between the highest point of the front wheelwell and the ground **(see illustration)** on both sides of the vehicle and record your measurements.
25  Compare your measurements with the dimensions listed in this Chapter's Specifications. If your measurements indicate the vehicle ride height is out of specification, adjust it as follows.
26  Raise the side of the vehicle to be adjusted and place it securely on jackstands.
27  Turn the torsion bar height adjusting nut clockwise (as viewed from below) to increase the height, or counterclockwise to decrease it **(see illustration)**. **Note:** *One complete turn of the nut will alter ride height by about 3/16-inch.*
28  After making an approximate adjustment, lower the vehicle, roll it back and forth about half a car's length, bounce the suspension, measure the ride height on both sides, compare your measurements to those listed at the front of this Chapter and, if necessary, raise it up and readjust the ride height nut(s).
29  Repeat this procedure until the ride height is correct.

### 7  Lower arm (Integra models) – removal and installation

Because of the special tools required to remove and install the lower arm, this procedure is beyond the scope of the average home mechanic.

**8.3  After removing the cotter pin, loosen this castle nut which secures the radius arm balljoint stud to the steering knuckle (don't remove the nut yet – it will prevent the components from separating violently)**

If you need to replace the lower arm, have it done by a dealer service department or other qualified repair shop.

### 8  Radius arm (Integra models) – removal and installation

### Removal

*Refer to illustrations 8.3, 8.4 and 8.7*

1  Loosen the front wheel lug nuts, raise the front of the vehicle and place it securely on jackstands. Remove the wheel.
2  Place a floor jack beneath the lower arm. **Warning:** *The jack must absorb the full torque reaction of the torsion bar when the balljoint is released, so make sure the jack is properly located.*
3  Remove the cotter pin from the castle nut on the balljoint stud of the radius arm and loosen the nut a few turns **(see illustration)**.
4  Using a balljoint separator, separate the balljoint stud from the steering knuckle **(see illustration)**. Remove the castle nut.
5  Unbolt the radius arm from the lower arm **(see illustration 6.3)**.
6  Detach the stabilizer bar from the radius arm (see Section 10).
7  Remove the self-locking nut from the radius arm stud which faces to the rear, through the radius arm bushing **(see illustration)**.
8  Carefully lower the floor jack to relieve the tension on the torsion bar/lower arm assembly and to provide enough clearance to remove the radius arm.
9  Rotate the radius arm down, slide it forward, out of its bushing and remove it.
10  Inspect the radius arm bushings. If they're worn, cracked or torn, replace them. Check the balljoint. If it's worn or damaged, replace the radius arm. The balljoint can't be replaced individually.

# Chapter 10  Suspension and steering systems

10 – 15

**8.4  Use a balljoint separator to separate the balljoint stud in the radius arm from the steering knuckle**

## Installation

11  Installation is the reverse of removal. Replace all self-locking nuts that don't fit tightly on their respective bolts. Tighten all fasteners to the torque listed in this Chapter's Specifications. Don't tighten the radius arm bushing nut to specification until the vehicle is resting on the ground.

## 9  Steering/suspension crossmember (Integra models) – removal and installation

*Refer to illustration 9.10*

1  Loosen the wheel lug nuts, raise the vehicle and place it securely on jackstands. Remove the wheels.

**8.7  Remove this self-locking nut from the radius arm stud**

2  Remove the steering gear (see Section 27).
3  Place a floor jack directly under the left radius arm balljoint **(see illustration 2.2)**. **Warning:** *The jack must absorb the full torque reaction of the torsion bar when the balljoint is released, so make sure the jack is properly located.*
4  Detach the radius arm balljoint from the steering knuckle (see Section 8).
5  Slowly lower the jack until the radius arm is no longer stressed.
6  Repeat Steps 3, 4 and 5 for the right-side radius arm balljoint.
7  Remove the bolts which secure the torque tube holders at the rear end of each torsion bar (see Section 6).
8  Support the engine with an engine hoist or with a floor jack under the oil pan (see Chapter 2). If a floor jack is used, be sure to place a wood block on the jack head to act as a cushion.
9  Remove the nuts and bolts which secure the engine/transaxle assembly to the rear mount (see Chapter 7).
10  Support the center of the steering/suspension crossmember with a floor jack and remove all mounting nuts and bolts **(see illustration)**.
11  Slowly lower the crossmember and remove it from under the vehicle.
12  Installation is the reverse of removal. Be sure to tighten all fasteners to the torque listed in this Chapter's Specifications.

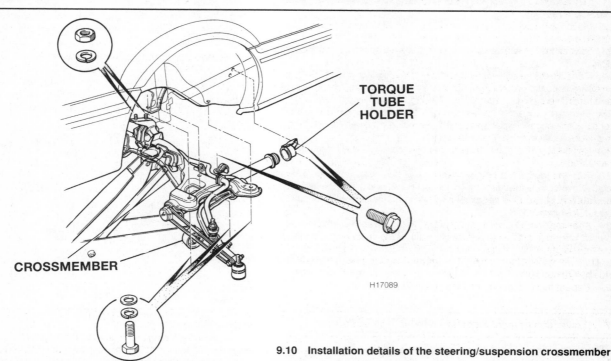

**9.10  Installation details of the steering/suspension crossmember**

# Chapter 10 Suspension and steering systems

**10.3 Stabilizer bar bracket on Legend models (driver's side bracket shown, passenger side bracket similar)**

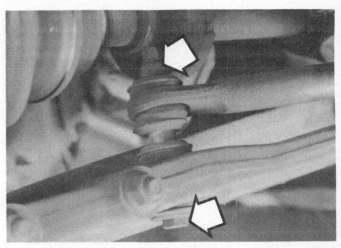

**10.4a Stabilizer bar link bolt assembly on Integra models**

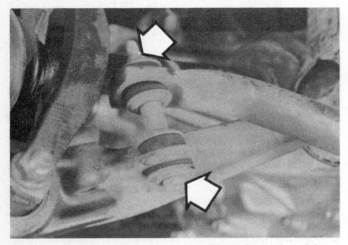

**10.4b Stabilizer bar link bolt assembly on Legend models**

**11.3 To detach the front end of the radius rod, remove the plug in the plastic splash shield, then remove this nut from the rod**

**11.4 To detach the rear end of the radius rod, remove the two bolts (arrows) which attach the radius rod to the lower arm**

## 10 Front stabilizer bar and bushings – removal and installation

*Refer to illustrations 10.3, 10.4a and 10.4b*

1 Apply the parking brake. Loosen the front wheel lug nuts, raise the front of the vehicle and support it securely on jackstands. Remove the wheels.

2 On Integra models, detach the steering/suspension crossmember from the underside of the vehicle and lower it to the ground (see Section 8). Remove the bolts which attach the stabilizer bar brackets to the crossmember **(see illustration 6.3)**.

3 On Legend models, remove the bolts which attach the stabilizer bar brackets to the underside of the vehicle **(see illustration)**.

4 Detach the stabilizer bar link bolts from the radius arms (Integra models) or the lower arms (Legend models) **(see illustrations)**. On all models, note the order in which the spacers, washers and bushings are arranged on the link bolt.

5 Remove the bar from under the vehicle.

6 Pull the brackets off the stabilizer bar and inspect the bushings for cracks, hardness and other signs of deterioration. If the bushings are damaged, replace them.

7 Installation is the reverse of removal.

## 11 Radius rod (Legend models) – removal and installation

*Refer to illustrations 11.3 and 11.4*

1 Loosen the wheel lug nuts, raise the front of the vehicle and place it securely on jackstands. Remove the wheel.

2 Remove the plug from the plastic splash shield **(see illustration 1.3)**.

# Chapter 10  Suspension and steering systems

12.5  Separate the lower arm from the steering knuckle balljoint with a two-jaw puller

12.6  To remove the lower arm, remove the pivot bolt from the inner end of the arm

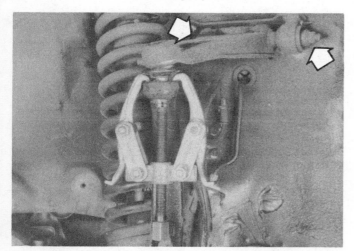

13.3  Use a two-jaw puller to separate the upper arm from the steering knuckle, then remove the pivot nuts and bolts (arrows)

3  Remove the nut from the front end of the radius rod in the front crossmember (see illustrations).
4  Remove the bolts which attach the rear end of the radius rod to the lower arm (see illustration) and remove the rod. If a new rod is being installed, measure the distance from the end of the rod to the adjuster nut, then remove the nut and install it on the new rod in the same position.
5  Installation is the reverse of removal. Be sure to tighten all fasteners to the torque values listed in this Chapter's Specifications.
6  Drive the vehicle to an alignment shop and have the front end alignment checked and, if necessary, adjusted.

## 12  Lower arm (Legend models) – removal and installation

*Refer to illustrations 12.5 and 12.6*
1  Loosen the front wheel lug nuts, raise the vehicle, place it securely on jackstands and remove the wheel.
2  Detach the damper fork from the strut assembly (see Section 3).
3  Detach the radius rod from the lower arm (see Section 11).
4  Detach the stabilizer bar from the lower arm (see Section 10).
5  Remove the cotter pin from the castle nut on the lower balljoint stud. Loosen the nut, but don't remove it yet. Using a two-jaw puller, separate the lower arm from the balljoint in the steering knuckle (see illustration). Remove the nut.
6  Remove the pivot bolt from the inner end of the lower arm (see illustration) and remove the arm.
7  Installation is the reverse of removal.

## 13  Upper arm (Legend models) – removal and installation

*Refer to illustration 13.3*
1  Loosen the front wheel lug nuts, raise the vehicle, place it securely on jackstands and remove the wheel.
2  Remove the cotter pin and loosen, but do not remove, the castle nut from the upper balljoint stud. The nut will prevent the upper arm and the steering knuckle from separating violently in the next step.
3  Separate the upper arm from the steering knuckle with a two-jaw puller (see illustration).
4  Remove the upper arm pivot nuts and bolts (see illustration 13.3) and the balljoint nut, then remove the upper arm. Note that the heads of the pivot bolts face toward each other – be sure to install them the same way.
5  Installation is the reverse of removal.

## 14  Balljoints – replacement

### Integra models
1  The balljoint in the radius arm isn't removable. If it's worn or damaged, replace the radius arm (see Section 8).

### Legend models
2  The front suspension uses two balljoints. The upper balljoint, located in the upper arm, can't be removed. If it's worn or damaged, replace the upper arm (see Section 13).
3  The lower balljoint, located in the steering knuckle, can be removed, but special tools are needed. If it's worn or damaged, remove the knuckle (see Section 5) and take it to a dealer service department or other repair shop to have it replaced.

## 15  Rear shock absorber/strut assembly – removal and installation

1  Loosen the rear wheel lug nuts, raise the vehicle, place it securely on jackstands and remove the rear wheels.

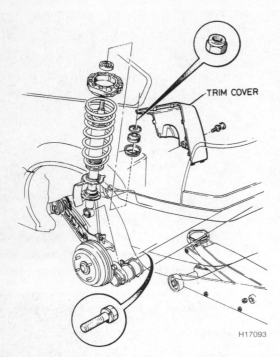

**15.2 Exploded view of the rear shock absorber assembly on Integra models – note the placement of the floor jack under the beam axle**

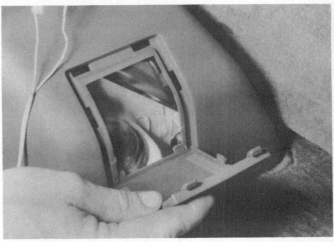

**15.3a To get to the upper mounting nut for a rear shock on an Integra, remove the trim panel . . .**

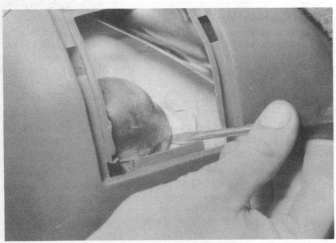

**15.3b . . . and the rubber cap**

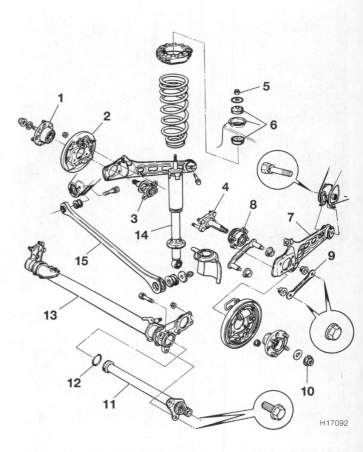

**15.4 An exploded view of the rear suspension on Integra models**

| | | | |
|---|---|---|---|
| 1 | Hub/bearing unit | 8 | Swing bearing unit |
| 2 | Backing plate | 9 | Control arm |
| 3 | Spindle | 10 | Hub unit nut |
| 4 | Spindle | 11 | Stabilizer bar assembly |
| 5 | Shock absorber upper mounting nut | 12 | Stabilizer bar seal |
| 6 | Upper mounting cushions | 13 | Rear axle beam |
| 7 | Trailing arm | 14 | Shock absorber |
| | | 15 | Panhard rod |

### Integra models

Refer to illustrations 15.2, 15.3a, 15.3b, 15.4 and 15.6

2   Place a floor jack under the rear axle beam (**see illustration**) and raise it slightly, so the weight of the axle is firmly supported. **Warning:** Once you loosen the shock absorber upper mounting nut and lower the floor jack, the axle beam will be under considerable force, so make sure it's positioned securely on the floor jack.

3   From inside the vehicle, remove the trim panel and protective cap (**see illustrations**).

4   Remove the shock absorber upper mounting nut (**see illustration**). You'll need to prevent the piston rod from turning with an Allen wrench while you're removing the nut. Also remove the big washer and rubber cushion.

5   Slowly lower the floor jack until tension is removed from the coil spring.

# Chapter 10  Suspension and steering systems

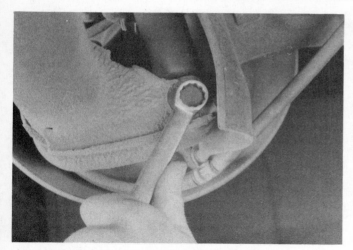

15.6  To detach the lower end of the shock absorber from the axle beam, remove this bolt

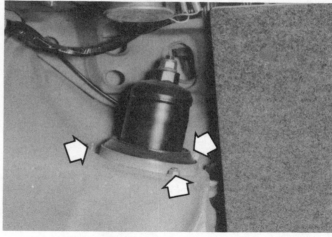

15.10  Remove the three strut upper mounting nuts (arrows) (1986 through 1988 Legend sedans)

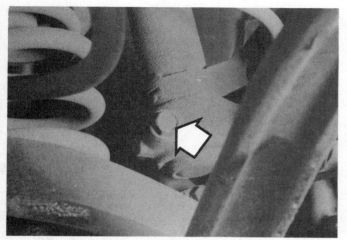

15.12  Remove the strut pinch bolt (arrow) and disconnect the strut from the hub carrier (1986 through 1988 Legend sedans)

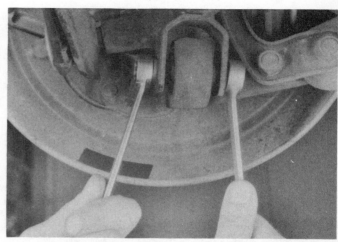

15.21  Remove the lower mounting bolt, pull the lower arms down and remove the shock absorber/coil spring assembly (1989 and 1990 sedans and all coupes)

6  Unbolt the shock absorber lower mounting bolt **(see illustration)**, lower the jack all the way and remove the shock absorber and spring assembly.
7  If you wish to inspect the shock absorber, refer to Section 4. Be sure to lay out the parts of the shock absorber assembly in order so they're reinstalled in exactly the same order in which they were removed.
8  Installation is the reverse of removal. Make sure the the smaller diameter coils of the spring are at the bottom, and tighten all fasteners to the torque listed in this Chapter's Specifications. **Note:** *Tighten the fasteners securely while the vehicle is raised but don't tighten them to the final torque until it's back on the ground.*

## Legend models

### 1986 through 1988 sedans

*Refer to illustrations 15.10 and 15.12*

9  Jack up the lower arm slightly.
10  From inside the trunk, peel back the trunk side carpet and remove the upper mounting nuts **(see illustration)**.
11  Lower the jack.
12  Remove the strut pinch bolt **(see illustration)**.
13  Detach the strut from the hub carrier.
14  To inspect the strut, see Section 4.
15  Installation is the reverse of removal.

### 1989 and 1990 sedans and all coupes

*Refer to illustration 15.21*

16  Remove the trunk side carpet.
17  Remove the shock absorber upper mounting nuts **(see illustration 15.10).**
18  Remove the clip for the parking brake cable from the trailing arm.
19  Disconnect the stabilizer bar from the trailing arm (see Section 21).
20  Remove the upper arm mounting bolts (see Section 23).
21  Remove the shock absorber lower mounting bolt **(see illustration)**.
22  Pull the lower arm down and remove the shock absorber/coil spring assembly.
23  To inspect or replace the shock absorber unit, see Section 4.
24  Installation is the reverse of removal. Be sure to tighten all fasteners to the torque values listed in this Chapter's Specifications.

## 16  Rear hub and bearing assembly – removal and installation

*Refer to illustrations 16.3a, 16.3b, 16.3c, 16.3d, 16.3e and 16.5*
**Note:** *The rear hub and bearing are a single assembly. The bearing is sealed for life and requires no lubrication or attention. If the bearing is worn or damaged, replace the entire hub and bearing assembly.*

# Chapter 10 Suspension and steering systems

**16.3a** Remove the rear hub nut and thrust washer, then pull off the hub unit (Integra models, 1988 and 1989 Legend sedans and all Legend coupes)

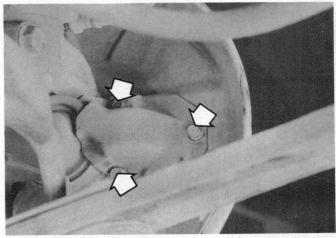

**16.3b** On 1986 through 1988 Legend sedans, remove the shield for the anti-lock brake sensor (if equipped), ...

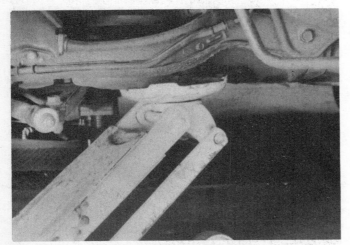

**16.3c** ... place a floor jack under the lower control arm and raise the arm until the hub carrier is at a right angle to the arm, ...

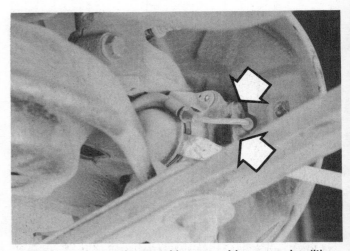

**16.3d** ... pry out the cap with a screwdriver or pry bar (it's easier to remove the cap if you remove the anti-lock brake sensor retaining bolts (arrows), detach the sensor, shove it through the space between hub carrier and lower arm and set it aside), ...

1  Loosen the rear wheel lug nuts, raise the vehicle, place it securely on jackstands and remove the rear wheel.
2  Remove the brake drum or caliper and disc (see Chapter 9).
3  Unstake the hub retaining nut, unscrew the nut and remove the thrust washer **(see illustration)**. On some models, you may have to bend down a lock tab to unscrew the nut. On Integra models, 1989 and 1990 Legend sedans and all coupes, the hub nut faces out. On 1986 through 1988 Legend sedans, the nut is on the inner side of the hub carrier. To remove the nut on these models, remove the shield for the anti-lock brake sensor (if equipped), place a floor jack under the lower arm then raise the arm until the hub carrier is at a right angle to the arm. Pry out the cap, unstake the nut and remove it **(see illustrations)**.
4  On Integra models, 1988 and 1989 Legend sedans and all coupes, pull the hub assembly off the spindle. If it's stuck, use a puller to get it off. On 1986 through 1988 Legend sedans, pull the hub/axleshaft assembly through the hub carrier.

## Installation

5  Install the new hub assembly and thrust washer, tighten the new nut to the torque listed in this Chapter's Specifications, then stake its edge into the groove in the spindle **(see illustration)** or bend up the lock tab against the nut.
6  The remainder of installation is the reverse of removal.

**16.3e** ... unstake the nut and remove it

# Chapter 10  Suspension and steering systems

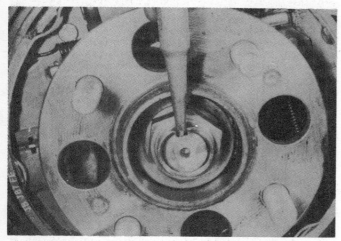

16.5  After tightening the new hub nut to the specified torque, stake it into the groove in the spindle with a hammer and punch

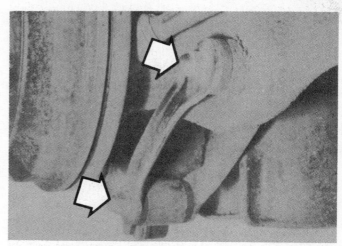

17.5  To disconnect the right trailing arm from the stabilizer assembly, remove these two nuts, pull out the bolts and detach the inner and outer stabilizer control plates

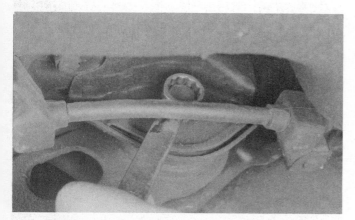

17.6  To disconnect the forward end of the trailing arm from the body, remove this pivot bolt

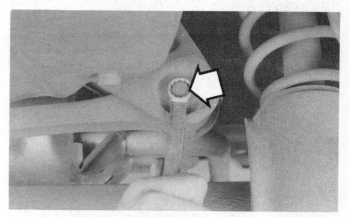

18.2  Remove this bolt (arrow) from the upper end of the Panhard rod

## 17  Trailing arm (Integra models) – removal and installation

*Refer to illustrations 17.5 and 17.6*

1  Loosen the rear wheel lug nuts, raise the rear of the vehicle, place it securely on jackstands and remove the wheel.
2  Remove the rear hub and bearing assembly (see Section 16).
3  Detach all brake hose clips from the trailing arm assembly. Disconnect the brake hose from the wheel cylinder or rear caliper and plug it to prevent leakage or contamination (see Chapter 9).
4  Remove the brake backing plate nuts and remove the backing plate **(see illustration 15.4)**.
5  If you're removing the right trailing arm, remove the two stabilizer control plate nuts and detach the inner and outer stabilizer control plates from the trailing arm and stabilizer bar assembly **(see illustration)**.
6  Remove the front trailing arm pivot bolt **(see illustration)** and remove the trailing arm from the vehicle.
7  Installation is the reverse of removal. Bleed the brake system (see Chapter 9).

## 18  Panhard rod (Integra models) – removal and installation

*Refer to illustrations 18.2 and 18.3*

1  Loosen the rear wheel lug nuts, raise the rear of the vehicle, place it securely on jackstands and remove the wheel.

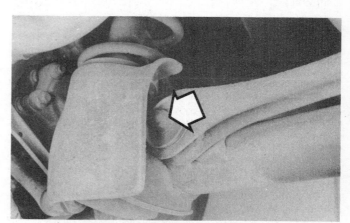

18.3  Remove this nut (arrow) from the Panhard rod's lower mounting stud and slide the rod off the stud

2  Remove the Panhard rod upper mounting bolt **(see illustration)**.
3  Remove the nut from the lower mounting stud **(see illustration)**, slide the rod off the stud and remove it.
4  Installation is the reverse of removal.

# Chapter 10 Suspension and steering systems

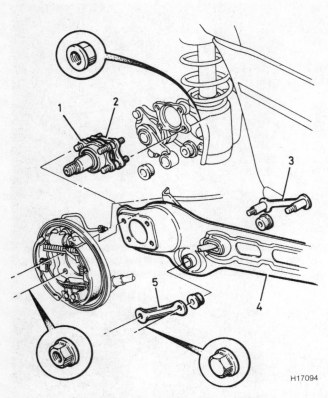

**19.5 Exploded view of the right trailing arm and related components**

1. Spindle
2. Swing bearing unit
3. Inner stabilizer control plate
4. Trailing arm
5. Outer stabilizer control plate

### 19 Rear axle beam (Integra models) – removal and installation

*Refer to illustration 19.5*

1. Loosen the rear wheel lug nuts, raise the rear of the vehicle, place it securely on jackstands and remove the wheel.
2. Detach the lower ends of both shock absorbers (see Section 15).
3. Detach the Panhard rod from the axle beam (see Section 18).
4. On the left side, remove the rear hub and bearing assembly (see Section 16) and remove the four nuts which attach the brake backing plate, trailing arm and spindle to the axle beam **(see illustration 15.4)**.
5. On the right side, remove the two stabilizer control plate nuts and remove the inner and outer stabilizer control plates from the trailing arm and stabilizer bar **(see illustration)**.
6. Remove the four bolts which secure the right spindle/swing bearing unit to the axle beam.
7. Support the axle beam with a floor jack and move it sideways to disengage the spindle studs, then lower the axle beam assembly to the ground.
8. Installation is the reverse of removal. Be sure to tighten all fasteners to the torque values listed in this Chapter's Specifications.

### 20 Spindle/swing bearing assembly (Integra models) – removal and installation

1. Loosen the rear wheel lug nuts, raise the rear of the vehicle, place it securely on jackstands and remove the rear wheels.

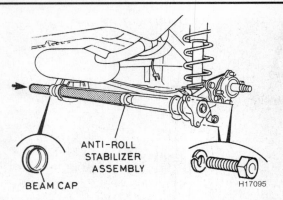

**21.5 Mounting details of the rear stabilizer bar assembly on Integra models – note the dowel inserted inside the left end of the axle beam to drive out the stabilizer bar**

#### Left spindle

2. Remove the rear axle beam (see Section 19).
3. Pull the spindle out of the brake backing plate and trailing arm **(see illustration 15.4)**. Installation is the reverse of removal.

#### Right spindle

4. Remove the right rear hub and bearing assembly (see Section 16).
5. Remove the four nuts which attach the brake backing plate and trailing arm to the swing bearing unit.
6. Remove the rear axle beam (see Section 19).
7. Pull the spindle/swing bearing assembly out of the brake backing plate and trailing arm **(see illustration 19.5)**.
8. Installation is the reverse of removal.

#### Swing bearing

**Note:** *In the process of removing the swing bearing, it will be destroyed. Therefore, do not have it removed unless you intend to replace it. If you are planning to install a new right spindle, have a new swing bearing pressed on too, because you won't be able to switch the bearing from the old spindle to the new one.*

9. Remove the right spindle/swing bearing assembly (see Steps 4 through 7).
10. Have the swing bearing pressed off the spindle by a dealer service department or other repair shop, and a new bearing pressed on.
11. Installation is the reverse of removal.

### 21 Rear stabilizer assembly – removal and installation

1. Loosen the rear wheel lug nuts, raise the rear of the vehicle, place it securely on jackstands and remove the rear wheels.

#### Integra models

*Refer to illustration 21.5*

2. Remove both rear hubs (see Section 16).
3. Remove the brake backing plates (see Chapter 9).
4. On the right side, remove the two nuts and remove the inner and outer control arms from the trailing arm and stabilizer bar **(see illustrations 15.4 and 19.5)**.
5. Extract the cap from the left end of the axle beam **(see illustration)**.
6. Remove the two bolts which attach the stabilizer to the axle beam.
7. Insert a long dowel, such as a broom stick, into the axle beam from the left end and tap out the stabilizer.
8. Installation is the reverse of removal. Be sure to tighten all fasteners to the torque listed in this Chapter's Specifications. Replace any self-locking type fasteners that no longer grip the threads. The two bolts which secure the stabilizer bar to the axle beam should not be tightened until after the control arms have been reinstalled.
9. When you're done, bleed the brakes (see Chapter 9).

# Chapter 10  Suspension and steering systems

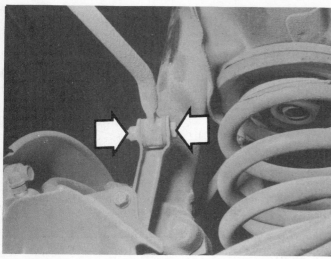

21.10a  To disconnect the stabilizer bar from the lower arm on 1986 through 1988 Legend sedans, remove the stabilizer bar-to-link bolt and nut (arrows)

21.10b  To disconnect the stabilizer bar from the trailing arm on 1989 and 1990 Legend sedans and all coupes, remove the stabilizer bar-to-link bolt and nut (arrow)

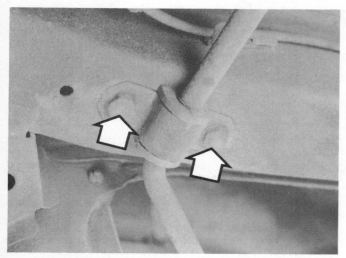

21.11a  To detach the stabilizer bar from the underside of 1986 through 1988 Legend sedans, remove these bolts (arrows) from both brackets

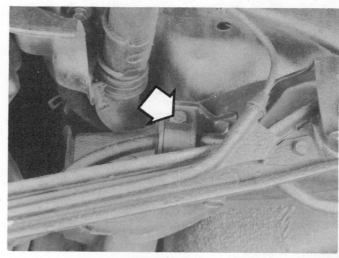

21.11b  To detach the stabilizer bar from the underside of 1989 and 1990 Legend sedans and all coupes, remove this bolt (arrow) from each bracket

### Legend models
*Refer to illustrations 21.10a, 21.10b, 21.11a and 21.11b*
10  Remove the stabilizer bar-to-link nuts and bolts **(see illustrations)**.
11  Remove the stabilizer-to-body clamp bolts **(see illustrations)** and remove the stabilizer bar.
12  Installation is the reverse of removal.

## 22  Rear suspension arms (1986 through 1988 Legend sedans) – removal and installation

1  Loosen the rear wheel lug nuts, raise the vehicle, place it securely on jackstands and remove the wheel.

### Trailing link/trailing link bracket
*Refer to illustrations 22.2, 22.3 and 22.4*
2  Remove the through bolt that attaches the trailing link bracket to the hub carrier **(see illustration)**.

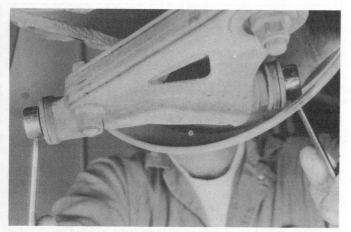

22.2  Remove the through bolt that attaches the trailing link bracket to the hub carrier

# 10-24  Chapter 10  Suspension and steering systems

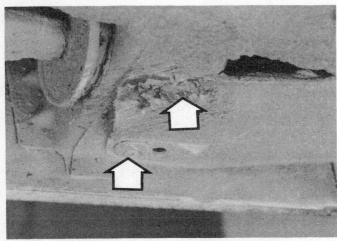

22.3  To get at the nut that attaches the forward end of the trailing link to the chassis, remove these cover screws (arrows) and remove the cover plate

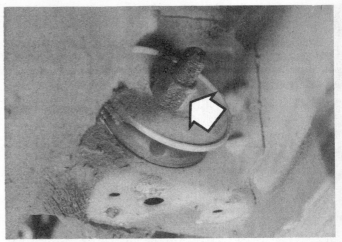

22.4  Remove this nut to disconnect the forward end of the trailing link from the chassis

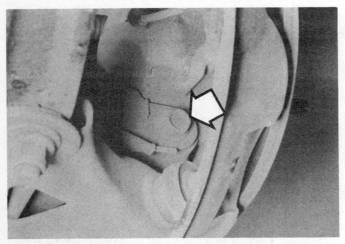

22.11  To disconnect the outer end of the lower arm from the hub carrier, put a backup wrench on the nut (not shown, on rear of hub carrier) and remove this bolt (arrow)

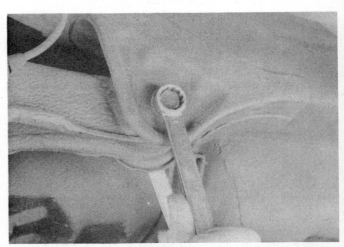

22.12  To disconnect the inner end of the lower arm from the chassis, remove this bolt

3  Remove the cover plate that protects the nut for the forward end of the trailing link (**see illustration**).
4  Remove the bolt that attaches the forward end of the trailing link to the chassis (**see illustration**).
5  Remove the trailing link.
6  Inspect the bushings at the forward end of the trailing link and at the bracket. If they're dry, lubricate them with silicone grease; if they're cracked, have them replaced by a dealer.
7  If you're replacing either the trailing link or its bracket, remove the bolts and nuts that attach the bracket to the trailing link and reattach the new trailing link to the old bracket, or the new bracket to the old trailing link. Tighten the bolts and nuts securely.
8  Installation is the reverse of removal. Be sure to tighten all fasteners securely.

## Lower arm

*Refer to illustrations 22.11 and 22.12*

9  Disconnect the stabilizer link from the lower arm (see Section 21).
10  Disconnect the trailing link bracket from the hub carrier (see above).
11  Remove the bolt that attaches the outer end of the lower arm to the hub carrier (**see illustration**).
12  Remove the bolt that attaches the inner end of the lower arm to the chassis (**see illustration**).

13  Remove the lower arm.
14  Inspect the bushings at the inner and outer ends of the lower arm. If they're dry, lubricate them with silicone grease; if they're cracked, have them replaced by a dealer.
15  Installation is the reverse of removal.
16  Install the wheel, hand tighten the wheel lug nuts, lower the vehicle and tighten the wheel lug nuts to the torque listed in the Chapter 1 Specifications.

## 23  Rear suspension arms (1989 and 1990 Legend sedans and all Legend coupes) – removal and installation

1  Loosen the rear wheel lug nuts, raise the vehicle, place it securely on jackstands and remove the wheel.

### Upper arm

*Refer to illustration 23.2*

2  To disconnect the outer end of the upper arm from the knuckle, remove the nut from the balljoint stud at the knuckle (**see illustration**), install a small two-jaw puller and separate the arm from the knuckle.
3  Remove the two mounting bolts that attach the inner end of the upper arm to the chassis.

# Chapter 10  Suspension and steering systems

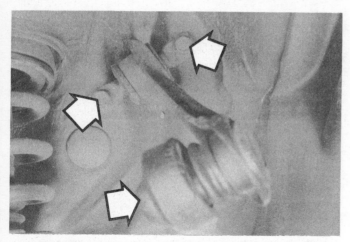

23.2  To remove the upper arm, remove the nut on the balljoint stud (lower arrow), separate the balljoint stud from the knuckle with a two-jaw puller and remove the two mounting bolts from the inner end of the arm (upper arrows)

23.7  To disconnect the outer ends of the lower arms from the knuckle, remove this through bolt – you'll need to use a backup wrench

23.8  To disconnect the inner ends of the lower arms from the chassis, remove the pivot bolts (bolt for rear arm shown)

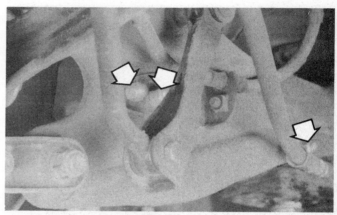

23.12  Detach the brake hose bracket (two upper arrows) and the parking brake cable bracket (lower arrow) from the trailing arm

6  Installation is the reverse of removal. Be sure to tighten all fasteners to the torque values listed in this Chapter's Specifications.

## Lower arms

*Refer to illustrations 23.7 and 23.8*

7  Remove the through bolt that attaches the lower arms to the knuckle **(see illustration)**.
8  Disconnect any brake hoses or cables from the lower arms. Remove the pivot bolts that attach the inner ends of the lower arms to the chassis **(see illustration)**.
9  Remove the lower arms.
10  Inspect the lower arm bushings for cracks and deterioration. If any of them are worn, have them replaced by a dealer.
11  Installation is the reverse of removal. Be sure to tighten all fasteners to the torque values listed in this Chapter's Specifications.

## Trailing arm

*Refer to illustrations 23.12 and 23.14*

12  Disconnect the brake hose and the parking brake cable brackets from the trailing arm **(see illustration)**.
13  Disconnect the stabilizer bar link from the trailing arm (see Section 21).
14  Remove the four bolts that attach the trailing arm bracket to the knuckle **(see illustration)**.
15  Remove the pivot bolt that attaches the forward end of the trailing arm to the chassis.
16  Remove the trailing arm.

23.14  To disconnect the rear end of the trailing arm and its bracket from the knuckle, remove these four bolts (arrows)

4  Remove the upper arm.
5  Inspect the bushing at the inner end of the arm for cracks. Inspect the balljoint at the outer end of the arm for excessive freeplay. If either is worn, have a dealer replace it.

# 10–26  Chapter 10  Suspension and steering systems

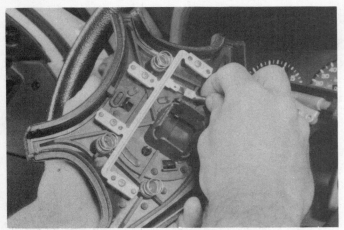

**24.2a** To get at the steering wheel retaining nut on Integra models, remove the four horn pad retaining screws from the side the steering wheel facing the dash, pull off the pad and disconnect the horn ground wire

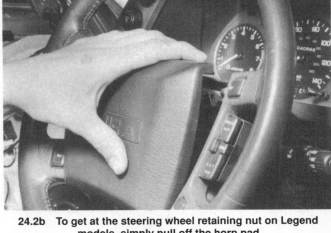

**24.2b** To get at the steering wheel retaining nut on Legend models, simply pull off the horn pad

**24.3** On some Legend models, you'll need to unplug this electrical connector before removing the steering wheel

**24.5** Paint or scribe an alignment mark between the steering shaft and the steering wheel before removing the wheel

17  Inspect the bushings at the forward and rear ends of the arm. If they're cracked or deteriorated, have them replaced by a dealer.
18  Installation is the reverse of removal. Be sure to tighten all fasteners to the torque listed in this Chapter's Specifications.

## 24  Steering wheel – removal and installation

Refer to illustrations 24.2a, 24.2b, 24.3 and 24.5
**Warning:** *Some 1988 and later Legend models are equipped with airbags. If your vehicle is equipped with an airbag, DO NOT attempt to remove the steering wheel. Have it removed by a dealer service department or other qualified repair shop..*
1  Disconnect the cable from the negative battery terminal.
2  On Integra models, remove the four horn pad retaining screws from the side of the steering wheel facing the dash, pull off the horn pad and disconnect the horn ground wire **(see illustration)**. On Legend models, simply pull off the horn pad **(see illustration)**.
3  On some Legend models, there's an electrical connector to the right of the steering wheel retaining nut. Unplug this connector **(see illustration)**.
4  Remove the retaining nut.
5  Paint or scribe a mark indicating the relationship of the steering shaft to the steering wheel hub **(see illustration)**.

6  Remove the steering wheel by pulling it straight off the shaft. The use of a steering wheel puller is unnecessary on most models.
7  Installation is the reverse of removal. Be sure to align the index mark on the steering wheel hub with the mark on the shaft when you slip the wheel onto the shaft. Install the mounting nut and tighten it to the torque listed in this Chapter's Specifications.
8  Connect the negative battery cable.

## 25  Tie-rod ends – removal and installation

### Removal
Refer to illustrations 25.2a, 25.2b and 25.4
1  Loosen the wheel lug nuts. Raise the front of the vehicle, support it securely on jackstands and remove the wheel.
2  Hold the tie-rod end with a backup wrench and loosen the jam nut enough to mark the position of the tie-rod end in relation to the threads **(see illustrations)**.
3  Remove the cotter pin and loosen the nut on the tie-rod end stud. Don't completely remove the nut.
4  Separate the tie-rod from the steering knuckle arm with a puller **(see illustrations)**. Remove the nut and detach the tie-rod.
5  Unscrew the tie-rod end from the tie-rod.

# Chapter 10  Suspension and steering systems

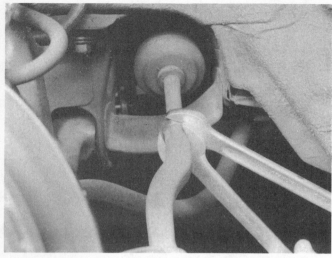

25.2a  Using a backup wrench to prevent the tie-rod end from turning, loosen the jam nut

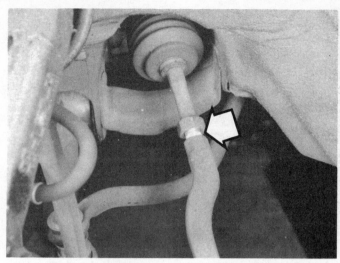

25.2b  Make an alignment mark on the exposed threads, along the edge of the tie-rod end, so the new tie-rod end will be screwed on to the exact same position

25.4  Use a two-jaw puller to separate the tie-rod end from the steering knuckle arm

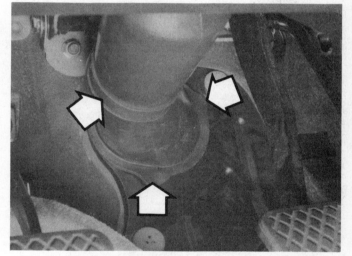

27.1  Pop off the upper and lower clamps (left arrow points to lower clamp; upper clamp not visible in this photo) and slide off the cover – when you put it back on, make sure the two nibs (arrows) on the flange at the lower end are pushed into the holes in the floor

## Installation

6  Thread the tie-rod end on to the marked position and insert the tie-rod stud into the steering knuckle arm. Don't tighten the jam nut yet.
7  Install the castellated nut on the stud and tighten it to the torque listed in this Chapter's Specifications. Install a new cotter pin.
8  Tighten the jam nut securely.
9  Install the wheel and lug nuts. Lower the vehicle and tighten the lug nuts to the torque listed in the Chapter 1 Specifications.
10  Have the alignment checked by a dealer service department or an alignment shop.

## 26  Steering gear boots – replacement

1  Loosen the lug nuts, raise the front of the vehicle and support it securely on jackstands. Remove the wheel.
2  Remove the tie-rod end and jam nut (see Section 25).
3  Remove the steering gear boot clamps and slide off the boot.
4  Before installing the new boot, wrap the threads and serrations on the end of the steering rod with a layer of tape so the small end of the new boot

isn't damaged.
5  Slide the new boot into position on the steering gear until it seats in the groove in the steering rod and install new clamps.
6  Remove the tape and install the tie-rod end (see Section 25).
7  Install the wheel and lug nuts. Lower the vehicle and tighten the lug nuts to the torque listed in the Chapter 1 Specifications.

## 27  Steering gear – removal and installation

*Refer to illustrations 27.1, 27.2, 27.4, 27.5a, 27.5b, 27.8a and 27.8b*
**Note:** *This procedure applies to both power and manual steering gear assemblies. When working on a vehicle equipped with a manual steering gear, simply ignore any references made to the power steering system.*

### Removal

1  Working under the dash, remove the steering joint cover **(see illustration)**.

# Chapter 10 Suspension and steering systems

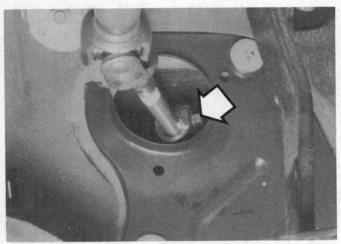

**27.2** Mark the relationship of the intermediate shaft to the steering gear input shaft and remove the pinch bolt (arrow)

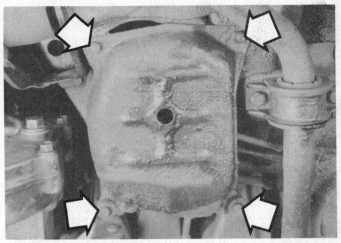

**27.4** On Legend models, remove these four retaining bolts (arrows) and this shield to get at the steering gearbox

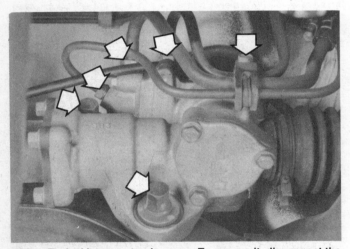

**27.5a** Typical Integra steering gear: To remove it, disconnect the power steering fluid lines (arrows) and plug them to prevent leakage and contamination, then remove the steering gear mounting bolts (arrows)

**27.5b** Typical Legend steering gear: To remove it, disconnect the power steering fluid lines (arrows) and plug them to prevent leakage and contamination, then remove the steering gear mounting bolts (arrows)

2   Mark the relationship of the intermediate shaft universal joint to the steering gear input shaft **(see illustration)** and remove the pinch bolt.
3   Raise the front of the vehicle and support it securely on jackstands. Apply the parking brake.
4   On Legend models, remove the steering gear shield **(see illustration)**.
5   Place a drain pan under the steering gear (power steering only). Disconnect the power steering fluid lines **(see illustrations)** and cap them to prevent contamination and loss of fluid.
6   Separate the tie-rod ends from the steering knuckle arms (see Section 25).
7   On Integra models, disconnect the shift linkage from the transaxle (see Chapter 7).
8   Support the steering gear and remove the right (passenger side) mounting bolts **(see illustrations)**. Lower the unit, separate the intermediate shaft from the steering gear input shaft and remove the steering gear from the vehicle.

## Installation

9   Raise the steering gear into position and connect the intermediate shaft, aligning the marks.
10   Install the steering gear mounting bolts and washers and tighten them securely.

**27.8a** Remove these bolts (arrows) from the right (passenger side) steering gear mounting clamp (Integra models) and remove the clamp

# Chapter 10  Suspension and steering systems

**27.8b  Remove these bolts (arrows) from the right (passenger side) steering gear mounting clamp (Legend models) and remove the clamp**

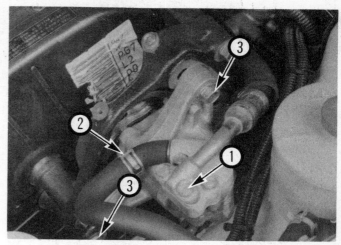

**28.1a  Power steering pump mounting details on Integra models**
1  High pressure line    2  Return line    3  Mounting bolts

**28.1b  Power steering pump mounting details on Legend models**
1  High pressure line    3  Upper mounting bolt
2  Return line               (lower bolt not visible)

**28.1c  Lower mounting bolt (arrow) for Legend power steering pump**

11  Connect the tie-rod ends to the steering knuckle arms (see Section 25).
12  Install the lower intermediate shaft pinch bolt and tighten it securely.
13  Install the steering joint cover and clamps. Make sure the two tabs on the flange at the lower end of the shield are aligned with the holes in the floor.
14  Connect the power steering hoses/lines to the steering gear and fill the power steering pump reservoir with the recommended fluid (see Chapter 1).
15  On Legend models, install the steering gear shield.
16  On Integra models, reconnect the shift linkage (see Chapter 7).
17  Lower the vehicle and bleed the steering system (see Section 29).

## 28  Power steering pump – removal and installation

*Refer to illustrations 28.1a, 28.1b and 28.1c*

1  Disconnect the fluid hoses at the pump **(see illustrations)**. Note the difference between the pressure and the return hoses. Cap both hoses to prevent leakage or contamination.
2  Remove the pump mounting bolts. If they're different diameters or lengths, note which hole they go to. Remove the pump.

3  Installation is the reverse of removal. Be sure to bleed the power steering system (see next Section) and adjust the drivebelt tension (see Chapter 1).

## 29  Power steering system – bleeding

1  Following any operation in which the power steering fluid lines have been disconnected, the power steering system must be bled to remove all air and obtain proper steering performance.
2  With the front wheels in the straight ahead position, check the power steering fluid level (see Chapter 1). If it's low, add fluid until it reaches the Cold mark on the dipstick (Legend models) or the lower mark on the reservoir (Integra models).
3  Start the engine and allow it to run at fast idle. Recheck the fluid level and add more if necessary to reach the Cold mark on the dipstick.
4  Bleed the system by turning the wheels from side-to-side, without hitting the stops. This will work the air out of the system. Keep the reservoir full of fluid as this is done.
5  When the air is worked out of the system, return the wheels to the straight ahead position and leave the vehicle running for several more minutes before shutting it off.
6  Road test the vehicle to be sure the steering system is functioning normally and noise free.

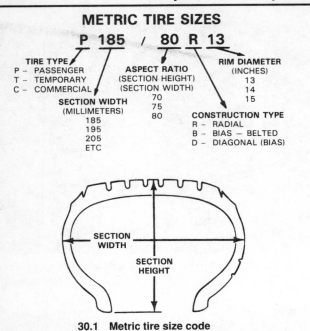

30.1  Metric tire size code

Getting the proper wheel alignment is a very exacting process, one in which complicated and expensive machines are necessary to perform the job properly. Because of this, you should have a technician with the proper equipment perform these tasks. We will, however, use this space to give you a basic idea of what is involved with wheel alignment so you can better understand the process and deal intelligently with the shop that does the work.

Toe-in is the turning in of the wheels. The purpose of a toe specification is to ensure parallel rolling of the wheels. In a vehicle with zero toe-in, the distance between the front edges of the wheels will be the same as the distance between the rear edges of the wheels. The actual amount of toe-in is normally only a fraction of an inch. At the front end, toe-in adjustment is controlled by the tie-rod end position on the tie-rod. At the rear (on Legend models), it is adjusted by moving the lower arm, in or out, within its bracket on the body. Incorrect toe-in will cause the tires to wear improperly by making them scrub against the road surface.

Camber is the tilting of the wheels from the vertical when viewed from the front or rear of the vehicle. When the wheels tilt out at the top, the camber is said to be positive (+). When the wheels tilt in at the top the camber is negative (-). The amount of tilt is measured in degrees from the vertical and this measurement is called the camber angle. This angle affects the amount of tire tread which contacts the road and compensates for changes in the suspension geometry when the vehicle is cornering or travelling over an undulating surface. Camber isn't adjustable on these vehicles.

Caster is the tilting of the top of the steering axis from the vertical. A tilt toward the rear is positive caster and a tilt toward the front is negative caster. On Legend models, caster on the front end is adjustable by lengthening or shortening the radius rod.

7  Recheck the fluid level to be sure it is up to the Hot mark on the dipstick while the engine is at normal operating temperature. Add fluid if necessary (see Chapter 1).

## 30  Wheels and tires – general information

*Refer to illustration 30.1*

All vehicles covered by this manual are equipped with metric-sized fiberglass or steel belted radial tires **(see illustration)**. Use of other size or type of tires may affect the ride and handling of the vehicle. Don't mix different types of tires, such as radials and bias belted, on the same vehicle as handling may be seriously affected. It's recommended that tires be replaced in pairs on the same axle, but if only one tire is being replaced, be sure it's the same size, structure and tread design as the other.

Because tire pressure has a substantial effect on handling and wear, the pressure on all tires should be checked at least once a month or before any extended trips (see Chapter 1).

Wheels must be replaced if they are bent, dented, leak air, have elongated bolt holes, are heavily rusted, out of vertical symmetry or if the lug nuts won't stay tight. Wheel repairs that use welding or peening are not recommended.

Tire and wheel balance is important to the overall handling, braking and performance of the vehicle. Unbalanced wheels can adversely affect handling and ride characteristics as well as tire life. Whenever a tire is installed on a wheel, the tire and wheel should be balanced by a shop with the proper equipment.

## 31  Wheel alignment – general information

*Refer to illustration 31.1*

A wheel alignment refers to the adjustments made to the wheels so they are in proper angular relationship to the suspension and the ground. Wheels that are out of proper alignment not only affect steering control, but also increase tire wear. On Legend models, toe-in and caster can be adjusted on the front wheels, and the rear toe-in can also be adjusted. On Integra models, the only adjustment possible is the toe-in of the front wheels. The front and rear camber angle, rear caster on Legend models, rear toe-in and caster (front and rear) on Integra models should be checked to determine if any of the suspension components are worn out or bent **(see illustration)**.

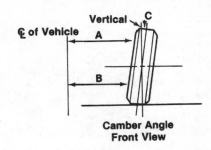
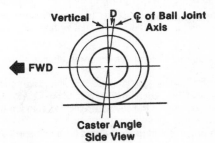
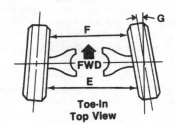

31.1  Front end alignment details
1  A minus B = C (degrees camber)
2  E minus F = toe-in (measured in inches)
3  G = toe-in (expressed in degrees)

# Chapter 11  Body

**Contents**

| | |
|---|---|
| Body – maintenance ............................................. 2 | Door window glass – removal and installation ............... 16 |
| Body repair – major damage .................................. 6 | Fixed glass – replacement ....................................... 8 |
| Body repair – minor damage ................................. 5 | General information .............................................. 1 |
| Bumpers – removal and installation ....................... 10 | Hinges and locks – maintenance ............................. 7 |
| Center console – removal and installation .............. 21 | Hood – removal, installation and adjustment .............. 9 |
| Door inside handle – removal and installation ......... 12 | Instrument cluster bezel – removal and installation ..... 22 |
| Door latch and lock cylinder – removal and installation ......... 14 | Liftgate – removal, installation and adjustment .......... 20 |
| Door outside handle – removal and installation ....... 13 | Mirrors – removal and installation .......................... 18 |
| Door – removal and installation ............................ 15 | Seat belt check ................................................... 23 |
| Door trim panel – removal and installation ............. 11 | Trunk lid – removal, installation and adjustment ........ 19 |
| Door window glass regulator assembly – removal, installation and adjustment ............................. 17 | Upholstery and carpets – maintenance ..................... 4 |
| | Vinyl trim – maintenance ....................................... 3 |

## 1  General information

These models feature a "unibody" layout, using a floor pan with front and rear frame side rails which support the body components, front and rear suspension systems and other mechanical components.

Certain components are particularly vulnerable to accident damage and can be unbolted and repaired or replaced. Among these parts are the body moldings, bumpers, the hood and trunk lid (or liftgate) and all glass.

Only general body maintenance practices and body panel repair procedures within the scope of the do-it-yourselfer are included in this Chapter.

## 2  Body – maintenance

1  The condition of your vehicle's body is very important, because the resale value depends a great deal on it. It's much more difficult to repair a neglected or damaged body than it is to repair mechanical components. The hidden areas of the body, such as the wheel wells, the frame and the engine compartment, are equally important, although they don't require as frequent attention as the rest of the body.

2  Once a year, or every 12,000 miles, it's a good idea to have the underside of the body steam cleaned. All traces of dirt and oil will be removed and the area can then be inspected carefully for rust, damaged brake lines, frayed electrical wires, damaged cables and other problems.

3  At the same time, clean the engine and the engine compartment with a steam cleaner or water soluble degreaser.

4  The wheel wells should be given close attention, since undercoating can peel away and stones and dirt thrown up by the tires can cause the paint to chip and flake, allowing rust to set in. If rust is found, clean down to the bare metal and apply an anti-rust paint.

5  The body should be washed about once a week. Wet the vehicle thoroughly to soften the dirt, then wash it down with a soft sponge and plenty of clean soapy water. If the surplus dirt is not washed off very carefully, it can wear down the paint.

6  Spots of tar or asphalt thrown up from the road should be removed with a cloth soaked in solvent.

7  Once every six months, wax the body and chrome trim. If a chrome cleaner is used to remove rust from any of the vehicle's plated parts, remember that the cleaner also removes part of the chrome, so use it sparingly.

## 3  Vinyl trim – maintenance

Don't clean vinyl trim with detergents, caustic soap or petroleum based cleaners. Plain soap and water works just fine, with a soft brush to clean dirt that may be ingrained. Wash the vinyl as frequently as the rest of the vehicle.

After cleaning, application of a high quality rubber and vinyl protectant will help prevent oxidation and cracks. The protectant can also be applied to weatherstripping, vacuum lines and rubber hoses, which often fail as a result of chemical degradation, and to the tires.

## 4  Upholstery and carpets – maintenance

1  Every three months remove the carpets or mats and clean the interior of the vehicle (more frequently if necessary). Vacuum the upholstery and carpets to remove loose dirt and dust.
2  Leather upholstery requires special care. Stains should be removed with warm water and a very mild soap solution. Use a clean, damp cloth to remove the soap, then wipe again with a dry cloth. Never use alcohol, gasoline, nail polish remover or thinner to clean leather upholstery.
3  After cleaning, regularly treat leather upholstery with a leather wax. Never use car wax on leather upholstery.
4  In areas where the interior of the vehicle is subject to bright sunlight, cover leather seats with a sheet if the vehicle is to be left out for any length of time.

## 5  Body repair – minor damage

*See photo sequence*

### Repair of minor scratches

1  If the scratch is superficial and does not penetrate to the metal of the body, repair is very simple. Lightly rub the scratched area with a fine rubbing compound to remove loose paint and built up wax. Rinse the area with clean water.
2  Apply touch-up paint to the scratch, using a small brush. Continue to apply thin layers of paint until the surface of the paint in the scratch is level with the surrounding paint. Allow the new paint at least two weeks to harden, then blend it into the surrounding paint by rubbing with a very fine rubbing compound. Finally, apply a coat of wax to the scratch area.
3  If the scratch has penetrated the paint and exposed the metal of the body, causing the metal to rust, a different repair technique is required. Remove all loose rust from the bottom of the scratch with a pocket knife, then apply rust inhibiting paint to prevent the formation of rust in the future. Using a rubber or nylon applicator, coat the scratched area with glaze-type filler. If required, the filler can be mixed with thinner to provide a very thin paste, which is ideal for filling narrow scratches. Before the glaze filler in the scratch hardens, wrap a piece of smooth cotton cloth around the tip of a finger. Dip the cloth in thinner and then quickly wipe it along the surface of the scratch. This will ensure that the surface of the filler is slightly hollow. The scratch can now be painted over as described earlier in this section.

### Repair of dents

4  When repairing dents, the first job is to pull the dent out until the affected area is as close as possible to its original shape. There is no point in trying to restore the original shape completely as the metal in the damaged area will have stretched on impact and cannot be restored to its original contours. It is better to bring the level of the dent up to a point which is about 1/8-inch below the level of the surrounding metal. In cases where the dent is very shallow, it is not worth trying to pull it out at all.
5  If the back side of the dent is accessible, it can be hammered out gently from behind using a soft-face hammer. While doing this, hold a block of wood firmly against the opposite side of the metal to absorb the hammer blows and prevent the metal from being stretched.
6  If the dent is in a section of the body which has double layers, or some other factor makes it inaccessible from behind, a different technique is required. Drill several small holes through the metal inside the damaged area, particularly in the deeper sections. Screw long, self tapping screws into the holes just enough for them to get a good grip in the metal. Now the dent can be pulled out by pulling on the protruding heads of the screws with locking pliers.
7  The next stage of repair is the removal of paint from the damaged area and from an inch or so of the surrounding metal. This is easily done with a wire brush or sanding disk in a drill motor, although it can be done just as effectively by hand with sandpaper. To complete the preparation for filling, score the surface of the bare metal with a screwdriver or the tang of a file or drill small holes in the affected area. This will provide a good grip for the filler material. To complete the repair, see the Section on filling and painting.

### Repair of rust holes or gashes

8  Remove all paint from the affected area and from an inch or so of the surrounding metal using a sanding disk or wire brush mounted in a drill motor. If these are not available, a few sheets of sandpaper will do the job just as effectively.
9  With the paint removed, you will be able to determine the severity of the corrosion and decide whether to replace the whole panel, if possible, or repair the affected area. New body panels are not as expensive as most people think and it is often quicker to install a new panel than to repair large areas of rust.
10  Remove all trim pieces from the affected area except those which will act as a guide to the original shape of the damaged body, such as headlight shells, etc. Using metal snips or a hacksaw blade, remove all loose metal and any other metal that is badly affected by rust. Hammer the edges of the hole inward to create a slight depression for the filler material.
11  Wire brush the affected area to remove the powdery rust from the surface of the metal. If the back of the rusted area is accessible, treat it with rust inhibiting paint.
12  Before filling is done, block the hole in some way. This can be done with sheet metal riveted or screwed into place, or by stuffing the hole with wire mesh.
13  Once the hole is blocked off, the affected area can be filled and painted. See the following subsection on filling and painting.

### Filling and painting

14  Many types of body fillers are available, but generally speaking, body repair kits which contain filler paste and a tube of resin hardener are best for this type of repair work. A wide, flexible plastic or nylon applicator will be necessary for imparting a smooth and contoured finish to the surface of the filler material. Mix up a small amount of filler on a clean piece of wood or cardboard (use the hardener sparingly). Follow the manufacturer's instructions on the package, otherwise the filler will set incorrectly.
15  Using the applicator, apply the filler paste to the prepared area. Draw the applicator across the surface of the filler to achieve the desired contour and to level the filler surface. As soon as a contour that approximates the original one is achieved, stop working the paste. If you continue, the paste will begin to stick to the applicator. Continue to add thin layers of paste at 20-minute intervals until the level of the filler is just above the surrounding metal.
16  Once the filler has hardened, the excess can be removed with a body file. From then on, progressively finer grades of sandpaper should be used, starting with a 180-grit paper and finishing with 600-grit wet-or-dry paper. Always wrap the sandpaper around a flat rubber or wooden block, otherwise the surface of the filler will not be completely flat. During the sanding of the filler surface, the wet-or-dry paper should be periodically rinsed in water. This will ensure that a very smooth finish is produced in the final stage.
17  At this point, the repair area should be surrounded by a ring of bare metal, which in turn should be encircled by the finely feathered edge of good paint. Rinse the repair area with clean water until all of the dust produced by the sanding operation is gone.
18  Spray the entire area with a light coat of primer. This will reveal any imperfections in the surface of the filler. Repair the imperfections with fresh filler paste or glaze filler and once more smooth the surface with sandpaper. Repeat this spray-and-repair procedure until you are satisfied

that the surface of the filler and the feathered edge of the paint are perfect. Rinse the area with clean water and allow it to dry completely.
19  The repair area is now ready for painting. Spray painting must be carried out in a warm, dry, windless and dust free atmosphere. These conditions can be created if you have access to a large indoor work area, but if you are forced to work in the open, you will have to pick the day very carefully. If you are working indoors, dousing the floor in the work area with water will help settle the dust which would otherwise be in the air. If the repair area is confined to one body panel, mask off the surrounding panels. This will help minimize the effects of a slight mismatch in paint color. Trim pieces such as chrome strips, door handles, etc., will also need to be masked off or removed. Use masking tape and several thicknesses of newspaper for the masking operations.
20  Before spraying, shake the paint can thoroughly, then spray a test area until the spray painting technique is mastered. Cover the repair area with a thick coat of primer. The thickness should be built up using several thin layers of primer rather than one thick one. Using 600-grit wet-or-dry sandpaper, rub down the surface of the primer until it is very smooth. While doing this, the work area should be thoroughly rinsed with water and the wet-or-dry sandpaper periodically rinsed as well. Allow the primer to dry before spraying additional coats.
21  Spray on the top coat, again building up the thickness by using several thin layers of paint. Begin spraying in the center of the repair area and then, using a circular motion, work out until the whole repair area and about two inches of the surrounding original paint is covered. Remove all masking material 10 to 15 minutes after spraying on the final coat of paint. Allow the new paint at least two weeks to harden, then use a very fine rubbing compound to blend the edges of the new paint into the existing paint. Finally, apply a coat of wax.

## 6  Body repair – major damage

1  Major damage must be repaired by an auto body shop specifically equipped to perform unibody repairs. These shops have the specialized equipment required to do the job properly.
2  If the damage is extensive, the body must be checked for proper alignment or the vehicle's handling characteristics may be adversely affected and other components may wear at an accelerated rate.
3  Due to the fact that all of the major body components (hood, fenders, etc.) are separate and replaceable units, any seriously damaged components should be replaced rather than repaired. Sometimes the components can be found in a wrecking yard that specializes in used vehicle components, often at considerable savings over the cost of new parts.

## 7  Hinges and locks – maintenance

Once every 3000 miles, or every three months, the hinges and latch assemblies on the doors, hood and trunk (or liftgate) should be given a few drops of light oil or lock lubricant. The door latch strikers should also be lubricated with a thin coat of grease to reduce wear and ensure free movement. Lubricate the door and trunk (or liftgate) locks with spray-on graphite lubricant.

## 8  Fixed glass – replacement

Replacement of the windshield and fixed glass requires the use of special fast-setting adhesive/caulk materials and some specialized tools and techniques. These operations should be left to a dealer service department or a shop specializing in glass work.

## 9  Hood – removal, installation and adjustment

*Refer to illustrations 9.2, 9.4, 9.10a and 9.10b*
**Note:** *The hood is heavy and somewhat awkward to remove and install – at least two people should perform this procedure.*

### Removal and installation

1  Use blankets or pads to cover the fenders and cowl areas. This will protect the body and paint as the hood is lifted off.
2  Scribe or draw alignment marks around the bolt heads to ensure proper alignment during installation **(see illustration)**.
3  Disconnect any cables or wire harnesses which will interfere with removal.
4  Have an assistant support the weight of the hood. On Legend models, remove the clips and detach the support strut **(see illustration)**. Remove the hinge-to-hood nuts or bolts and shims.
5  Lift off the hood.
6  Installation is the reverse of removal. Be sure to reinstall any shims in their original locations.

### Adjustment

7  Fore-and-aft adjustment of the hood is done by moving the hood after loosening the hinge-to-body bolts.
8  Scribe a line around the entire hinge plate so you can judge the amount of movement.

9.2  Scribe or mark around the hood hinge bolts (arrows) before loosening them

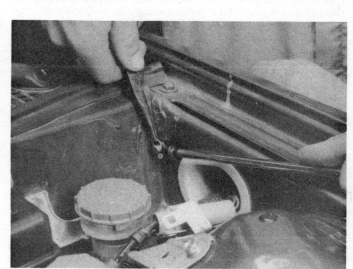

9.4  On Legend models, use needle-nose pliers to detach the clip when disconnecting the lower end of the support struts

These photos illustrate a method of repairing simple dents. They are intended to supplement *Body repair - minor damage* in this Chapter and should not be used as the sole instructions for body repair on these vehicles.

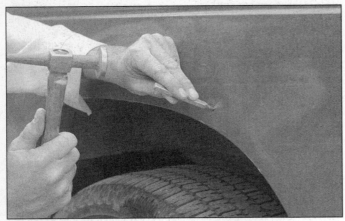

1  If you can't access the backside of the body panel to hammer out the dent, pull it out with a slide-hammer-type dent puller. In the deepest portion of the dent or along the crease line, drill or punch hole(s) at least one inch apart . . .

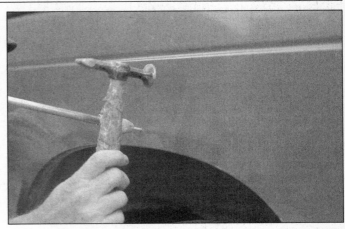

2  . . . then screw the slide-hammer into the hole and operate it. Tap with a hammer near the edge of the dent to help 'pop' the metal back to its original shape. When you're finished, the dent area should be close to its original contour and about 1/8-inch below the surface of the surrounding metal

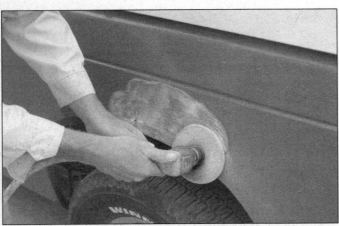

3  Using coarse-grit sandpaper, remove the paint down to the bare metal. Hand sanding works fine, but the disc sander shown here makes the job faster. Use finer (about 320-grit) sandpaper to feather-edge the paint at least one inch around the dent area

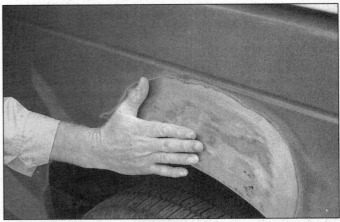

4  When the paint is removed, touch will probably be more helpful than sight for telling if the metal is straight. Hammer down the high spots or raise the low spots as necessary. Clean the repair area with wax/silicone remover

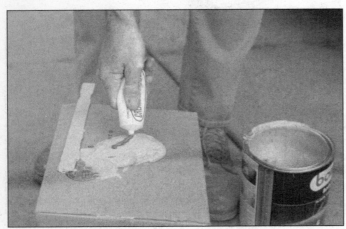

5  Following label instructions, mix up a batch of plastic filler and hardener. The ratio of filler to hardener is critical, and, if you mix it incorrectly, it will either not cure properly or cure too quickly (you won't have time to file and sand it into shape)

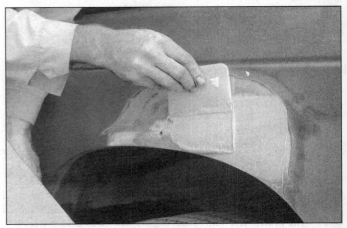

6  Working quickly so the filler doesn't harden, use a plastic applicator to press the body filler firmly into the metal, assuring it bonds completely. Work the filler until it matches the original contour and is slightly above the surrounding metal

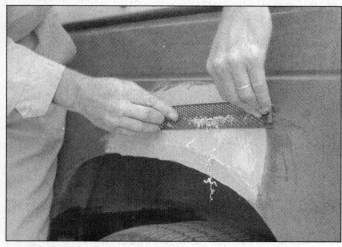

7  Let the filler harden until you can just dent it with your fingernail. Use a body file or Surform tool (shown here) to rough-shape the filler

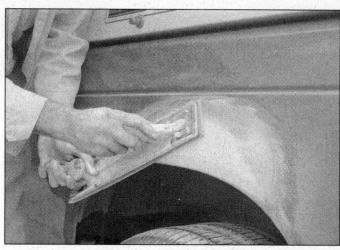

8  Use coarse-grit sandpaper and a sanding board or block to work the filler down until it's smooth and even. Work down to finer grits of sandpaper - always using a board or block - ending up with 360 or 400 grit

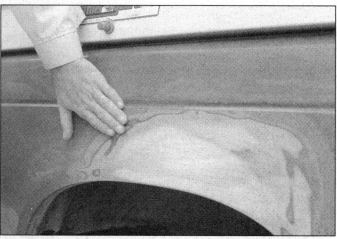

9  You shouldn't be able to feel any ridge at the transition from the filler to the bare metal or from the bare metal to the old paint. As soon as the repair is flat and uniform, remove the dust and mask off the adjacent panels or trim pieces

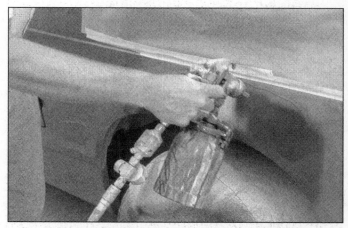

10  Apply several layers of primer to the area. Don't spray the primer on too heavy, so it sags or runs, and make sure each coat is dry before you spray on the next one. A professional-type spray gun is being used here, but aerosol spray primer is available inexpensively from auto parts stores

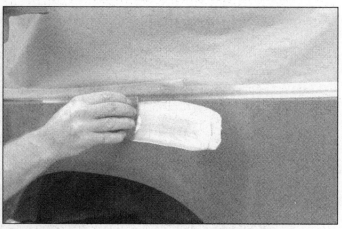

11  The primer will help reveal imperfections or scratches. Fill these with glazing compound. Follow the label instructions and sand it with 360 or 400-grit sandpaper until it's smooth. Repeat the glazing, sanding and respraying until the primer reveals a perfectly smooth surface

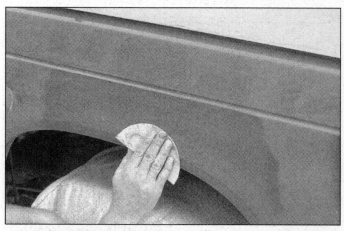

12  Finish sand the primer with very fine sandpaper (400 or 600-grit) to remove the primer overspray. Clean the area with water and allow it to dry. Use a tack rag to remove any dust, then apply the finish coat. Don't attempt to rub out or wax the repair area until the paint has dried completely (at least two weeks)

## Chapter 11 Body

9.10a  On Integra models, loosen the hood latch bolts (arrows) and move the latch to adjust the hood position

9.10b  On Legend models, the hood position can be adjusted after loosening the striker bolts (arrows) – adjust how flush the hood is with the fenders by turning the adjusting screw (A) in-or-out after loosening the locknut at the base

9   Loosen the bolts and move the hood into correct alignment. Move it only a little at a time. Tighten the hinge bolts or nuts and carefully lower the hood to check the alignment.
10   If necessary after installation, the entire hood latch assembly can be adjusted up-and-down as well as from side-to-side on the radiator support (Integra) or the hood strikers (Legend) so the hood closes securely and is flush with the fenders **(see illustrations)**. To do this, scribe a line around the hood latch mounting bolts to provide a reference point. Then loosen the bolts and reposition the latch assembly or strikers as necessary. Following adjustment, retighten the mounting bolts.
11   Finally, adjust the hood cushions on the hood (Integra) or the striker adjusting screws (Legend) so the hood, when closed, is flush with the fenders **(see illustration 9.10b)**. Adjust the rear edge of the hood until it's flush with the fenders using shims under the hinge plates.
12   The hood latch assembly, as well as the hinges, should be periodically lubricated with white lithium-base grease to prevent sticking and wear.

### 10   Bumpers – removal and installation

1   Detach the bumper cover.
2   Disconnect any wiring or other components that would interfere with bumper removal.
3   Support the bumper with a jack or jackstand. Alternatively, have an assistant support the bumper as the bolts are removed.
4   Remove the mounting bolts and detach the bumper.
5   Installation is the reverse of removal.
6   Tighten the retaining bolts securely.
7   Install the bumper cover and any other components that were removed.

### 11   Door trim panel – removal and installation

*Refer to illustrations 11.2a, 11.2b, 11.3, 11.5a, 11.5b, 11.6 and 11.8*
1   Disconnect the negative cable from the battery.
2   Remove all door trim panel retaining screws and door pull/armrest assemblies **(see illustrations)**.
3   On manual window regulator equipped models, remove the window crank handle **(see illustration)**. On power regulator models, pry out the control switch assembly and unplug it.
4   Insert a putty knife between the trim panel and the door and disengage the retaining clips. Work around the outer edge until the panel is free.
5   Once all of the clips are disengaged, detach the trim panel, unplug

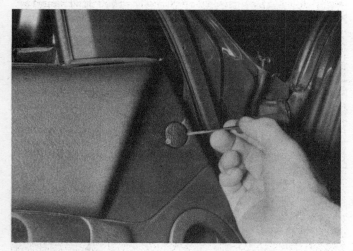

11.2a  Some trim panel screws are hidden under covers

11.2b  Pry the inner door handle trim cover off for access to the screw

# Chapter 11  Body

11.5a  Grasp the panel securely and pull up to remove it

any wire harness connectors and remove the trim panel from the vehicle **(see illustrations)**.

6  For access to the inner door, carefully peel back the plastic shield **(see illustration)**.

7  Prior to installation of the door panel, be sure to reinstall any clips in the panel which may have come out during the removal procedure and remain in the door itself.

8  Plug in the wire harness connectors and place the panel in position in the door. Press the door panel into place until the clips are seated and install the armrest/door pulls. Install the manual regulator crank handle **(see illustration)** or power window switch assembly.

## 12  Door inside handle – removal and installation

*Refer to illustration 12.2*

1  Remove the door trim panel and plastic shield (see Section 11).

2  Remove the inside door handle retaining screws, disconnect the latch rod and rotate the handle out of the door **(see illustration)**.

3  Installation is the reverse of removal.

11.3  Use a wire hook to remove the clip from the manual window crank handle

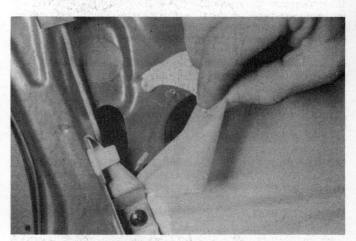

11.6  Pull the plastic shield off carefully so you don't distort or tear it

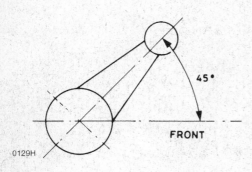

11.5b  On some Legend models, it will be necessary to detach the top of the trim panel by prying with a screwdriver

11.8  Install the crank handle so it's at a 45-degree angle with the window closed

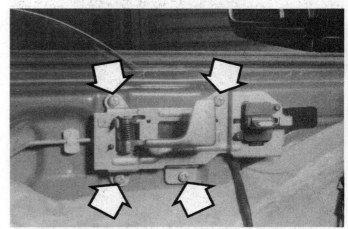

12.2  Remove the inner door handle screws (arrows)

**13.3 Apply tape around the door handle opening to protect the paint from damage when the handle is withdrawn from the door**

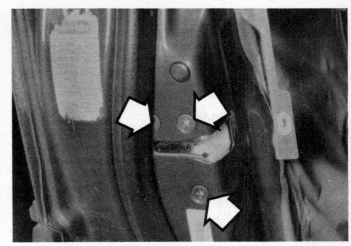

**14.4 Remove the three screws (arrows) and detach the latch from the door**

## 13 Door outside handle – removal and installation

*Refer to illustration 13.3*

1  Remove the door trim panel and plastic shield (see Section 11).
2  Disconnect the operating rods attached to the outside door handle.
3  Prior to removing the handle, apply tape around the handle opening to protect the paint **(see illustration)**.
4  Remove the door handle retaining nuts or bolts and detach the handle from the door.
5  Installation is the reverse of removal.

## 14 Door latch and lock cylinder – removal and installation

1  Remove the door trim panel and plastic shield (see Section 11).

### Latch

*Refer to illustration 14.4*

2  Remove the inside door handle (see Section 12).
3  Disconnect the operating rods from the latch.
4  Remove the three latch retaining screws located in the end of the door **(see illustration)**.

5  Detach the latch assembly from the door.
6  Installation is the reverse of removal.

### Lock cylinder

7  Remove the outside door handle (see Section 13).
8  Remove the retaining clip and withdraw the lock cylinder from the handle.
9  Installation is the reverse of removal.

## 15 Door – removal and installation

*Refer to illustrations 15.4a and 15.4b*

1  Remove the door trim panel (see Section 11). Disconnect any wire harness connectors and push them through the door opening so they won't interfere with door removal.
2  Place a jack or jackstand under the door or have an assistant on hand to support it when the hinge bolts are removed. **Note:** *If a jack or jackstand is used, place a rag between it and the door to protect the door's painted surfaces.*
3  Scribe around the door hinges.
4  Remove the check strap pin and hinge-to-door bolts, then carefully lift off the door **(see illustrations)**.

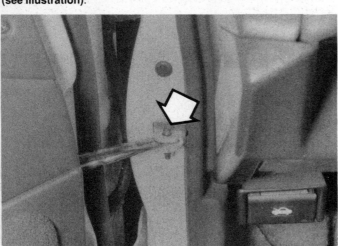

**15.4a Detach the check strap by driving out the pin in the direction shown with a hammer and punch, then, with an assistant supporting the door, . . .**

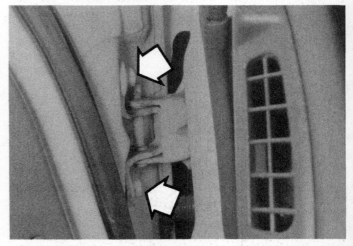

**15.4b . . . remove the hinge bolts (arrows) from the upper and lower hinges**

# Chapter 11  Body

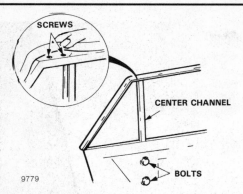

16.4  On rear doors, remove the screws and bolts retaining the center channel, detach the channel and remove the stationary glass

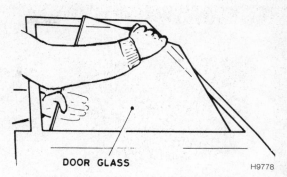

16.5  Lift the glass up out of the door while tilting it

5  Installation is the reverse of removal.
6  Following installation of the door, check the alignment and adjust it if necessary as follows:
  a) Up-and-down and forward-and-backward adjustments are made by loosening the hinge-to-body bolts and moving the door as necessary.
  b) The door lock striker can also be adjusted both up-and-down and sideways to provide positive engagement with the lock mechanism. This is done by loosening the mounting bolts and moving the striker as necessary.

## 16  Door window glass – removal and installation

*Refer to illustrations 16.4 and 16.5*

1  Remove the door trim panel and plastic shield (see Section 11).
2  Remove the inside door handle (see Section 12).
3  Lower the window so the mounting bolts can be reached through the access hole in the door, then remove the bolts **(see illustration 17.4)**.
4  On rear doors, remove the retaining screws and bolts and detach the center channel and stationary glass **(see illustration)**.
5  Lift the door glass up and out of the door window slot, then tilt it and remove it from the door **(see illustration)**.
6  Installation is the reverse of removal.

## 17  Door window glass regulator assembly – removal, installation and adjustment

*Refer to illustrations 17.4*

1  Remove the door trim panel and plastic shield (see Section 11).
2  Remove the inside door handle (see Section 12).
3  Remove the door window glass (see Section 16).
4  Remove the retaining bolts and withdraw the regulator mechanism through the access hole in the door **(see illustration)**. On power window models, unplug the electrical connector.
5  Prior to installation, lubricate all contact surfaces with multi-purpose grease. Installation is the reverse of removal.
6  To adjust the glass position evenly in the opening, loosen the roller guide or motor mounting bolts. Raise the window as far as possible, making sure it's centered in its channel, then tighten the roller guide or motor mounting bolts securely.

## 18  Mirrors – removal and installation

### Interior mirror

*Refer to illustrations 18.1, 18.2 and 18.3*

1  Pry off the mirror base cover with a screwdriver **(see illustration)**.
2  Remove the three retaining screws, then detach the mirror **(see illustration)**.

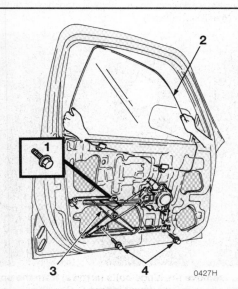

17.4  Typical front door regulator installation details

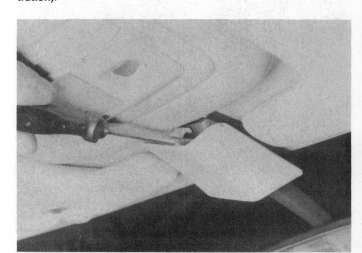

18.1  Use a screwdriver to pry off the mirror base cover

18.2  Use a Phillips head screwdriver to remove the screws

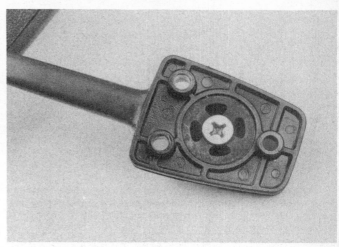

18.3  Remove the large Phillips head screw and detach the base from the old mirror

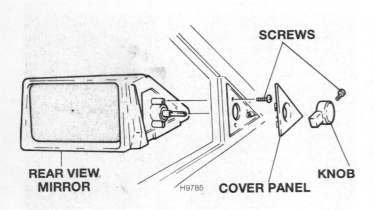

18.6a  Manual mirror details

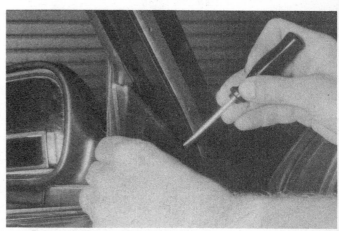

18.6b  On power mirrors, pry the cover off with a screwdriver

18.7a  Remove the mirror screws (arrows)

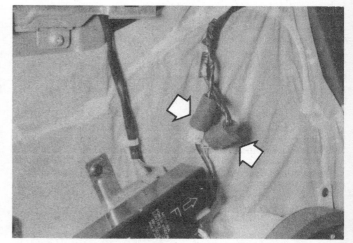

18.7b  Unplug the power mirror at the connectors (arrows)

3  If you are installing a new mirror, remove the large phillips head retaining screw and detach the base from the old mirror (see illustration). Transfer the base to the new mirror.
4  Place the mirror in position and install the screws
5  Snap the cover onto the mirror base.

### Exterior mirror
*Refer to illustrations 18.6a, 18.6b, 18.7a and 18.7b*
6  On manual mirrors, remove the screw and pull off the knob (see illustration). Pry off the cover (see illustration). On power mirrors, disconnect the negative battery cable, then remove the door trim panel (see

# Chapter 11  Body

19.3  Mark the position of the trunk lid bolts with a scribe or a permanent marker

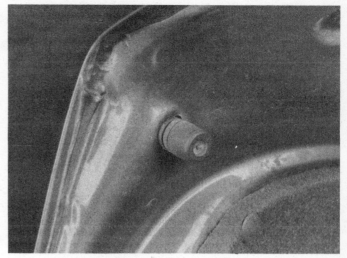

19.7  Adjust the trunk vertically on Legend models using the rubber bumpers

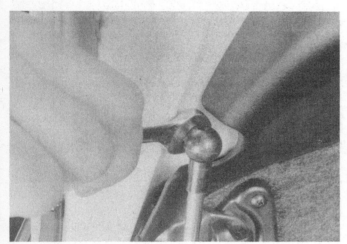

20.4a  Use a small wrench to loosen the upper . . .

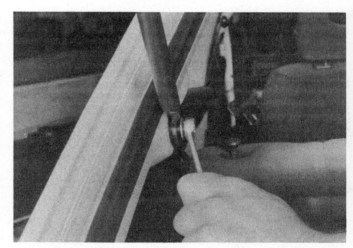

20.4b  . . . and lower nuts, then detach the support struts

Section 11) for access.

7   Remove the three retaining screws and lift the mirror off **(see illustration)**. On models equipped with power mirrors, unplug the electrical connector **(see illustration)**.

8   Installation is the reverse of removal.

## 19  Trunk lid – removal, installation and adjustment

*Refer to illustrations 19.3 and 19.7*

1   Open the trunk lid and cover the edges of the trunk compartment with pads or cloths to protect the painted surfaces when the lid is removed.

2   Disconnect any cables or wire harness connectors attached to the trunk lid that would interfere with removal.

3   Scribe or draw alignment marks around the hinge-to-trunk lid bolt heads **(see illustration)**.

4   While an assistant supports the trunk lid, remove the hinge-to-trunk lid bolts from both sides and lift off the trunk lid.

5   Installation is the reverse of removal. **Note:** *When reinstalling the trunk lid, align the hinge bolt flanges with the marks made during removal.*

6   After installation, close the lid and see if it's in proper alignment with the surrounding panels. Fore-and-aft and side-to-side adjustments of the lid are controlled by the position of the hinge bolts in the slots. To adjust it, loosen the hinge bolts, reposition the lid and retighten the bolts.

7   The height of the rear of the lid in relation to the surrounding body panels when closed can be adjusted by loosening the lock striker bolts and adding or removing adjusting shims (available from a dealer) between the striker and the body and retightening the bolts. On Legend models, make a final height adjustment by screwing the rubber bumpers in-or-out **(see illustration)**.

## 20  Liftgate – removal, installation and adjustment

*Refer to illustrations 20.4a and 20.4b*

1   Open the liftgate and cover the upper body area around the opening with pads or cloths to protect the painted surfaces when the liftgate is removed.

2   Disconnect all cables and wire harness connectors that would interfere with removal of the liftgate (tie string or wire to the cables so they can be pulled back into the body when the liftgate is reinstalled). Detach the headliner for access to the tailgate hinge retaining nuts.

3   Paint or scribe around the hinge flanges.

4   While an assistant supports the liftgate, detach the support struts **(see illustrations)**.

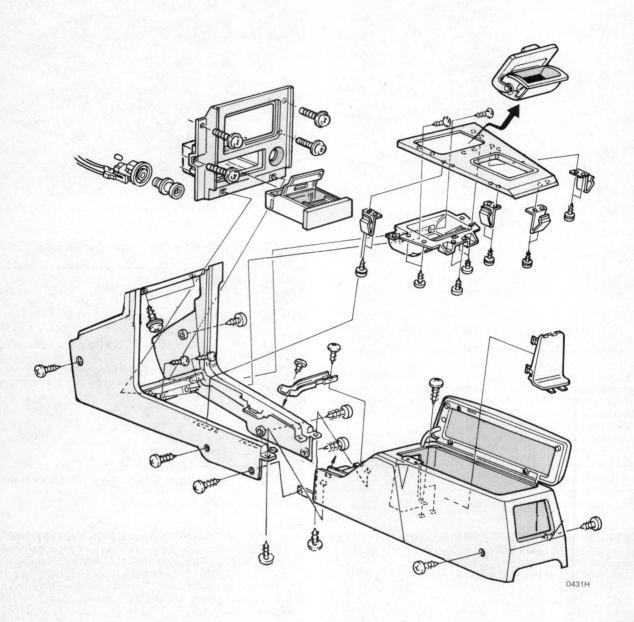

**21.3a Legend center console – exploded view**

5  Remove the hinge nuts and detach the liftgate from the vehicle.
6  Installation is the reverse of removal.
7  After installation, close the liftgate and make sure it's in proper alignment with the surrounding body panels. Adjustments are made by moving the position of the hinge studs in the slots. To adjust it, loosen the hinge nuts and reposition the hinges either side-to-side or fore-and-aft the desired amount and retighten the nuts.

8  The engagement of the liftgate can be adjusted by loosening the lock striker bolts, repositioning the striker and retightening the bolts.

## 21  Center console – removal and installation

*Refer to illustrations 21.3a and 21.3b*

1  Disconnect the negative battery cable.

# Chapter 11  Body

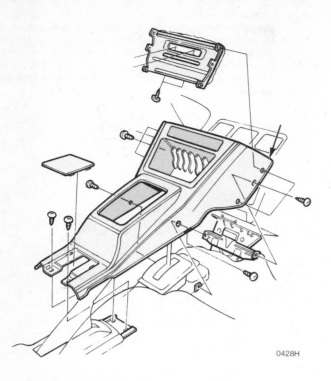

**21.3b  Integra center console details**

2   Unscrew the shift handle or knob (manual transaxle models).
3   Remove the retaining screws or bolts. Some screws are hidden under the cigar lighter, ashtray or under a center lid at the back of the console **(see illustrations)**.
4   Detach the console and lift it out of the vehicle.
5   Installation is the reverse of removal.

## 22  Instrument cluster bezel – removal and installation

1   Disconnect the negative battery cable.

**22.4  Integra instrument cluster details**

### Integra
*Refer to illustration 22.4*

2   Pry the right and left side switches out of the bezel, then unplug and remove the switches.
3   Remove the bolts and lower the steering column.
4   Remove the screws and detach the bezel from the dashboard **(see illustration)**
5   Installation is the reverse of removal.

### Legend
*Refer to illustrations 22.6 and 22.7*

6   Remove the screws **(see illustration)**.
7   Grasp the bezel securely and pull it straight back **(see illustration)**.
8   Unplug the electrical connectors and remove the bezel.
9   Installation is the reverse of removal.

**22.6  Use a Phillips head screwdriver to remove the cluster retaining screws (arrows) – Legend models**

**22.7  Pull the bezel back for access, then unplug the electrical connectors (Legend models)**

## 23 Seat belt check

1 Check the seatbelts, buckles, latch plates and guide loops for obvious damage and signs of wear.
2 Check that the seatbelt reminder light comes on when the key is turned to the Run and Start positions. A chime should also sound.
3 The seatbelts are designed to lock up during a sudden stop or impact, yet allow free movement during normal driving. Check that the retractors return the belt against your chest while driving and rewind the belt completely when the buckle is unlatched.
4 If any of the above checks reveal problems with the seatbelts, replace parts as necessary.

# Chapter 12 Chassis electrical system

## Contents

| | |
|---|---|
| Antenna – removal and installation | 9 |
| Battery check and maintenance | See Chapter 1 |
| Battery – removal and installation | See Chapter 5 |
| Bulb replacement | 13 |
| Brake light switch – removal, installation and adjustment | See Chapter 9 |
| Circuit breakers – general information | 4 |
| Combination switches – check and replacement | 7 |
| Cruise control system – description and check | 19 |
| Electrical troubleshooting – general information | 2 |
| Fuses – general information | 3 |
| General information | 1 |
| Hazard/turn-signal flashers – check and replacement | 6 |
| Headlights – adjustment | 12 |
| Headlights – removal and installation | 11 |
| Ignition switch and lock cylinder – check and replacement | 10 |
| Instrument cluster – removal and installation | 15 |
| Instrument panel – removal and installation | 16 |
| Neutral start switch – check and replacement | See Chapter 7B |
| Power door lock system – description and check | 17 |
| Power window system – description and check | 18 |
| Radio – removal and installation | 8 |
| Relays – general information | 5 |
| Windshield wiper motor – removal and installation | 14 |
| Wiring diagrams – general information | 20 |

## 1 General information

The electrical system is a 12-volt, negative ground type. Power for the lights and all electrical accessories is supplied by a lead/acid-type battery which is charged by the alternator.

This Chapter covers repair and service procedures for the various electrical components not associated with the engine. Information on the battery, alternator, distributor and starter motor can be found in Chapter 5.

It should be noted that when portions of the electrical system are serviced, the negative battery cable should be disconnected from the battery to prevent electrical shorts and/or fires.

## 2 Electrical troubleshooting – general information

**Warning:** *Some later Legend models are equipped with an air bag or Supplemental Restrain System (SRS). Do not test any circuits involved with this system to avoid possible damage. Yellow insulation is used on all SRS wiring harnesses to make them readily identifiable.*

A typical electrical circuit consists of an electrical component, any switches, relays, motors, fuses, fusible links or circuit breakers related to that component and the wiring and connectors that link the component to both the battery and the chassis. To help you pinpoint an electrical circuit problem, wiring diagrams are included at the end of this book.

Before tackling any troublesome electrical circuit, first study the appropriate wiring diagrams to get a complete understanding of what makes up that individual circuit. Trouble spots, for instance, can often be narrowed down by noting if other components related to the circuit are operating properly. If several components or circuits fail at one time, chances are the problem is in a fuse or ground connection, because several circuits are often routed through the same fuse and ground connections.

Electrical problems usually stem from simple causes, such as loose or corroded connections, a blown fuse, a melted fusible link or a bad relay. Visually inspect the condition of all fuses, wires and connections in a problem circuit before troubleshooting it.

If testing instruments are going to be utilized, use the diagrams to plan ahead of time where you will make the necessary connections in order to accurately pinpoint the trouble spot.

The basic tools needed for electrical troubleshooting include a circuit tester or voltmeter (a 12-volt bulb with a set of test leads can also be used), a continuity tester, which includes a bulb, battery and set of test leads, and a jumper wire, preferably with a circuit breaker incorporated, which can be used to bypass electrical components. Before attempting to locate a problem with test instruments, use the wiring diagram(s) to decide where to make the connections.

### Voltage checks

Voltage checks should be performed if a circuit is not functioning properly. Connect one lead of a circuit tester to either the negative battery terminal or a known good ground. Connect the other lead to a connector in the circuit being tested, preferably nearest to the battery or fuse. If the bulb of the tester lights, voltage is present, which means that the part of the circuit between the connector and the battery is problem free. Continue checking the rest of the circuit in the same fashion. When you reach a point at which no voltage is present, the problem lies between that point and the last test point with voltage. Most of the time the problem can be traced to a loose connection. **Note:** *Keep in mind that some circuits receive voltage only when the ignition key is in the Accessory or Run position.*

### Finding a short

One method of finding shorts in a circuit is to remove the fuse and connect a test light or voltmeter in its place to the fuse terminals. There should be no voltage present in the circuit. Move the wiring harness from side-to-side while watching the test light. If the bulb goes on, there is a short to ground somewhere in that area, probably where the insulation has rubbed through. The same test can be performed on each component in the circuit, even a switch.

### Ground check

Perform a ground test to check whether a component is properly grounded. Disconnect the battery and connect one lead of a self-powered test light, known as a continuity tester, to a known good ground. Connect the other lead to the wire or ground connection being tested. If the bulb goes on, the ground is good. If the bulb does not go on, the ground is not good.

### Continuity check

A continuity check is done to determine if there are any breaks in a circuit – if it is passing electricity properly. With the circuit off (no power in the circuit), a self-powered continuity tester can be used to check the circuit. Connect the test leads to both ends of the circuit (or to the "power" end and a good ground), and if the test light comes on the circuit is passing current properly. If the light doesn't come on, there is a break somewhere in the circuit. The same procedure can be used to test a switch, by connecting the continuity tester to the switch terminals. With the switch turned On, the test light should come on.

### Finding an open circuit

When diagnosing for possible open circuits, it is often difficult to locate them by sight because oxidation or terminal misalignment are hidden by the connectors. Merely wiggling a connector on a sensor or in the wiring harness may correct the open circuit condition. Remember this when an open circuit is indicated when troubleshooting a circuit. Intermittent problems may also be caused by oxidized or loose connections.

Electrical troubleshooting is simple if you keep in mind that all electrical circuits are basically electricity running from the battery, through the wires, switches, relays, fuses and fusible links to each electrical component (light

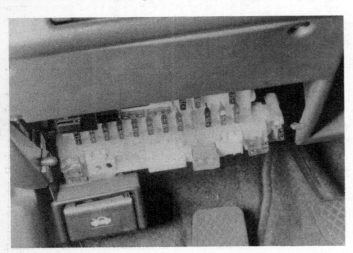

3.1a  On Integra models, the interior fuse block is located at the lower left corner of the dashboard, under a cover

3.1b  The interior fuse block is located under the driver's side kick panel, under a hinged cover on Legend models

3.1c  The engine compartment fuse block is located on the passenger's side of the firewall on Integra models

3.1d  The Legend engine compartment fuse block is near the battery on the driver's side

# Chapter 12 Chassis electrical system

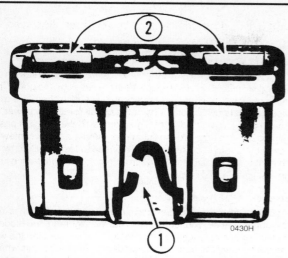

**3.3 To test for a blown fuse, pull it out and inspect it for an open (1), then, with the fuse installed and the circuit activated, connect a test light across the terminals (2)**

**3.6 On later models, the main fuses are located in the engine compartment fuse block (Integra shown)**

bulb, motor, etc.) and to ground, from which it is passed back to the battery. Any electrical problem is an interruption in the flow of electricity to and from the battery.

## 3  Fuses – general information

*Refer to illustrations 3.1a, 3.1b, 3.1c, 3.1d, 3.3 and 3.6*

The electrical circuits of the vehicle are protected by a combination of fuses, circuit breakers and fusible links. The two fuse blocks are located under the instrument panel on the left side of the dashboard and in the engine compartment **(see illustrations)**.

Each of the fuses is designed to protect a specific circuit, and the various circuits are identified on the fuse panel itself.

Miniaturized fuses are employed in the fuse block. These compact fuses, with blade terminal design, allow fingertip removal and replacement. If an electrical component fails, always check the fuse first. A blown fuse is easily identified through the clear plastic body. Visually inspect the element for evidence of damage **(see illustration)**. If a continuity check is called for, the blade terminal tips are exposed in the fuse body.

Be sure to replace blown fuses with the correct type. Fuses of different ratings are physically interchangeable, but only fuses of the proper rating should be used. Replacing a fuse with one of a higher or lower value than specified is not recommended. Each electrical circuit needs a specific amount of protection. The amperage value of each fuse is molded into the fuse body.

If the replacement fuse immediately fails, don't replace it again until the cause of the problem is isolated and corrected. In most cases, the cause will be a short circuit in the wiring caused by a broken or deteriorated wire.

These models are equipped with main fuses (also known as a fusible links) which protect all the circuits coming from the battery. If these circuits are overloaded, the main fuse blows, preventing damage to the main wiring harness. On 1986 and 1987 Integra models, the main fuses consist of metal strips which will be visibly melted when overloaded. On these models, disconnect the battery before replacing a main fuse with a new one (available from your dealer). On all other models, the main fuse (or fuses, depending on model) are located in the engine compartment fuse block **(see illustration)**. They are very similar in appearance to standard fuses and are replaced in the same way.

## 4  Circuit breakers – general information

Circuit breakers protect components such as power windows, power door locks and headlights.

On some models the circuit breaker resets itself automatically, so an electrical overload in a circuit breaker protected system will cause the circuit to fail momentarily, then come back on. If the circuit does not come back on, check it immediately. Once the condition is corrected, the circuit breaker will resume its normal function. Some circuit breakers must be reset manually.

## 5  Relays – general information

Several electrical accessories in the vehicle use relays to transmit the electrical signal to the component. If the relay is defective, that component will not operate properly.

The various relays are grouped together in several locations.

If a faulty relay is suspected, it can be removed and tested by a dealer service department or a repair shop. Defective relays must be replaced as a unit.

## 6  Hazard/turn-signal flashers – check and replacement

1  The hazard/turn-signal flasher is a small canister-shaped unit located in the interior fuse block.
2  When the flasher unit is functioning properly, an audible click can be heard during its operation. If the turn signals fail on one side or the other and the flasher unit does not make its characteristic clicking sound, a faulty turn-signal bulb is indicated.
3  If both turn signals fail to blink, the problem may be due to a blown fuse, a faulty flasher unit, a broken switch or a loose or open connection. If a quick check of the fuse box indicates that the turn-signal fuse has blown, check the wiring for a short before installing a new fuse.
4  To replace the flasher, simply pull it out of the fuse block or wiring harness.
5  Make sure that the replacement unit is identical to the original. Compare the old one to the new one before installing it.
6  Installation is the reverse of removal.

## 7  Combination switch(es) – check and replacement

**Warning:** *Some later model Legends are equipped with an air bag or Supplemental Restraint System. To avoid possible damage to this system, the manufacturer recommends that, on these models, the following procedures should be left to a dealer service department because of the special tools and techniques required.*

**Note:** *On Legend models, the following procedure covers both the headlight/turn signal and wiper/washer combination switches*

## Windshield Wiper/Washer Switch

| POSITION | A1 | A2 | A3 | A4 | A6 | A7 | A8 |
|---|---|---|---|---|---|---|---|
| Off | O—|—O | | | | | | |
| INT | | | O—|—O | | | |
| LO | O—|—|—|—|—O | | |
| HI | | | | | O—|—O | |
| Mist Switch ON | | | | | O—|—O | |
| Washer Switch ON | | | | O—|—|—|—|—O |

## Rear Window Wiper/Washer Switch

| POSITION | B1 | B2 | B3 | B4 | B6 |
|---|---|---|---|---|---|
| Washer Switch ON | O—|—|—O | | |
| Off | O—|—O | | | |
| On | | | | O—|—O |
| Washer Switch ON | O—|—|—O | | |

## Lighting/Dimmer/Passing Switch

| Lighting Switch | POSITION | C1 | C3 | C4 | C5 | C7 | C8 |
|---|---|---|---|---|---|---|---|
| • | | O—|—|—O | | | |
| • | Dimmer Switch Low | O—|—|—|—|—|—O—|—O |
| • | Dimmer Switch High | O—|—|—|—O—|—|—O |
| | Passing Switch On | | O—|—O | | | |

## Hazard Switch

| POSITION | D1 | D2 | D3 | D4 | D5 | D6 |
|---|---|---|---|---|---|---|
| Off | | | | | O—|—O | |
| On | O—|—|—O | O—|—|—|—|—O |

## Turn Signal Switch

| POSITION | E1 | E2 | E3 |
|---|---|---|---|
| R | O—|—|—O |
| Neutral | | | |
| L | O—|—O | |

7.2a Make the continuity checks indicated in these charts – the switch must be in the indicated position for each check (Integra models)

# Chapter 12 Chassis electrical system

12 – 5

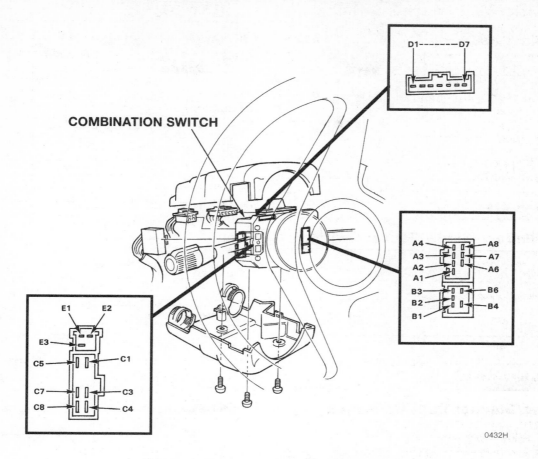

7.2b Integra combination switch connector details

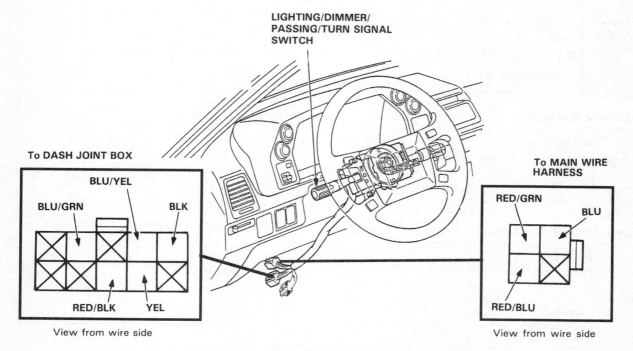

7.2c Legend combination headlight/turn-signal switch connector details

12-6 Chapter 12 Chassis electrical system

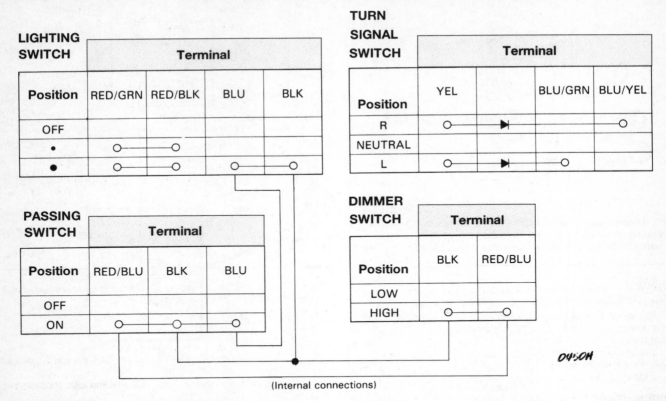

7.2d Make the continuity checks for the headlight/turn-signal switch with the switch in the indicated positions (Legend models)

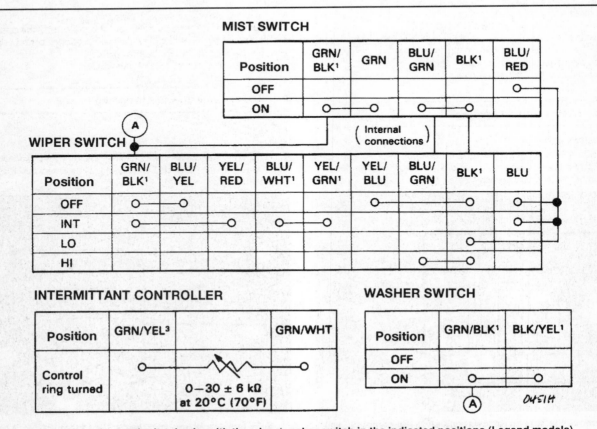

7.2e Make the continuity checks with the wiper/washer switch in the indicated positions (Legend models)

# Chapter 12  Chassis electrical system

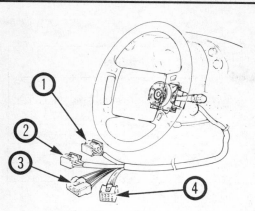
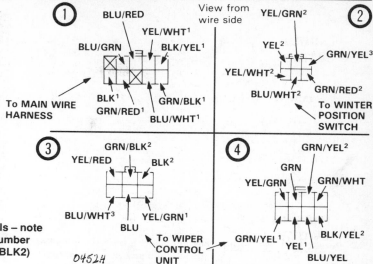

**7.2f  Legend combination wiper/washer switch connector details – note that several different wires have the same color and have a number suffix to distinguish them (GRN/BLK1 isn't the same as GRN/BLK2)**

## Check

Refer to illustrations 7.2a, 7.2b, 7.2c, 7.2d, 7.2e and 7.2f

1  Remove the steering column upper and lower covers, then unplug the switch electrical connectors. On Legend models, remove the lower dash panel for access.
2  Check the connector terminals for continuity with the switch in each position **(see illustrations)**.
3  If the continuity is not as specified, replace the switch.

## Replacement

Refer to illustration 7.6

4  On most Legend models, the turn-signal and wiper/washer switches can be removed separately.
5  Remove the steering wheel (see Chapter 10).
6  Remove the retaining screws and lift the switch off **(see illustration)**.
7  Detach the switch electrical connector from the wiring harness and remove the combination switch from the vehicle.
8  Installation is the reverse of removal.

---

## 8  Radio – removal and installation

Refer to illustration 8.7

**Warning:** *Some later model Legends are equipped with an air bag or Supplemental Restraint System. To avoid possible damage to this system, the manufacturer recommends that, on these models, the following procedures should be left to a dealer service department because of the special tools and techniques required.*

1  Disconnect the negative cable from the battery.

### Integra

2  Remove the ashtray and ashtray holder.
3  From under the radio, remove the two screws, then push the radio out from behind.
4  Unplug the electrical connector and antenna lead, then remove the radio.
5  Installation is the reverse of removal.

### Legend

6  Remove the center console (see Chapter 11).
7  Remove the mounting screws, pull the radio out of the dashboard for access to the back, then unplug the electrical connector and antenna lead and remove the radio **(see illustration)**.
8  Installation is the reverse of removal.

---

## 9  Antenna – removal and installation

### Integra

Refer to illustration 9.2

1  On manual antenna models, remove the radio and disconnect the antenna lead (see the previous Section). On power antenna models, disconnect the cable at the motor.

**7.6  Remove the screws (arrows) and slide the combination switch off the steering shaft**

**8.7  Remove the screws (arrows) and pull the radio out for access to the electrical connectors (Legend models)**

12-8  Chapter 12  Chassis electrical system

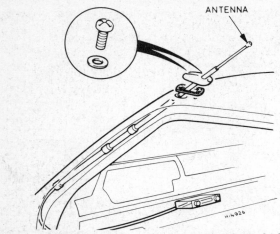

9.2 After attaching a string or wire to the lead (to make installing the new one easier), remove the antenna screws and pull the assembly out of the body pillar (Integra models)

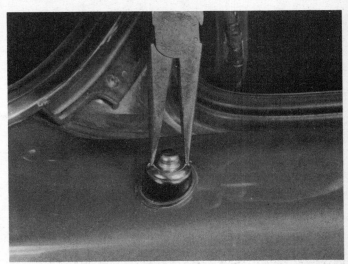

9.5 Use needle-nose pliers to unscrew the antenna cap nut

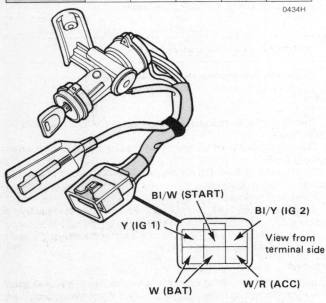

10.3a Integra ignition switch continuity check details

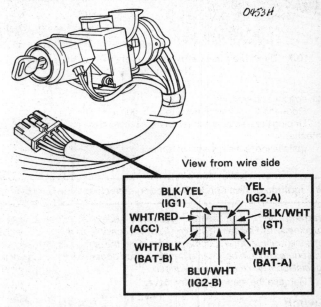

10.3b Check the Legend ignition switch continuity with the switch in the indicated positions

2  Connect a piece of string or thin wire to the antenna lead (at the radio end) or, on power antenna models, the cable (at the motor end). Remove the mounting screws and pull the antenna and lead out of the body pillar **(see illustration)**.

3  Fasten the wire or string to the lead or cable of the new antenna. Lower the antenna into place while pulling the new lead or cable into the pillar with the wire or string.

4  Disconnect the string or wire and connect the antenna lead or cable to the radio or motor. Install the antenna retaining screws.

### Legend

*Refer to illustration 9.5*

5  Lower the antenna, then unscrew the cap nut securing the antenna to fender, using a pair of needle-nose pliers **(see illustration)**. Remove the

# Chapter 12  Chassis electrical system

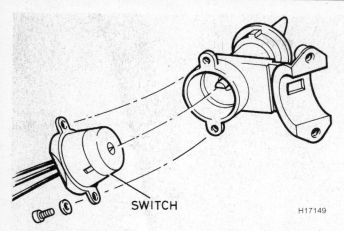

10.4  The ignition switch is held to the housing by two screws

10.6  On Legend models, lift the key light off the lock cylinder

10.7  Push the pin in while pulling the lock cylinder out of the housing

nut, spacer and bushing.
6   Open the trunk and remove the left side inner trim panel.
7   Unplug the electrical connector, remove the two nuts and remove the antenna.
8   Installation is the reverse of removal.

## 10  Ignition switch and lock cylinder – check and replacement

**Warning:** *Some later model Legends are equipped with an air bag or Supplemental Restraint System. To avoid possible damage to this system, the manufacturer recommends that, on these models, the following procedures should be left to a dealer service department because of the special tools and techniques required.*

1   Remove the steering column cover.

### Switch

*Refer to illustrations 10.3a, 10.3b and 10.4*

2   Trace the wire harness for the ignition switch/key lock cylinder assembly to the fuse box, then unplug the connector from the fuse box.
3   Check the connector for continuity with the key in each position **(see illustrations)**.
4   If the continuity is not as specified, replace the switch. Do this by placing the key in the Off (Lock) position, removing the two screws and lifting the switch off the steering column **(see illustration)**. Make sure the recess in the switch is aligned with the projection on the lock when installing the new switch. Install the steering column cover.

### Lock cylinder

*Refer to illustrations 10.6 and 10.7*

5   Check the lock cylinder in each position to make sure it isn't worn or loose and that the key position corresponds to the markings on the housing. Replace the lock cylinder if it's worn or damaged.
6   Remove the ignition key light (Legend models) **(see illustration)**.
7   With the key in the Accessory position, insert a pin or thin tool into the recess, then depress the pin while pulling the lock cylinder out of the housing **(see illustration)**.
8   When installing, turn the key to the Lock position and line up the lock cylinder with the housing. Turn the key almost to the Lock position, insert the lock cylinder until the pin touches the housing, then turn the key all the way to Lock as you push in and insert the lock cylinder into the housing until it clicks. Install the steering column cover.

## 11  Headlights – removal and installation

1   Disconnect the negative cable from the battery.

### Integra

2   Raise the headlights.
3   Remove the lower front grille and the headlight bezel screws, then detach the bezel.
4   Remove the headlight retainer screws, taking care not to disturb the adjusting screws.
5   Remove the retainer and pull the headlight out enough to allow the connector to be unplugged.
6   Remove the headlight.
7   To install the headlight, plug the connector in, place the headlight in position and install the retainer and screws. Tighten the screws securely.
8   Place the headlight bezel in position and install the retaining screws. Re-connect the negative battery cable.

### Legend

**Warning:** *Halogen gas-filled bulbs are under pressure and may shatter if the surface is scratched or the bulb is dropped. Wear eye protection and handle the bulbs carefully, grasping only the base whenever possible. Do not touch the surface of the bulb with your fingers because the oil from your skin could cause it to overheat and fail prematurely. If you do touch the bulb surface, clean it with rubbing alcohol.*

9   Open the hood.
10  Reach behind the headlight assembly, unplug the electrical connector, grasp the bulb holder retaining collar, turn it counterclockwise to remove it, then pull the bulb assembly out.
11  Insert the new bulb assembly into the housing, place the retaining collar in position and rotate it clockwise, then plug in the electrical connector. Re-connect the negative battery cable

# 12-10 Chapter 12 Chassis electrical system

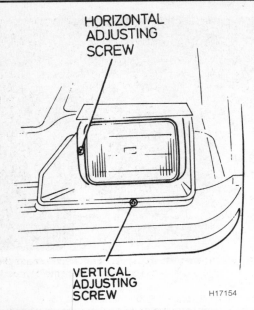

**12.1a  Integra adjusting screws are accessible after removing the headlight bezel – later models have two horizontal adjusting screws**

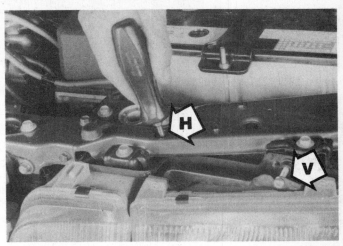

**12.1b  On Legend models, the horizontal adjusting screw (H) is accessible using a Phillips screwdriver behind the housing, while the vertical screw (V) is at the top**

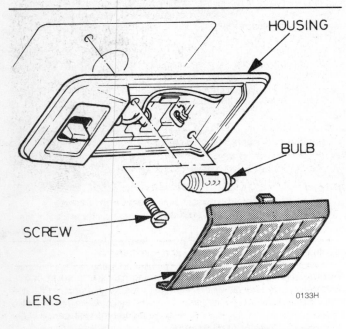

**13.2  Pry off the interior light lens for access to the bulb**

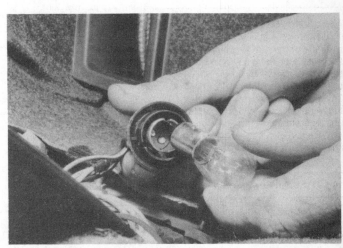

**13.3  Remove most bulbs by pushing in and turning the bulb counterclockwise**

## 12  Headlights – adjustment

*Refer to illustrations 12.1a and 12.1b*

**Note:** *The headlights must be aimed correctly. If adjusted incorrectly they could blind the driver of an oncoming vehicle and cause a serious accident or seriously reduce your ability to see the road. The headlights should be checked for proper aim every 12 months and any time a new headlight is installed or front end body work is performed. It should be emphasized that the following procedure is only an interim step which will provide temporary adjustment until the headlights can be adjusted by a properly equipped shop.*

1  Headlights have spring-loaded adjusting screws for controlling up-and-down and left-and-right movement **(see illustrations)**.

2  There are several methods of adjusting the headlights. The simplest method requires a blank wall 25 feet in front of the vehicle and a level floor.
3  Position masking tape vertically on the wall in reference to the vehicle centerline and the centerlines of both headlights.
4  Position a horizontal tape line in reference to the centerline of all the headlights. **Note:** *It may be easier to position the tape on the wall with the vehicle parked only a few inches away.*
5  Adjustment should be made with the vehicle sitting level, the gas tank half-full and no unusually heavy load in the vehicle.
6  Starting with the low beam adjustment, position the high intensity zone so it is two inches below the horizontal line and two inches to the right of the headlight vertical line. Adjustment is made by turning the vertical adjusting screw. The horizontal adjusting screw should be used to move the beam left or right.
7  With the high beams on, the high intensity zone should be vertically centered with the exact center just below the horizontal line. **Note:** *It may not be possible to position the headlight aim exactly for both high and low beams. If a compromise must be made, keep in mind that the low beams are the most used and have the greatest effect on driver safety.*
8  Have the headlights adjusted by a dealer service department or service station at the earliest opportunity.

# Chapter 12  Chassis electrical system

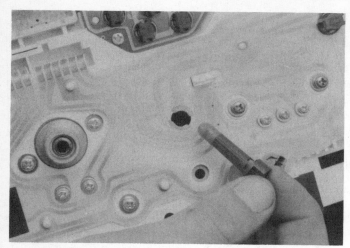

**13.4  After removing the instrument cluster, flip it over and replace the bulbs by rotating them and pulling them out of the housing**

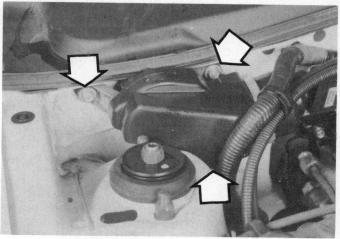

**14.5  After unplugging the electrical connector, remove the bolts (arrows) and lift the wiper motor from the engine compartment**

## 13  Bulb replacement

*Refer to illustrations 13.2, 13.3 and 13.4*

1  The lenses of many lights are held in place by screws, which makes it a simple procedure to gain access to the bulbs.
2  On some lights the lenses are held in place by clips. The lenses can be removed either by unsnapping them or by using a small screwdriver to pry them off **(see illustration)**.
3  Several types of bulbs are used. Some are removed by pushing in and turning them counterclockwise **(see illustration)**. Others can simply be unclipped from the terminals or pulled straight out of the socket.
4  To gain access to the instrument panel lights **(see illustration)**, the instrument cluster will have to be removed first (see Section 15).

## 14  Windshield wiper motor – removal and installation

*Refer to illustration 14.5*

1  Disconnect the negative cable from the battery.
2  Detach the wiper arms.
3  Remove the scoop and hood seal from the cowl.
4  Use a screwdriver to detach the wiper linkage balljoint from the motor arm.
5  Unplug the electrical connector, remove the retaining bolts and lift the wiper motor from the engine compartment **(see illustration)**.
6  Prior to installation, lubricate the contact points of the wiper linkage with multi-purpose grease. Installation is the reverse of removal.

## 15  Instrument cluster – removal and installation

**Warning:** *Some later model Legends are equipped with an air bag or Supplemental Restraint System. To avoid possible damage to this system, the manufacturer recommends that, on these models, the following procedures should be left to a dealer service department because of the special tools and techniques required.*

*Refer to illustration 15.5*

1  Disconnect the negative cable from the battery.
2  Remove the steering wheel (see Chapter 10).
3  Remove the steering column upper and lower covers.
4  Remove the instrument cluster bezel (see Chapter 11).
5  Remove the instrument cluster screws **(see illustration)**.
6  Pull the cluster out, unplug the wiring connectors and, on Integra models, disconnect the speedometer cable from the instrument cluster.
7  Installation is the reverse of removal.

**15.5  Instrument cluster screw locations (arrows) (Legend shown)**

## 16  Instrument panel – removal and installation

**Warning:** *Some later model Legends are equipped with an air bag or Supplemental Restraint System. To avoid possible damage to this system, the manufacturer recommends that, on these models, the following procedures should be left to a dealer service department because of the special tools and techniques required.*

1  Disconnect the negative cable from the battery.
2  Remove the steering wheel (see Chapter 10) and the steering column upper and lower covers.
3  Remove the steering column mounting bracket bolts and lower the steering column.
4  On Legend models, remove the defroster garnishes, the left air duct and the hood opener.
5  Remove the lower dash panels and the fuse box cover. Disconnect the wiring harness from the fuse box.
6  Remove the air conditioner and heater control assembly (see Chapter 3).
7  Disconnect the ground cable from the steering column.
8  Remove the retaining bolts, nuts and screws and unplug any electrical connectors. Support the instrument panel, pull it out and remove it from the vehicle.
9  Installation is the reverse of removal.

## 17  Power door lock system – description and check

The power door lock system operates the door lock actuators mounted in each door. The system consists of the switches, actuators and associated wiring. Since special tools and techniques are required to diagnose the system, it should be left to a dealer service department or a repair shop. However, it is possible for the home mechanic to make simple checks of the wiring connections and actuators for minor faults which can be easily repaired. These include:
a) Check the system fuse and/or circuit breaker.
b) Check the switch wires for damage and loose connections. Check the switches for continuity.
c) Remove the door panel(s) and check the actuator wiring connections to see if they're loose or damaged. Inspect the actuator rods (if equipped) to make sure they aren't bent or damaged. Inspect the actuator wiring for damaged or loose connections. The actuator can be checked by applying battery power momentarily. A discernible click indicates that the solenoid is operating properly.

## 18  Power window system – description and check

The power window system operates the electric motors mounted in the doors which lower and raise the windows. The system consists of the control switches, the motors (regulators), glass mechanisms and associated wiring.

Because of the complexity of the power window system and the special tools and techniques required for diagnosis, repair should be left to a dealer service department or a repair shop. However, it is possible for the home mechanic to make simple checks of the wiring connections and motors for minor faults which can be easily repaired. These include:
a) Inspect the power window actuating switches for broken wires and loose connections.
b) Check the power window fuse and/or circuit breaker.
c) Remove the door panel(s) and check the power window motor wires to see if they're loose or damaged. Inspect the glass mechanisms for damage which could cause binding.

## 19  Cruise control system – description and check

The cruise control system maintains vehicle speed with a vacuum actuated servo motor located in the engine compartment, which is connected to the throttle linkage by a cable. The system consists of the servo motor, clutch switch, brake switch, control switches, a relay and associated vacuum hoses.

Because of the complexity of the cruise control system and the special tools and techniques required for diagnosis, repair should be left to a dealer service department or a repair shop. However, it is possible for the home mechanic to make simple checks of the wiring and vacuum connections for minor faults which can be easily repaired. These include:
a) Inspect the cruise control actuating switches for broken wires and loose connections.
b) Check the cruise control fuse.
c) The cruise control system is operated by vacuum so it's critical that all vacuum switches, hoses and connections are secure. Check the hoses in the engine compartment for tight connections, cracks and obvious vacuum leaks.

## 20  Wiring diagrams – general information

*Refer to illustration 20.4*

Since it isn't possible to include all wiring diagrams for every year covered by this manual, the following diagrams are those that are typical and most commonly needed.

Prior to troubleshooting any circuits, check the fuse and circuit breakers (if equipped) to make sure they're in good condition. Make sure the battery is properly charged and check the cable connections (Chapter 1).

When checking a circuit, make sure that all connectors are clean, with no broken or loose terminals. When unplugging a connector, do not pull on the wires. Pull only on the connector housings themselves.

Refer to the accompanying table for the wire color codes applicable to your vehicle.

| | | |
|---|---|---|
| WHT | .......... | White |
| YEL | .......... | Yellow |
| BLK | .......... | Black |
| BLU | .......... | Blue |
| GRN | .......... | Green |
| RED | .......... | Red |
| ORN | .......... | Orange |
| PNK | .......... | Pink |
| BRN | .......... | Brown |
| GRY | .......... | Gray |
| LT BLU | ....... | Light Blue |
| LT GRN | ....... | Light Green |

**20.4  Wiring diagram color codes**

# Chapter 12  Chassis electrical system

12 – 13

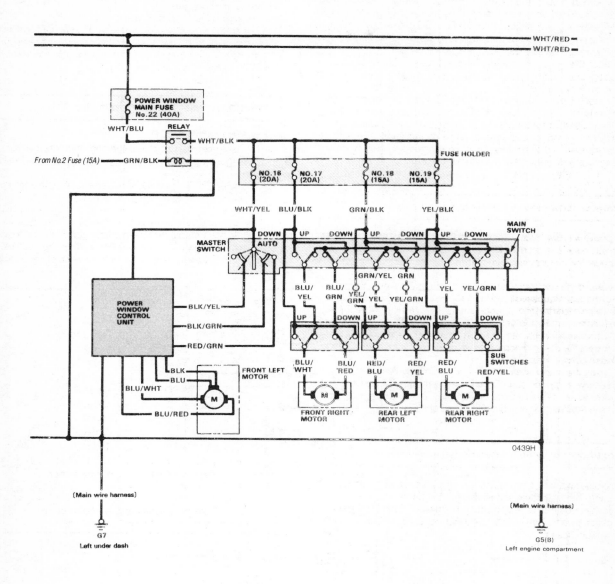

Integra power window circuit

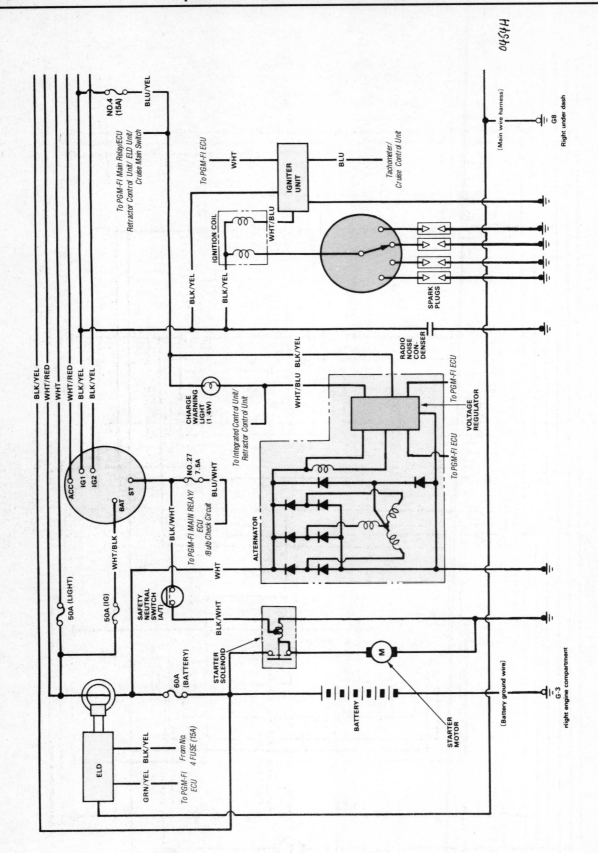

# Chapter 12 Chassis electrical system

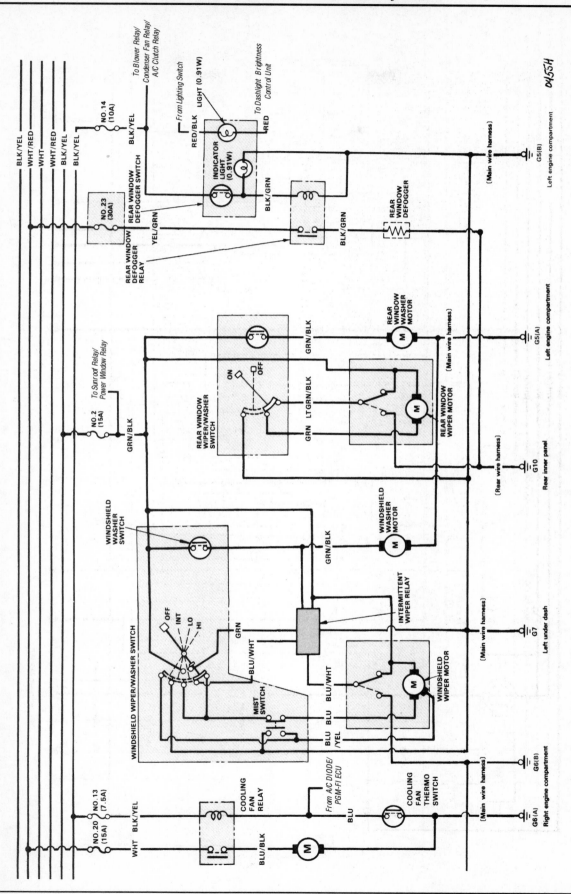

Integra engine wiring diagram (2 of 2)

# Chapter 12 Chassis electrical system

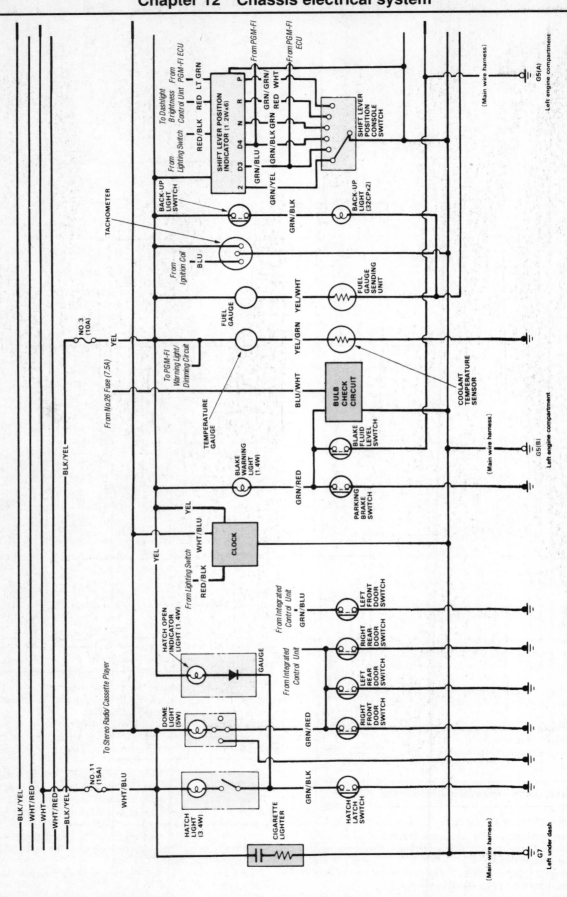

Integra engine wiring diagram (1 of 2)

# Chapter 12  Chassis electrical system

12 – 17

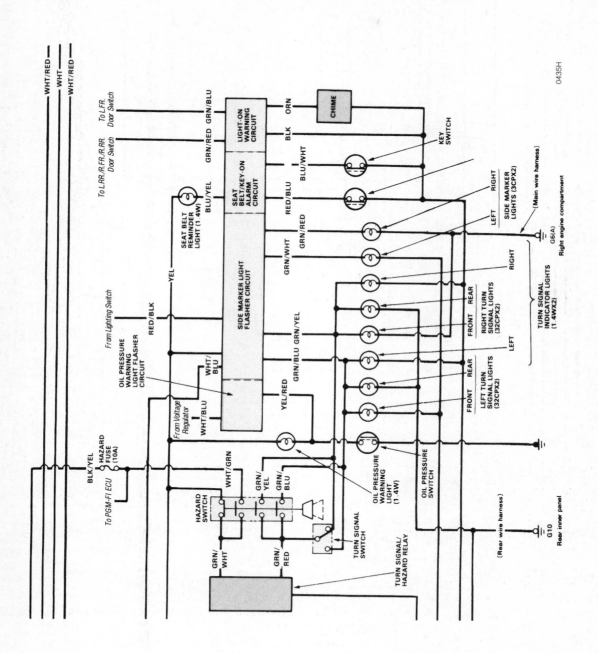

Integra engine wiring diagram (2 of 2)

## 12-18 Chapter 12 Chassis electrical system

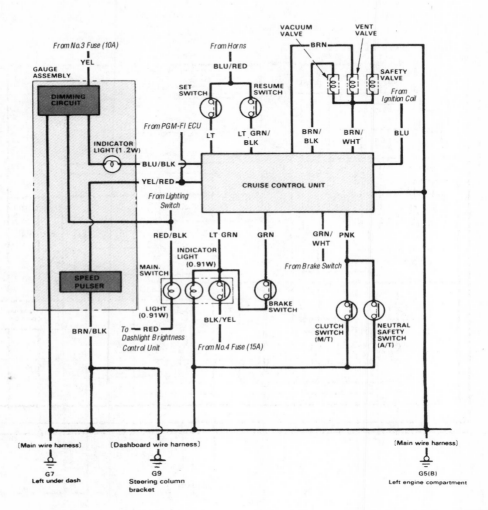

Integra cruise control circuit

# Chapter 12 Chassis electrical system

12 – 19

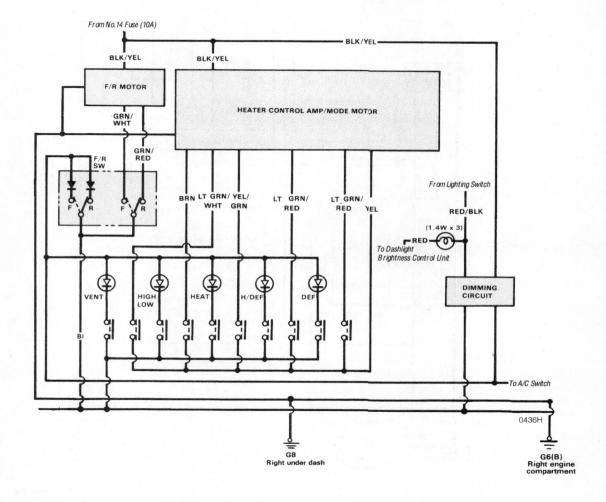

Integra heater control circuit

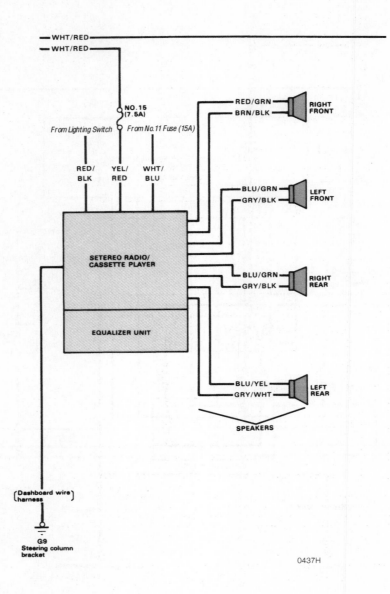

Integra stereo system

# Chapter 12  Chassis electrical system

12 – 21

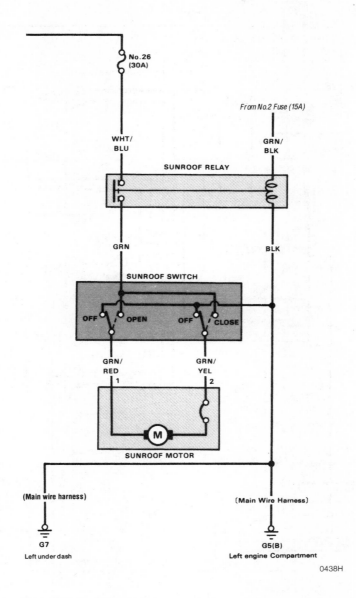

Integra power sunroof

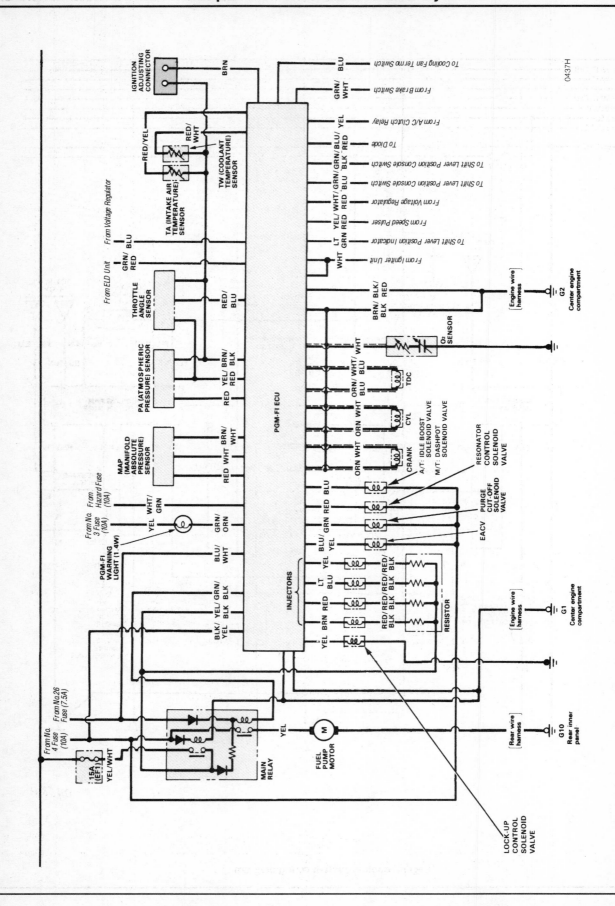

Integra PGM-FI Electronic control unit wiring

# Chapter 12  Chassis electrical system

12 – 23

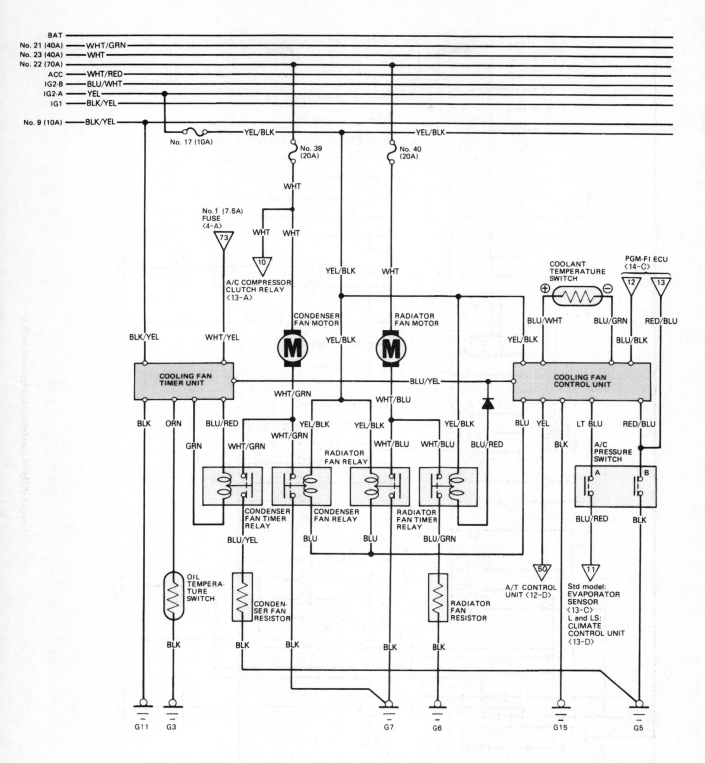

Legend cooling system wiring diagram

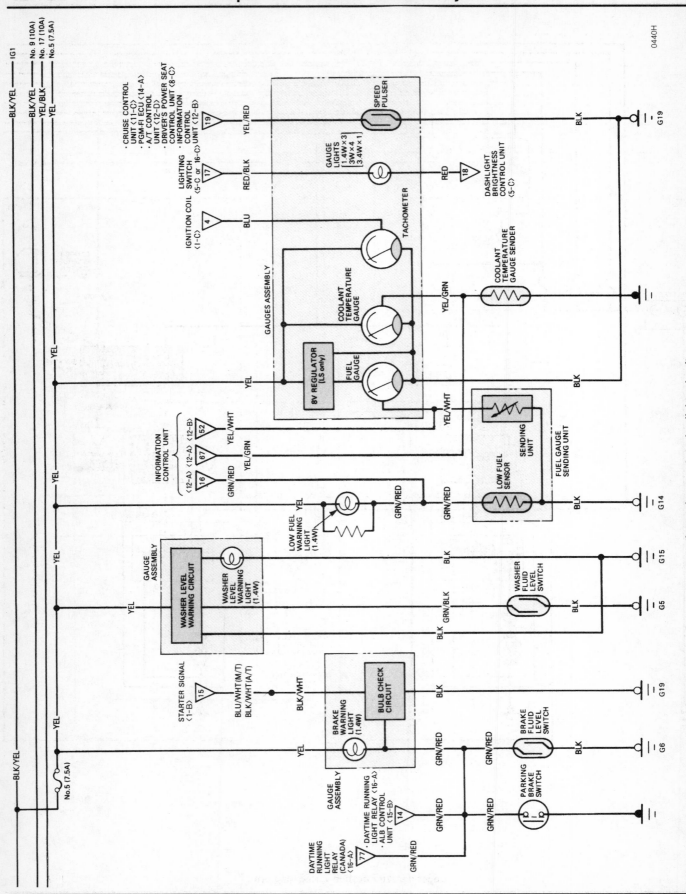

Legend gauges/warning lights circuit

# Chapter 12  Chassis electrical system

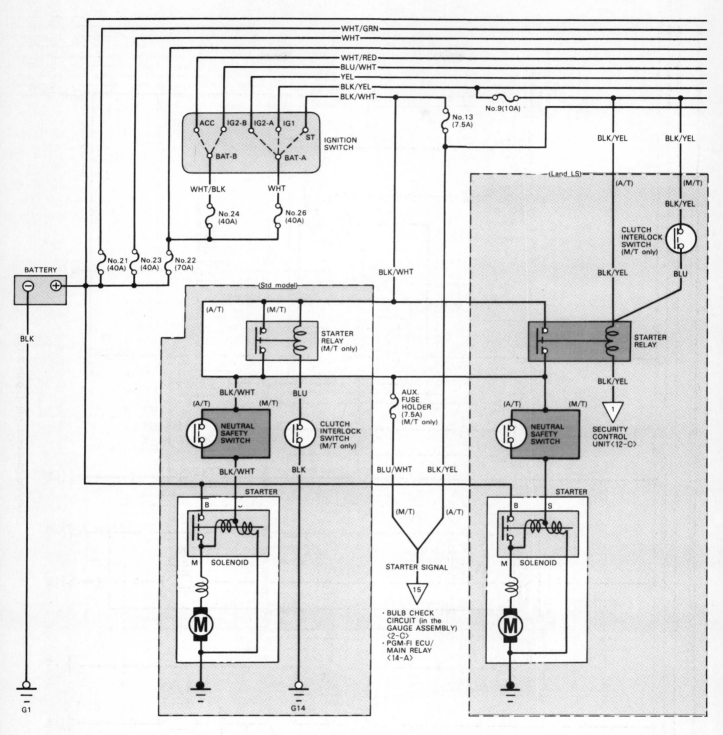

Legend starter system wiring diagram

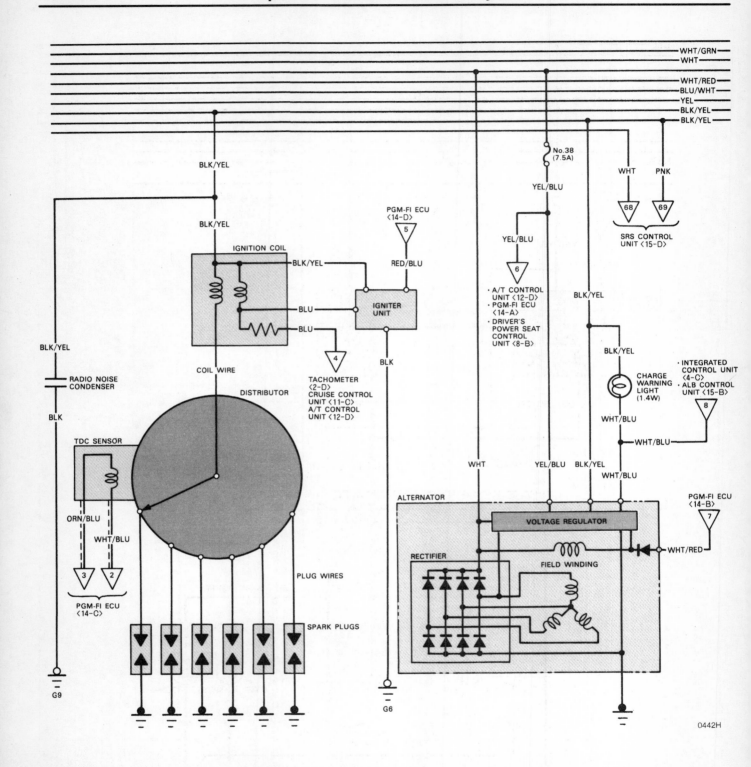

Legend ignition system wiring diagram

# Chapter 12 Chassis electrical system

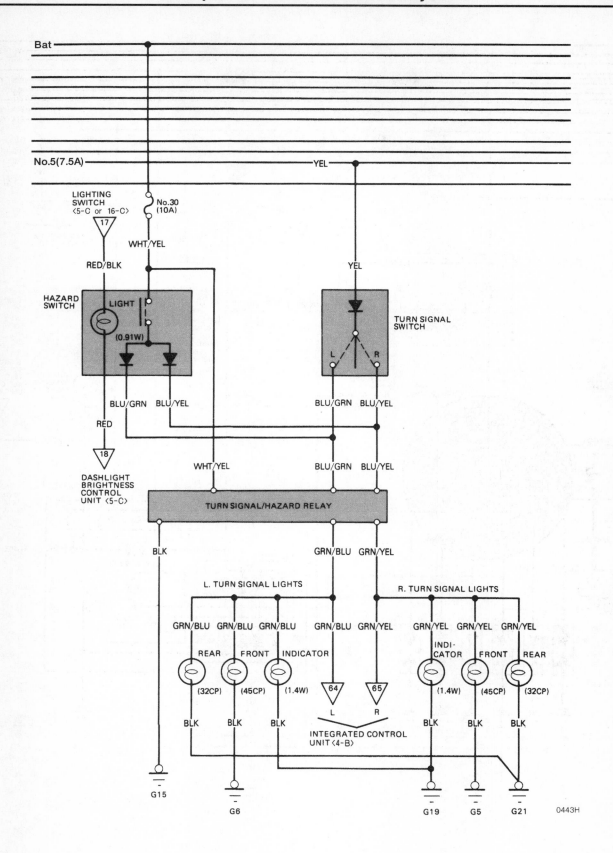

**Legend turn signal/hazard warning lights**

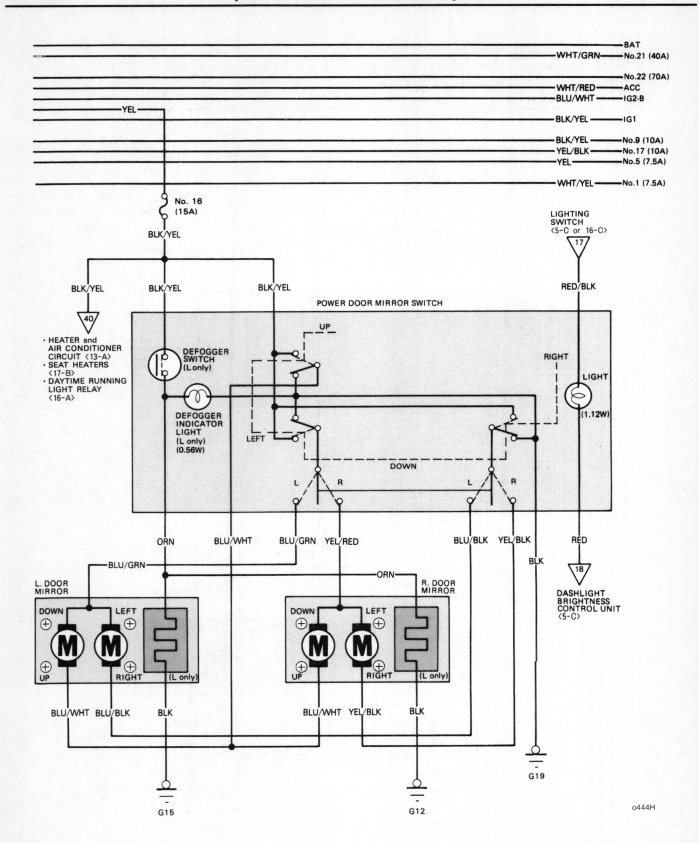

Legend power mirror circuit

# Chapter 12  Chassis electrical system

12 – 29

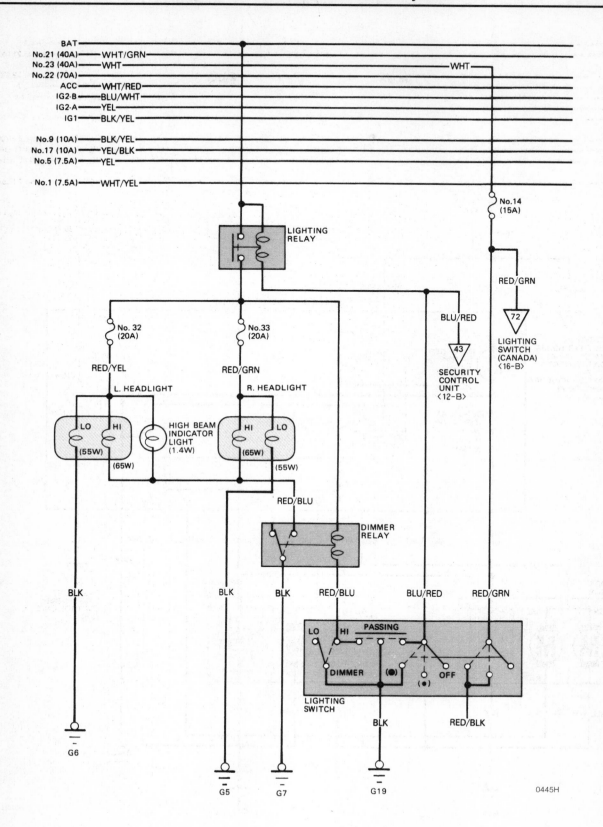

Legend headlight circuit

# Index

## A

**About this manual:** 0-6
**Accelerator, linkage, inspection:** 1-34
**Accelerator cable, replacement:** 4-5
**Air cleaner, housing, removal and installation:** 4-5
**Air conditioning**
    blower motor, removal and installation: 3-9
    compressor, removal and installation: 3-13
    condenser, removal and installation: 3-13
    control assembly, removal and installation: 3-10
    receiver/drier, removal and installation: 3-12
    system, general information: 3-2
**Air conditioning system:** 3-1 through 14
    check and maintenance: 3-11
**Air filter, replacement:** 1-25
**Air injection system:** 6-11
**ALB (Anti-Lock Brake) system, general information:** 9-26
**Alternator, removal and installation:** 5-8
**Antenna, removal and installation:** 12-7
**Anti-Lock Brake (ALB) system, general information:** 9-26
**Antifreeze**
    draining and refilling: 1-31
    general information: 3-2
**Automatic transaxle:** 7B-1 through 6
    diagnosis: 7B-2
    filter, change: 1-33
    fluid, change: 1-33
    fluid level, check: 1-15
    general information: 7B-1
    removal and installation: 7B-5
    shift linkage, adjustment: 7B-2
**Axle beam (rear), removal and installation:** 10-22

## B

**Back-up light bulb, replacement:** 12-11
**Balljoint, replacement:** 10-17
**Battery**
    cables, check and replacement: 5-2
    charging: 1-18
    check and maintenance: 1-18
    emergency jump starting: 5-2
    removal and installation: 5-2
**Battery jump starting:** 0-14
**Bearing (main, engine), oil clearance check:** 2C-24
**Bearings, main and connecting rod, inspection:** 2C-20
**Belt, timing:** 2A-4
    removal, inspection and installation: 2B-7
**Belts (engine), check, adjustment and replacement:** 1-20
**Bleeding**
    brake system: 9-23
    power steering system: 10-29
**Block (engine)**
    cleaning: 2C-16
    honing: 2C-17
    inspection: 2C-16
**Blower motor, heater and air conditioner, removal and installation:** 3-9
**Body:** 11-1 through 14
    general information: 11-1
    maintenance: 11-1
**Body repair**
    major damage: 11-3
    minor damage: 11-2
**Booster (brake), inspection, removal and installation:** 9-23
**Booster battery (jump) starting:** 0-14

# Index

**Boots**
    driveaxle, check: 1-30
    steering gear, replacement: 10-27
**Brake**
    booster, inspection, removal and installation: 9-23
    cables, removal and installation: 9-24
    caliper, removal, overhaul and installation: 9-6, 12
    fluid
        bleeding: 9-23
        level check: 1-10
    hoses and lines, inspection and replacement: 9-22
    light bulb, replacement: 12-11
    light switch, check, replacement and adjustment: 9-25
    master cylinder, removal, overhaul and installation: 9-19
    pads, replacement: 9-2, 8
    parking, adjustment: 9-24
    proportioning valve, general information, removal and installation: 9-22
    system, check: 1-23
    system bleeding: 9-23
    wheel cylinder, removal, overhaul and installation: 9-18
**Brake disc, inspection, removal and installation: 9-12**
**Brake shoes, replacement: 9-14**
**Brakes: 9-1 through 26**
    general information: 9-2
**Break-in (engine): 2C-28**
**Brushes (alternator), replacement: 5-9**
**Bulb, replacement: 12-11**
**Bumpers, removal and installation: 11-6**
**Buying parts: 0-7**
**Bypass control system, general information and check: 4-15**

## C

**Cable**
    battery, check and replacement: 5-2
    parking brake, removal and installation: 9-24
**Caliper (brake), removal, overhaul and installation: 9-6, 12**
**Camshaft**
    oil seal, replacement: 2A-7
    oil seals, replacement: 2B-10
    removal, inspection and installation: 2A-7
**Camshafts and valve components, removal, inspection and installation: 2B-10**
**Carpets, maintenance: 11-2**
**Catalytic converter: 6-11**
**Center console, removal and installation: 11-12**
**Centrifugal advance mechanism, check: 5-7**
**Charging, battery: 1-18**
**Charging system**
    check: 5-8
    general information and precautions: 5-8
**Chassis electrical system: 12-1 through 12**
**Check Engine light: 6-5**
**Chemicals and lubricants: 0-16**
**Circuit breakers, general information: 12-3**
**Clutch**
    cable, removal, installation and adjustment: 8-3
    components, removal, inspection and installation: 8-8
    description and check: 8-2
    fluid, level check: 1-10
    general information: 8-1
    hydraulic system, bleeding: 8-6
    interlock switch, check, replacement and adjustment: 8-10
    master cylinder, removal, overhaul and installation: 8-4
    pedal, adjustment: 8-7
    release arm freeplay check and adjustment: 1-25
    release bearing and fork, removal, inspection and installation: 8-7
    release cylinder, removal, overhaul and installation: 8-5
**Clutch and driveaxles: 8-1 through 14**
**Coil, ignition, removal and installation: 5-3**
**Combination switch(es), check and replacement: 12-3**
**Compression check: 2C-6**
**Compressor, removal and installation: 3-13**
**Condenser, removal and installation: 3-13**
**Connecting rods**
    bearing inspection: 2C-20
    inspection: 2C-18
    installation and oil clearance check: 2C-27
    removal: 2C-14
**Console (center), removal and installation: 11-12**
**Constant Velocity (CV) joint**
    boot replacement: 8-12
    overhaul: 8-12
**Converter (catalytic): 6-11**
**Coolant**
    general information: 3-2
    level check: 1-10
    reservoir, removal and installation: 3-6
    temperature sending unit, check and replacement: 3-9
**Cooler (oil), removal and installation: 3-7**
**Cooling fan, check and replacement: 3-3**
**Cooling system: 3-1 through 14**
    check: 1-22
    coolant, general information: 3-2
    general information: 3-2
    servicing: 1-31
**Core (heater), removal and installation: 3-10**
**Crank angle sensor: 6-5**
**Crankshaft**
    front oil seal, replacement: 2A-6; 2B-9
    inspection: 2C-19
    installation and main bearing oil clearance check: 2C-24
    rear oil seal, replacement: 2A-11; 2B-17
    removal: 2C-15
**Crossmember (steering and suspension), removal and installation: 10-15**
**Cruise control, description and check: 12-12**
**CV (Constant Velocity) joint**
    boot replacement: 8-12
    overhaul: 8-12
**Cylinder block**
    cleaning: 2C-16
    honing: 2C-17
    inspection: 2C-16
**Cylinder compression check: 2C-6**
**Cylinder head**
    cleaning and inspection: 2C-11
    disassembly: 2C-10
    reassembly: 2C-13
    removal and installation: 2A-10; 2B-13
    valves, servicing: 2C-13
**Cylinder head covers, removal and installation: 2B-3**
**Cylinder honing: 2C-17**

## D

**Dent repair**
    major damage: 11-3
    minor damage: 11-2
**Diagnosis: 0-19**
**Disc (brake), inspection, removal and installation: 9-12**

# Index

**Disc brake**
    caliper, removal, overhaul and installation: 9-6, 12
    pads, replacement: 9-2, 8
**Distributor**
    cap, check and replacement: 1-27
    removal and installation: 5-5
**Distributor rotor, check and replacement: 1-27**
**Door**
    glass regulator, removal and installation: 11-9
    handles, removal and installation: 11-7, 8
    lock cylinder, removal and installation: 11-8
    lock system (power), description and check: 12-12
    removal and installation: 11-8
    trim panel, removal and installation: 11-6
    window glass, removal and installation: 11-9
**Door latch, removal and installation: 11-8**
**Driveaxle, boot, check: 1-30**
**Driveaxle boot replacement: 8-12**
**Driveaxle intermediate shaft, removal and installation: 8-11**
**Driveaxles: 8-1 through 14**
    general information: 8-1
    removal and installation: 8-10
**Drivebelt, check, adjustment and replacement: 1-20**
**Driveplate, removal and installation: 2A-11; 2B-16**
**Drum brake wheel cylinder, removal, overhaul and installation: 9-18**

## E

**EACV (Electronic Air Control Valve), check and replacement: 6-10**
**EGR (Exhaust Gas Recirculation) system: 6-11**
**EGR system, check: 1-35**
**EICV (Electronic Idle Control Valve), check and replacement: 6-10**
**Electric**
    door lock system, description and check: 12-12
    windows, description and check: 12-12
**Electrical**
    circuit diagrams: 12-12
    system, general information: 12-1
    troubleshooting, general information: 12-1
**Electrical system (chassis): 12-1 through 12**
**Electrical systems (engine): 5-1 through 10**
**Electronic Air Control Valve (EACV), check and replacement: 6-10**
**Electronic Idle Control Valve (EICV), check and replacement: 6-10**
**Emergency battery jump starting: 0-14**
**Emissions control systems: 6-1 through 12**
    general information: 6-1
**Engine**
    block
        cleaning: 2C-16
        inspection: 2C-16
    coolant
        general information: 3-2
        level check: 1-10
    cooling fan, check and replacement: 3-3
    cylinder head, removal and installation: 2A-10; 2B-13
    cylinder head covers, removal and installation: 2B-3
    drivebelts, check, adjustment and replacement: 1-20
    electrical systems, general information: 5-1
    Integra: 2A-1
        general information: 2A-2
    Legend: 2B-1
        general information: 2B-2
    mount, check and replacement: 2A-12; 2B-17
    oil
        change: 1-15
        level check: 1-10
    oil seal, replacement: 2A-6, 7, 11; 2B-9, 10, 17
    overhaul: 2C-1 through 28
        disassembly sequence: 2C-9
        general information: 2C-5
        reassembly sequence: 2C-23
    overhaul alternatives: 2C-9
    rebuilding, alternatives: 2C-9
    removal, methods and precautions: 2C-7
    removal and installation: 2C-7
    repair operations possible with engine in vehicle: 2A-2; 2B-2
    tune-up: 1-1 through 36
**Engine electrical systems: 5-1 through 10**
**Engine emissions control systems: 6-1 through 12**
**Engine fuel and exhaust systems: 4-1 through 18**
**EVAP system: 6-10**
    check: 1-35
**Exhaust, manifold, removal and installation: 2A-4; 2B-5**
**Exhaust Gas Recirculation (EGR) system: 6-11**
    check: 1-35
**Exhaust system: 4-1 through 18**
    check: 1-32
    servicing, general information: 4-17

## F

**Fan, engine cooling, check and replacement: 3-3**
**Fault-finding: 0-19**
**Filter**
    automatic transaxle, replacement: 1-33
    fuel, replacement: 1-35
    oil, change: 1-15
**Fluid**
    automatic transaxle
        change: 1-33
        checking: 1-15
    level checks: 1-10
    manual transaxle
        change: 1-33
        checking: 1-28
    power steering, checking: 1-14
**Flywheel, removal and installation: 2A-11; 2B-16**
**Four-cylinder engine: 2A-1**
**Front stabilizer bar and bushings, removal and installation: 10-16**
**Front strut assembly, removal and installation: 10-8**
**Front-end alignment, general information: 10-30**
**Fuel**
    filter, replacement: 1-35
    pressure, check: 4-2
    pump
        check: 4-2
        removal and installation: 4-3
    system, check: 1-28
    tank, cleaning and repair: 4-5
**Fuel and exhaust systems, general information: 4-1**
**Fuel evaporative control system: 6-10**
**Fuel injection system**
    check: 4-10
    general information: 4-7
    injector resistor, check and replacement: 4-15
**Fuel injectors, check and replacement: 4-13**
**Fuel pressure regulator, check and replacement: 4-11**
**Fuel pressure relief procedure: 4-1**
**Fuel system: 4-1 through 18**
**Fuel tank, removal and installation: 4-4**
**Fuses, general information: 12-3**

# Index

## G

Gauge cluster, removal and installation: 12-11
Gear (steering), removal and installation: 10-27
General information
    Anti-Lock Brake (ALB) system: 9-26
    automatic transaxle: 7B-1
    body: 11-1
    brakes: 9-2
    cooling, heating and air conditioning systems: 3-2
    electrical system: 12-1
    emissions control systems: 6-1
    engine electrical systems: 5-1
    engine overhaul: 2C-5
    exhaust system: 4-1
    fuel injection system: 4-7
    fuel system: 4-1
    ignition system: 5-3
    Legend engine: 2B-2
    manual transaxle: 7A-1
    steering and suspension systems: 10-2
    tune-up: 1-10
    wheel alignment: 10-30
    wheels and tires: 10-30
Generator (alternator), removal and installation: 5-8

## H

Hand brake, adjustment: 9-24
Hazard, flasher, check and replacement: 12-3
Head, cylinder, removal and installation: 2A-10; 2B-13
Head (cylinder)
    cleaning and inspection: 2C-11
    disassembly: 2C-10
    reassembly: 2C-13
Headlight
    adjustment: 12-10
    removal and installation: 12-9
Heater
    blower motor, removal and installation: 3-9
    control assembly, removal and installation: 3-10
    core, removal and installation: 3-10
Heating system: 3-1 through 14
    check and maintenance: 3-11
    general information: 3-2
Hinges, maintenance: 11-3
Honing, cylinder: 2C-17
Hood
    adjustment: 11-3
    removal and installation: 11-3
Hoses
    brake, inspection and replacement: 9-22
    check and replacement: 1-21
Hub (front), removal and installation: 10-11
Hub and bearing (rear), removal and installation: 10-19
Hydraulic system (clutch), bleeding: 8-6

## I

Idle speed check and adjustment: 1-34
Igniter, check: 5-5
Ignition
    coil, removal and installation: 5-3
    distributor, removal and installation: 5-5
    key lock cylinder, replacement: 12-9
    system
        check: 5-3
        general information: 5-3
Ignition switch, check and replacement: 12-9
Ignition timing, check and adjustment: 1-33
Information sensors: 6-5
Initial start-up and break-in after overhaul: 2C-28
Injector resistor, check and replacement: 4-15
Instrument cluster, removal and installation: 12-11
Instrument cluster bezel, removal and installation: 11-13
Instrument panel, removal and installation: 12-11
Intake air temperature (TA) sensor: 6-5
Intake manifold, removal and installation: 2A-3; 2B-4
Integra engine: 2A-1
    general information: 2A-2
Intermediate shaft (driveaxle), removal and installation: 8-11
Introduction to the Acura Integra and Legend: 0-6

## J

Jacking: 0-15
Jump starting: 0-14

## K

Key lock cylinder, replacement: 12-9
Knuckle and hub, removal and installation: 10-11

## L

Legend engine: 2B-1
    general information: 2B-2
Liftgate, removal, installation and adjustment: 11-11
Linings (brake), check: 1-23
Lock (key), replacement: 12-9
Locks
    maintenance: 11-3
    power door, description and check: 12-12
Lower arm, removal and installation: 10-14, 17
Lubricants: 0-16

## M

Main and connecting rod bearings, inspection: 2C-20
Main bearing, oil clearance, check: 2C-24
Main bearing oil seal, installation: 2C-26
Maintenance
    introduction: 1-2
    techniques: 0-7
Maintenance schedule: 1-9
Manifold
    exhaust, removal and installation: 2A-4; 2B-5
    intake, removal and installation: 2A-3; 2B-4
Manifold Absolute Pressure (MAP) sensor: 6-5
Manual transaxle: 7A-1 through 6
    general information: 7A-1
    lubricant
        change: 1-33
        check: 1-28
    oil seal, replacement: 7A-1
    overhaul, general information: 7A-4
    removal and installation: 7A-2

# Index

MAP sensor: 6-5
Master cylinder, removal, overhaul and installation: 9-19
Mirrors, removal and installation: 11-9
Mount
    engine, check and replacement: 2A-12; 2B-17
    transaxle, check and replacement: 7A-2
Muffler, check: 1-32

## N

Neutral start switch, check and replacement: 7B-3

## O

Oil
    change: 1-15
    filter, change: 1-15
    level check: 1-10
    pan, removal and installation: 2A-10; 2B-14
    pump, removal and installation: 2A-11; 2B-15
    seal
        engine: 2A-6, 7, 11; 2B-9, 10, 17
        transaxle: 7A-1
Oil cooler, removal and installation: 3-7
Oil seal, main bearing, installation: 2C-26
Overhaul
    engine: 2C-1 through 28
        general information: 2C-5
    manual transaxle, general information: 7A-4
Overhaul (engine)
    alternatives: 2C-9
    disassembly sequence: 2C-9
    initial start-up and break-in: 2C-28
    reassembly sequence: 2C-23
Owner maintenance: 1-1
Oxygen sensor: 6-5

## P

Pads (brake), replacement: 9-2, 8
Pan, oil, removal and installation: 2A-10; 2B-14
Panhard rod, removal and installation: 10-21
Parking brake
    adjustment: 9-24
    cables, removal and installation: 9-24
Parking light bulb, replacement: 12-11
Parts, replacement: 0-7
PCV (Positive Crankcase Ventilation) system: 6-10
PCV system, valve, check and replacement: 1-31
PGM-FI system, general information: 6-2
Piston
    installation: 2C-27
    rings, installation: 2C-23
Pistons and connecting rods
    inspection: 2C-18
    installation: 2C-27
    installation and rod bearing oil clearance check: 2C-27
    removal: 2C-14
Positive Crankcase Ventilation (PCV) system: 6-10
    valve, check and replacement: 1-31
Power
    brake booster, inspection, removal and installation: 9-23
    door lock system, description and check: 12-12
    window system, description and check: 12-12
Power steering
    bleeding: 10-29
    fluid level, check: 1-14
    pump, removal and installation: 10-29
Pressure (tire), checking: 1-13
Programmed Fuel Injection (PGM-FI) system, general information: 6-2
Proportioning valve, general information, removal and installation: 9-22
Pump
    oil, removal and installation: 2A-11; 2B-15
    power steering, removal and installation: 10-29
    water, removal and installation: 3-8
Pump (fuel), removal and installation: 4-3
Pump (water), check: 3-8

## R

Radiator
    coolant, general information: 3-2
    draining, flushing and refilling: 1-31
    removal and installation: 3-5
Radio
    antenna, removal and installation: 12-7
    removal and installation: 12-7
Radio condenser, check: 5-5
Radius arm, removal and installation: 10-14
Radius rod, removal and installation: 10-16
Rear
    hub and wheel bearing assembly, removal and installation: 10-19
    main oil seal, installation: 2C-26
    shock absorber, removal and installation: 10-17
Rear axle beam, removal and installation: 10-22
Rear stabilizer assembly, removal and installation: 10-22
Rear suspension arms, removal and installation: 10-23, 24
Receiver/drier, removal and installation: 3-12
Relays, general information: 12-3
Release bearing and fork (clutch), removal, inspection and installation: 8-7
Reluctor, replacement: 5-7
Reluctor air gap, check and adjustment: 5-6
Reservoir (coolant), removal and installation: 3-6
Rings (piston), installation: 2C-23
Rocker arms, removal, inspection and installation: 2A-7
Rods (connecting)
    inspection: 2C-18
    installation: 2C-27
    removal: 2C-14
Rotation (tire): 1-22
Rotor, check and replacement: 1-27
Rotor (disc brake), inspection, removal and installation: 9-12
Routine maintenance: 1-1 through 36

## S

Safe automotive repair practices: 0-17
Safety: 0-17
Scheduled maintenance: 1-9
Seat belt check: 11-14
Self-diagnosis system, general information and code access: 6-5
Sending unit, cooling system temperature, check and replacement: 3-9
Shift assembly (manual transaxle), removal and installation: 7A-2
Shift linkage (automatic transaxle), adjustment: 7B-2

# Index

**Shock absorber, front, removal and installation:** 10-8
**Shock absorber (rear), removal and installation:** 10-17
**Shoes (brake)**
   check: 1-23
   replacement: 9-14
**Smog control systems:** 6-1 through 12
**Solenoid (starter), removal and installation:** 5-10
**Spark plug**
   replacement: 1-25
   wires: 1-27
**Speed control system, description and check:** 12-12
**Spindle/swing bearing assembly, removal and installation:** 10-22
**Stabilizer assembly (rear), removal and installation:** 10-22
**Stabilizer bar, front, removal and installation:** 10-16
**Starter**
   motor, removal and installation: 5-10
   neutral start switch, check and replacement: 7B-3
   solenoid, removal and installation: 5-10
**Starter motor, in-vehicle check:** 5-10
**Starter/clutch interlock switch, check, replacement and adjustment:** 8-10
**Starting system, general information and precautions:** 5-10
**Steering**
   gear
      boots, replacement: 10-27
      removal and installation: 10-27
   knuckle and hub, removal and installation: 10-11
   pump (power), removal and installation: 10-29
   system check: 1-28
   wheel, removal and installation: 10-26
**Steering and suspension systems, general information:** 10-2
**Steering system:** 10-1 through 30
**Steering/suspension crossmember, removal and installation:** 10-15
**Stereo system, removal and installation:** 12-7
**Strut assembly, removal and installation:** 10-8
**Strut/shock absorber assembly, replacement:** 10-10
**Suspension**
   balljoint, replacement: 10-17
   system check: 1-28
**Suspension arms (rear), removal and installation:** 10-23, 24
**Suspension system:** 10-1 through 30
   general information: 10-2
**Swing bearing/spindle assembly, removal and installation:** 10-22
**Switch, brake lights, check, replacement and adjustment:** 9-25

## T

**TA (intake air temperature) sensor:** 6-5
**Taillight bulb, replacement:** 12-11
**Tank (fuel)**
   cleaning and repair: 4-5
   removal and installation: 4-4
**TDC (top dead center), locating:** 2A-2; 2B-2
**Temperature sending unit, check and replacement:** 3-9
**Thermostat, check and replacement:** 3-2
**Throttle**
   linkage, inspection: 1-34
   valve, cable, adjustment: 7B-4
**Throttle angle sensor:** 6-5
**Throttle body, component check and replacement:** 4-10
**Throttle cable, replacement:** 4-5
**Throwout bearing (clutch), removal, inspection and installation:** 8-7
**Tie-rod ends, removal and installation:** 10-26
**Timing, ignition:** 1-33
**Timing belt, removal, inspection and installation:** 2A-4

**Timing belt and sprockets, removal, inspection and installation:** 2B-7
**Tire**
   checking: 1-13
   rotation: 1-22
**Tires and wheels, general information:** 10-30
**Tools:** 0-7
**Top Dead Center (TDC), locating:** 2A-2; 2B-2
**Torsion bar, removal, installation and adjustment:** 10-12
**Towing:** 0-15
**Trailing arm, removal and installation:** 10-21
**Transaxle (automatic):** 7B-1 through 6
   diagnosis: 7B-2
   fluid level, check: 1-15
   general information: 7B-1
   removal and installation: 7B-5
**Transaxle (manual):** 7A-1 through 6
   general information: 7A-1
   overhaul, general information: 7A-4
   removal and installation: 7A-2
**Transaxle mount, check and replacement:** 7A-2
**Trouble codes:** 6-5
**Troubleshooting:** 0-19
   electrical, general information: 12-1
**Trunk lid, removal, installation and adjustment:** 11-11
**Tune-up**
   general information: 1-10
   introduction: 1-2
**Tune-up and routine maintenance:** 1-1 through 36
**Turn signal**
   bulb, replacement: 12-11
   flasher, check and replacement: 12-3
**TV (Throttle Valve) cable, adjustment:** 7B-4

## U

**Underhood hose check and replacement:** 1-21
**Upholstery, maintenance:** 11-2
**Upper arm, removal and installation:** 10-17
**Using this manual:** 0-6

## V

**Vacuum advance mechanism, check and replacement:** 5-7
**Valve**
   clearance check and adjustment: 1-28
   lash check and adjustment: 1-28
   seals, replacement: 2A-9
   springs, retainers and seals – replacement: 2A-9
**Valve components, removal, inspection and installation:** 2B-10
**Valve cover, removal and installation:** 2A-3
**Valve job:** 2C-13
**Valves, servicing:** 2C-13
**Vehicle identification numbers:** 0-7
**Vinyl trim, maintenance:** 11-2
**Voltage regulator and alternator brushes, replacement:** 5-9

## W

**Water pump**
   check: 3-8
   removal and installation: 3-8

# Index

**Wheel**
    bearing (rear), removal and installation: 10-19
    cylinder, removal, overhaul and installation: 9-18
    steering, removal and installation: 10-26
**Wheel alignment, general information: 10-30**
**Wheels and tires, general information: 10-30**
**Window, power, description and check: 12-12**

**Windshield**
    and fixed glass, replacement: 11-3
    washer fluid, level check: 1-10
    wiper blades, inspection and replacement: 1-17
    wiper motor, removal and installation: 12-11
**Wiring diagrams: 12-13 through 12-30**
    general information: 12-12
**Working facilities: 0-7**

# Haynes Automotive Manuals

NOTE: New manuals are added to this list on a periodic basis. If you do not see a listing for your vehicle, consult your local Haynes dealer for the latest product information.

## ACURA
- *12020 Integra '86 thru '89 & Legend '86 thru '90

## AMC
- Jeep CJ - see JEEP (50020)
- 14020 Mid-size models, Concord, Hornet, Gremlin & Spirit '70 thru '83
- 14025 (Renault) Alliance & Encore '83 thru '87

## AUDI
- 15020 4000 all models '80 thru '87
- 15025 5000 all models '77 thru '83
- 15026 5000 all models '84 thru '88

## AUSTIN-HEALEY
- Sprite - see MG Midget (66015)

## BMW
- *18020 3/5 Series not including diesel or all-wheel drive models '82 thru '92
- *18021 3 Series except 325iX models '92 thru '97
- 18025 320i all 4 cyl models '75 thru '83
- 18035 528i & 530i all models '75 thru '80
- 18050 1500 thru 2002 except Turbo '59 thru '77

## BUICK
- Century (front wheel drive) - see GM (829)
- *19020 Buick, Oldsmobile & Pontiac Full-size (Front wheel drive) all models '85 thru '98
  Buick Electra, LeSabre and Park Avenue; Oldsmobile Delta 88 Royale, Ninety Eight and Regency; Pontiac Bonneville
- 19025 Buick Oldsmobile & Pontiac Full-size (Rear wheel drive)
  Buick Estate '70 thru '90, Electra '70 thru '84, LeSabre '70 thru '85, Limited '74 thru '79
  Oldsmobile Custom Cruiser '70 thru '90, Delta 88 '70 thru '85, Ninety-eight '70 thru '84
  Pontiac Bonneville '70 thru '81, Catalina '70 thru '81, Grandville '70 thru '75, Parisienne '83 thru '86
- 19030 Mid-size Regal & Century all rear-drive models with V6, V8 and Turbo '74 thru '87
  Regal - see GENERAL MOTORS (38010)
  Riviera - see GENERAL MOTORS (38030)
  Roadmaster - see CHEVROLET (24046)
  Skyhawk - see GENERAL MOTORS (38015)
  Skylark '80 thru '85 - see GM (38020)
  Skylark '86 on - see GM (38025)
  Somerset - see GENERAL MOTORS (38025)

## CADILLAC
- *21030 Cadillac Rear Wheel Drive all gasoline models '70 thru '93
  Cimarron - see GENERAL MOTORS (38015)
  Eldorado - see GENERAL MOTORS (38030)
  Seville '80 thru '85 - see GM (38030)

## CHEVROLET
- *24010 Astro & GMC Safari Mini-vans '85 thru '93
- 24015 Camaro V8 all models '70 thru '81
- 24016 Camaro all models '82 thru '92
  Cavalier - see GENERAL MOTORS (38015)
  Celebrity - see GENERAL MOTORS (38005)
- 24017 Camaro & Firebird '93 thru '97
- 24020 Chevelle, Malibu & El Camino '69 thru '87
- 24024 Chevette & Pontiac T1000 '76 thru '87
  Citation - see GENERAL MOTORS (38020)
- *24032 Corsica/Beretta all models '87 thru '96
- 24040 Corvette all V8 models '68 thru '82
- *24041 Corvette all models '84 thru '96
- 10305 Chevrolet Engine Overhaul Manual
- 24045 Full-size Sedans Caprice, Impala, Biscayne, Bel Air & Wagons '69 thru '90
- 24046 Impala SS & Caprice and Buick Roadmaster '91 thru '96
  Lumina - see GENERAL MOTORS (38010)
- 24048 Lumina & Monte Carlo '95 thru '98
  Lumina APV - see GM (38035)
- 24050 Luv Pick-up all 2WD & 4WD '72 thru '82
- *24055 Monte Carlo all models '70 thru '88
  Monte Carlo '95 thru '98 - see LUMINA (24048)
- 24059 Nova all V8 models '69 thru '79
- *24060 Nova and Geo Prizm '85 thru '92
- 24064 Pick-ups '67 thru '87 - Chevrolet & GMC, all V8 & in-line 6 cyl, 2WD & 4WD '67 thru '87; Suburbans, Blazers & Jimmys '67 thru '91
- *24065 Pick-ups '88 thru '98 - Chevrolet & GMC, all full-size pick-ups, '88 thru '98; Blazer & Jimmy '92 thru '94; Suburban '92 thru '98; Tahoe & Yukon '98
- 24070 S-10 & S-15 Pick-ups '82 thru '93, Blazer & Jimmy '83 thru '94
- *24071 S-10 & S-15 Pick-ups '94 thru '96 Blazer & Jimmy '95 thru '96
- *24075 Sprint & Geo Metro '85 thru '94
- *24080 Vans - Chevrolet & GMC, V8 & in-line 6 cylinder models '68 thru '96

## CHRYSLER
- 25015 Chrysler Cirrus, Dodge Stratus, Plymouth Breeze '95 thru '98
- 25025 Chrysler Concorde, New Yorker & LHS, Dodge Intrepid, Eagle Vision, '93 thru '97
- 10310 Chrysler Engine Overhaul Manual
- *25020 Full-size Front-Wheel Drive '88 thru '93
  K-Cars - see DODGE Aries (30008)
  Laser - see DODGE Daytona (30030)
- *25030 Chrysler & Plymouth Mid-size front wheel drive '82 thru '95
  Rear-wheel Drive - see Dodge (30050)

## DATSUN
- 28005 200SX all models '80 thru '83
- 28007 B-210 all models '73 thru '78
- 28009 210 all models '79 thru '82
- 28012 240Z, 260Z & 280Z Coupe '70 thru '78
- 28014 280ZX Coupe & 2+2 '79 thru '83
  300ZX - see NISSAN (72010)
- 28016 310 all models '78 thru '82
- 28018 510 and PL521 Pick-up '68 thru '73
- 28020 510 all models '78 thru '81
- 28022 620 Series Pick-up all models '73 thru '79
  720 Series Pick-up - see NISSAN (72030)
- 28025 810/Maxima all gasoline models, '77 thru '84

## DODGE
- 400 & 600 - see CHRYSLER (25030)
- *30008 Aries & Plymouth Reliant '81 thru '89
- 30010 Caravan & Plymouth Voyager Mini-Vans all models '84 thru '95
- *30011 Caravan & Plymouth Voyager Mini-Vans all models '96 thru '98
- 30012 Challenger/Plymouth Saporro '78 thru '83
- 30016 Colt & Plymouth Champ (front wheel drive) all models '78 thru '87
- *30020 Dakota Pick-ups all models '87 thru '96
- 30025 Dart, Demon, Plymouth Barracuda, Duster & Valiant 6 cyl models '67 thru '76
- *30030 Daytona & Chrysler Laser '84 thru '89
  Intrepid - see CHRYSLER (25025)
- *30034 Neon all models '95 thru '97
- *30035 Omni & Plymouth Horizon '78 thru '90
- *30040 Pick-ups all full-size models '74 thru '93
- *30041 Pick-ups all full-size models '94 thru '96
- *30045 Ram 50/D50 Pick-ups & Raider and Plymouth Arrow Pick-ups '79 thru '93
- 30050 Dodge/Plymouth/Chrysler rear wheel drive '71 thru '89
- *30055 Shadow & Plymouth Sundance '87 thru '94
- *30060 Spirit & Plymouth Acclaim '89 thru '95
- *30065 Vans - Dodge & Plymouth '71 thru '96

## EAGLE
- Talon - see Mitsubishi Eclipse (68030)
- Vision - see CHRYSLER (25025)

## FIAT
- 34010 124 Sport Coupe & Spider '68 thru '78
- 34025 X1/9 all models '74 thru '80

## FORD
- 10355 Ford Automatic Transmission Overhaul
- *36004 Aerostar Mini-vans all models '86 thru '96
- *36006 Contour & Mercury Mystique '95 thru '98
- 36008 Courier Pick-up all models '72 thru '82
- 36012 Crown Victoria & Mercury Grand Marquis '88 thru '96
- 10320 Ford Engine Overhaul Manual
- 36016 Escort/Mercury Lynx all models '81 thru '90
- *36020 Escort/Mercury Tracer '91 thru '96
- *36024 Explorer & Mazda Navajo '91 thru '95
- 36028 Fairmont & Mercury Zephyr '78 thru '83
- 36030 Festiva & Aspire '88 thru '97
- 36032 Fiesta all models '77 thru '80
- 36036 Ford & Mercury Full-size,
  Ford LTD & Mercury Marquis ('75 thru '82);
  Ford Custom 500, Country Squire, Crown Victoria & Mercury Colony Park ('75 thru '87);
  Ford LTD Crown Victoria & Mercury Gran Marquis ('83 thru '87)
- 36040 Granada & Mercury Monarch '75 thru '80
- 36044 Ford & Mercury Mid-size,
  Ford Thunderbird & Mercury Cougar ('75 thru '82);
  Ford LTD & Mercury Marquis ('83 thru '86);
  Ford Torino, Gran Torino, Ranchero pick-up, LTD II, Mercury Montego, Comet, XR-7 & Lincoln Versailles ('75 thru '86)
- 36048 Mustang V8 all models '64-1/2 thru '73
- 36049 Mustang II 4 cyl, V6 & V8 models '74 thru '78
- 36050 Mustang & Mercury Capri all models Mustang, '79 thru '93; Capri, '79 thru '86
- *36051 Mustang all models '94 thru '97
- 36054 Pick-ups & Bronco '73 thru '79
- 36058 Pick-ups & Bronco '80 thru '96
- 36059 Pick-ups, Expedition & Mercury Navigator '97 thru '98
- 36062 Pinto & Mercury Bobcat '75 thru '80
- 36066 Probe all models '89 thru '92
- 36070 Ranger/Bronco II gasoline models '83 thru '92
- *36071 Ranger '93 thru '97 & Mazda Pick-ups '94 thru '97
- 36074 Taurus & Mercury Sable '86 thru '95
- *36075 Taurus & Mercury Sable '96 thru '98
- *36078 Tempo & Mercury Topaz '84 thru '94
- *36082 Thunderbird/Mercury Cougar '83 thru '88
- *36086 Thunderbird/Mercury Cougar '89 and '97
- 36090 Vans all V8 Econoline models '69 thru '91
- *36094 Vans full size '92-'95
- *36097 Windstar Mini-van '95-'98

## GENERAL MOTORS
- *10360 GM Automatic Transmission Overhaul
- *38005 Buick Century, Chevrolet Celebrity, Oldsmobile Cutlass Ciera & Pontiac 6000 all models '82 thru '96
- *38010 Buick Regal, Chevrolet Lumina, Oldsmobile Cutlass Supreme & Pontiac Grand Prix front-wheel drive models '88 thru '95
- *38015 Buick Skyhawk, Cadillac Cimarron, Chevrolet Cavalier, Oldsmobile Firenza & Pontiac J-2000 & Sunbird '82 thru '94
- *38016 Chevrolet Cavalier & Pontiac Sunfire '95 thru '98
- 38020 Buick Skylark, Chevrolet Citation, Olds Omega, Pontiac Phoenix '80 thru '85
- 38025 Buick Skylark & Somerset, Oldsmobile Achieva & Calais and Pontiac Grand Am models '85 thru '95
- 38030 Cadillac Eldorado '71 thru '85, Seville '80 thru '85, Oldsmobile Toronado '71 thru '85 & Buick Riviera '79 thru '85
- *38035 Chevrolet Lumina APV, Olds Silhouette & Pontiac Trans Sport all models '90 thru '95
  General Motors Full-size Rear-wheel Drive - see BUICK (19025)

*(Continued on other side)*

* Listings shown with an asterisk (*) indicate model coverage as of this printing. These titles will be periodically updated to include later model years - consult your Haynes dealer for more information.

**Haynes North America, Inc., 861 Lawrence Drive, Newbury Park, CA 91320-1514 • (805) 498-6703**

# Haynes Automotive Manuals (continued)

*NOTE: New manuals are added to this list on a periodic basis. If you do not see a listing for your vehicle, consult your local Haynes dealer for the latest product information.*

## GEO
- Metro - see CHEVROLET Sprint (24075)
- Prizm - '85 thru '92 see CHEVY (24060), '93 thru '96 see TOYOTA Corolla (92036)
- *40030 Storm all models '90 thru '93
- Tracker - see SUZUKI Samurai (90010)

## GMC
- Safari - see CHEVROLET ASTRO (24010)
- Vans & Pick-ups - see CHEVROLET

## HONDA
- 42010 Accord CVCC all models '76 thru '83
- 42011 Accord all models '84 thru '89
- 42012 Accord all models '90 thru '93
- 42013 Accord all models '94 thru '95
- 42020 Civic 1200 all models '73 thru '79
- 42021 Civic 1300 & 1500 CVCC '80 thru '83
- 42022 Civic 1500 CVCC all models '75 thru '79
- 42023 Civic all models '84 thru '91
- *42024 Civic & del Sol '92 thru '95
- *42040 Prelude CVCC all models '79 thru '89

## HYUNDAI
- *43015 Excel all models '86 thru '94

## ISUZU
- Hombre - see CHEVROLET S-10 (24071)
- *47017 Rodeo '91 thru '97; Amigo '89 thru '94; Honda Passport '95 thru '97
- *47020 Trooper & Pick-up, all gasoline models Pick-up, '81 thru '93; Trooper, '84 thru '91

## JAGUAR
- *49010 XJ6 all 6 cyl models '68 thru '86
- *49011 XJ6 all models '88 thru '94
- *49015 XJ12 & XJS all 12 cyl models '72 thru '85

## JEEP
- *50010 Cherokee, Comanche & Wagoneer Limited all models '84 thru '96
- 50020 CJ all models '49 thru '86
- *50025 Grand Cherokee all models '93 thru '98
- 50029 Grand Wagoneer & Pick-up '72 thru '91 Grand Wagoneer '84 thru '91, Cherokee & Wagoneer '72 thru '83, Pick-up '72 thru '88
- *50030 Wrangler all models '87 thru '95

## LINCOLN
- Navigator - see FORD Pick-up (36059)
- 59010 Rear Wheel Drive all models '70 thru '96

## MAZDA
- 61010 GLC Hatchback (rear wheel drive) '77 thru '83
- 61011 GLC (front wheel drive) '81 thru '85
- *61015 323 & Protogé '90 thru '97
- *61016 MX-5 Miata '90 thru '97
- *61020 MPV all models '89 thru '94
- Navajo - see Ford Explorer (36024)
- 61030 Pick-ups '72 thru '93 Pick-ups '94 thru '96 - see Ford Ranger (36071)
- 61035 RX-7 all models '79 thru '85
- *61036 RX-7 all models '86 thru '91
- 61040 626 (rear wheel drive) all models '79 thru '82
- *61041 626/MX-6 (front wheel drive) '83 thru '91

## MERCEDES-BENZ
- 63012 123 Series Diesel '76 thru '85
- *63015 190 Series four-cyl gas models, '84 thru '88
- 63020 230/250/280 6 cyl sohc models '68 thru '72
- 63025 280 123 Series gasoline models '77 thru '81
- 63030 350 & 450 all models '71 thru '80

## MERCURY
- See FORD Listing.

## MG
- 66010 MGB Roadster & GT Coupe '62 thru '80
- 66015 MG Midget, Austin Healey Sprite '58 thru '80

## MITSUBISHI
- *68020 Cordia, Tredia, Galant, Precis & Mirage '83 thru '93
- *68030 Eclipse, Eagle Talon & Ply. Laser '90 thru '94
- *68040 Pick-up '83 thru '96 & Montero '83 thru '93

## NISSAN
- 72010 300ZX all models including Turbo '84 thru '89
- *72015 Altima all models '93 thru '97
- *72020 Maxima all models '85 thru '91
- 72030 Pick-ups '80 thru '96 Pathfinder '87 thru '95
- 72040 Pulsar all models '83 thru '86
- 72050 Sentra all models '82 thru '94
- *72051 Sentra & 200SX all models '95 thru '98
- *72060 Stanza all models '82 thru '90

## OLDSMOBILE
- *73015 Cutlass V6 & V8 gas models '74 thru '88
- For other OLDSMOBILE titles, see BUICK, CHEVROLET or GENERAL MOTORS listing.

## PLYMOUTH
- For PLYMOUTH titles, see DODGE listing.

## PONTIAC
- 79008 Fiero all models '84 thru '88
- 79018 Firebird V8 models except Turbo '70 thru '81
- 79019 Firebird all models '82 thru '92
- For other PONTIAC titles, see BUICK, CHEVROLET or GENERAL MOTORS listing.

## PORSCHE
- *80020 911 except Turbo & Carrera 4 '65 thru '89
- 80025 914 all 4 cyl models '69 thru '76
- 80030 924 all models including Turbo '76 thru '82
- *80035 944 all models including Turbo '83 thru '89

## RENAULT
- Alliance & Encore - see AMC (14020)

## SAAB
- *84010 900 all models including Turbo '79 thru '88

## SATURN
- 87010 Saturn all models '91 thru '96

## SUBARU
- 89002 1100, 1300, 1400 & 1600 '71 thru '79
- *89003 1600 & 1800 2WD & 4WD '80 thru '94

## SUZUKI
- *90010 Samurai/Sidekick & Geo Tracker '86 thru '96

## TOYOTA
- 92005 Camry all models '83 thru '91
- 92006 Camry all models '92 thru '96
- 92015 Celica Rear Wheel Drive '71 thru '85
- *92020 Celica Front Wheel Drive '86 thru '93
- 92025 Celica Supra all models '79 thru '92
- 92030 Corolla all models '75 thru '79
- 92032 Corolla all rear wheel drive models '80 thru '87
- 92035 Corolla all front wheel drive models '84 thru '92
- *92036 Corolla & Geo Prizm '93 thru '97
- 92040 Corolla Tercel all models '80 thru '82
- 92045 Corona all models '74 thru '82
- 92050 Cressida all models '78 thru '82
- 92055 Land Cruiser FJ40, 43, 45, 55 '68 thru '82
- 92056 Land Cruiser FJ60, 62, 80, FZJ80 '80 thru '96
- *92065 MR2 all models '85 thru '87
- 92070 Pick-up all models '69 thru '78
- *92075 Pick-up all models '79 thru '95
- *92076 Tacoma '95 thru '98, 4Runner '96 thru '98, & T100 '93 thru '98
- *92080 Previa all models '91 thru '95
- 92085 Tercel all models '87 thru '94

## TRIUMPH
- 94007 Spitfire all models '62 thru '81
- 94010 TR7 all models '75 thru '81

## VW
- 96008 Beetle & Karmann Ghia '54 thru '79
- 96012 Dasher all gasoline models '74 thru '81
- *96016 Rabbit, Jetta, Scirocco, & Pick-up gas models '74 thru '91 & Convertible '80 thru '92
- 96017 Golf & Jetta all models '93 thru '97
- 96020 Rabbit, Jetta & Pick-up diesel '77 thru '84
- 96030 Transporter 1600 all models '68 thru '79
- 96035 Transporter 1700, 1800 & 2000 '72 thru '79
- 96040 Type 3 1500 & 1600 all models '63 thru '73
- 96045 Vanagon all air-cooled models '80 thru '83

## VOLVO
- 97010 120, 130 Series & 1800 Sports '61 thru '73
- 97015 140 Series all models '66 thru '74
- *97020 240 Series all models '76 thru '93
- 97025 260 Series all models '75 thru '82
- *97040 740 & 760 Series all models '82 thru '88

## TECHBOOK MANUALS
- 10205 Automotive Computer Codes
- 10210 Automotive Emissions Control Manual
- 10215 Fuel Injection Manual, 1978 thru 1985
- 10220 Fuel Injection Manual, 1986 thru 1996
- 10225 Holley Carburetor Manual
- 10230 Rochester Carburetor Manual
- 10240 Weber/Zenith/Stromberg/SU Carburetors
- 10305 Chevrolet Engine Overhaul Manual
- 10310 Chrysler Engine Overhaul Manual
- 10320 Ford Engine Overhaul Manual
- 10330 GM and Ford Diesel Engine Repair Manual
- 10340 Small Engine Repair Manual
- 10345 Suspension, Steering & Driveline Manual
- 10355 Ford Automatic Transmission Overhaul
- 10360 GM Automatic Transmission Overhaul
- 10405 Automotive Body Repair & Painting
- 10410 Automotive Brake Manual
- 10415 Automotive Detailing Manual
- 10420 Automotive Eelectrical Manual
- 10425 Automotive Heating & Air Conditioning
- 10430 Automotive Reference Manual & Dictionary
- 10435 Automotive Tools Manual
- 10440 Used Car Buying Guide
- 10445 Welding Manual
- 10450 ATV Basics

## SPANISH MANUALS
- 98903 Reparación de Carrocería & Pintura
- 98905 Códigos Automotrices de la Computadora
- 98910 Frenos Automotriz
- 98915 Inyección de Combustible 1986 al 1994
- 99040 Chevrolet & GMC Camionetas '67 al '87 Incluye Suburban, Blazer & Jimmy '67 al '91
- 99041 Chevrolet & GMC Camionetas '88 al '95 Incluye Suburban '92 al '95, Blazer & Jimmy '92 al '94, Tahoe y Yukon '95
- 99042 Chevrolet & GMC Camionetas Cerradas '68 al '95
- 99055 Dodge Caravan & Plymouth Voyager '84 al '95
- 99075 Ford Camionetas y Bronco '80 al '94
- 99077 Ford Camionetas Cerradas '69 al '91
- 99083 Ford Modelos de Tamaño Grande '75 al '87
- 99088 Ford Modelos de Tamaño Mediano '75 al '86
- 99091 Ford Taurus & Mercury Sable '86 al '95
- 99095 GM Modelos de Tamaño Grande '70 al '90
- 99100 GM Modelos de Tamaño Mediano '70 al '88
- 99110 Nissan Camionetas '80 al '96, Pathfinder '87 al '95
- 99118 Nissan Sentra '82 al '94
- 99125 Toyota Camionetas y 4Runner '79 al '95

Over 100 Haynes motorcycle manuals also available

*Listings shown with an asterisk (*) indicate model coverage as of this printing. These titles will be periodically updated to include later model years - consult your Haynes dealer for more information.

**Haynes North America, Inc., 861 Lawrence Drive, Newbury Park, CA 91320-1514 • (805) 498-6703**